De Gruyter Graduate

Aresta, Dibenedetto, Dumeignil (Eds.) • Biorefineries: An Introdu

Also of interest

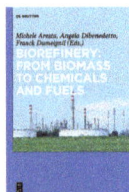

Biorefinery: From Biomass to Chemicals and Fuels
Aresta, Dibenedetto, Dumeignil *(Eds); 2012*
ISBN 978-3-11-026023-6, e-ISBN 978-3-11-026028-1

Chemical Reaction Technology
Murzin; 2015
ISBN 978-3-11-033643-6, e-ISBN 978-3-11-033644-3

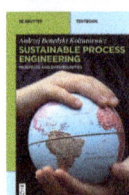

Sustainable Process Engineering: Prospects and Opportunities
Koltuniewicz; 2014
ISBN 978-3-11-030875-4, e-ISBN 978-3-11-030876-1

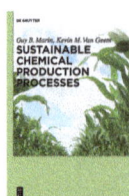

Sustainable Chemical Production Processes
Marin, Van Geem; 2016
ISBN 978-3-11-026975-8, e-ISBN 978-3-11-026992-5

Biohydrogen
Rögner (Ed.); 2015
ISBN 978-3-11-033645-0, e-ISBN 978-3-11-033673-3

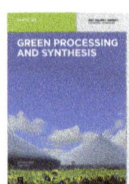

Green Processing and Synthesis
Hessel, Volker (Editor-in-Chief)
ISSN 2191-9542, e-ISSN 2191-9550

Biorefineries

An Introduction

Edited by
Michele Aresta, Angela Dibenedetto and Franck Dumeignil

DE GRUYTER

Editors

Prof. Michele Aresta
Chemical and Biomolecular Engineering Department, NUS
4, Engineering Drive 4, Singapore
CIRCC
Via Celso Ulpiani, 27
Bari 70126 Bari, Italy
cheam@nus.edu.sg

Prof. Angela Dibenedetto
University of Bari and CIRCC
Department of Chemistry
Via Celso Ulpiani 27
70176 Bari, Italy
angela.dibenedetto@uniba.it

Prof. Franck Dumeignil
IUF, FRSC
Université de Lille
UCCS, UMR CNRS 8181
59655 Villeneuve d'Ascq Cedex, France
Institut Universitaire de France

ISBN 978-3-11-033153-0
e-ISBN (PDF) 978-3-11-033158-5
e-ISBN (EPUB) 978-3-11-038999-9

Library of Congress Cataloging-in-Publication Data
A CIP catalog record for this book has been applied for at the Library of Congress.

Bibliographic information published by the Deutsche Nationalbibliothek
The Deutsche Nationalbibliothek lists this publication in the Deutsche Nationalbibliografie;
detailed bibliographic data are available on the Internet at http://dnb.dnb.de.

© 2015 Walter de Gruyter GmbH, Berlin/Boston
Typesetting: Compuscript Ltd., Shannon, Ireland
Printing and binding: CPI books GmbH, Leck
Cover image: Jim Parkin/iStock/thinkstock
∞ Printed on acid-free paper
Printed in Germany

www.degruyter.com

Preface

Biomass is a complex substance that may be used for the extraction or production of various chemicals and materials or for the production of heat used in the chemical and energy sectors. The latter represents the most primitive use (since man developed the capacity to manage a fire) and is the less economically rewarding one. The use of biomass for producing heat is still practiced today, but, more frequently, residual refractory biomass, more than primary biomass, is used to this end. In fact, biomass is an extremely rich raw substance that can be used as a source of many different molecular compounds and polymeric materials, increasing the profit and reducing the environmental burden of its use. Recently, the conversion of primary biomass into syngas ($CO+H_2$) has been pursued as an objective, which is affected by a drawback: complex matter is converted into a C1 molecule (CO), which is then used to build more complex C_n molecules. Such huge change in entropy in using biomass is energetically questionable. The direct use of the complexity of biomass is certainly a much clever approach to its exploitation.

Biorefinery is a recently implemented concept for making the most economically profitable and environmentally friendly use of biomass.

This book aims at giving a quite comprehensive view of the opportunities offered by the implementation of the concept of biorefinery to biomass utilization.

The introduction, written by Dr. Maria Georgiadou of the EU Commission, sets the political basis for the technical chapters that follow.

In Chapter 1, a few basic concepts are recalled that may be useful to non-technical readers. Chapter 2 presents the production of terrestrial biomass and Chapter 3 introduces the aquatic biomass, which has quite different properties compared to the former. Chapter 4 deals with the bioconversion of biomass and downstream technologies. Chapter 5 illustrates the key issues relevant to strain engineering for the production of second-generation bioethanol, starting from cellulose and tackling the food/non-food competition. Chapter 6 gives a panoramic view of homogeneous and heterogeneous catalytic processes for biomass conversion into chemicals, fuels, or materials. Chapter 7 completes the insight into polymeric materials sourced from biomass. Chapter 8 presents the route to fuels from biomass derived intermediate molecules. Chapter 9 makes the analysis of the use of biomass (residual) sourced syngas. Chapter 10 is an industrial trip into oil chemistry, with an interesting analysis of economics via value chains. Chapter 11 deals with the most difficult component of biomass: lignin. Routes to its valorization are discussed. Building-up a biorefinery can be a very costly operation. Therefore, the retrofit into existing assets may be an advantageous solution, which is discussed in Chapter 12.

As reported above, biorefineries aim at reducing the environmental impact of the use of biomass. Chapters 1–12 show the way the biomass components can be used. To close the cycle, waste streams need to be also treated or used. This is done by heat production of refractory solids or by anaerobic digestion of wet residues to produce biogas (CH_4+CO_2), which can be separated into its components with methane utilization, e.g., as a source of energy in the above processes. CO_2 can also find use, even in combination with waste biomass, but this is another fascinating story, which goes in the direction of man-directed photosynthesis!

We wish readers will go through this book and find answers to their questions. The book has been written in a simple manner, so that beginners may have access to basic concepts, and contains enough updated information so that experts may find a good literature summary.

Michele Aresta, Angela Dibenedetto, and Franck Dumeignil

Contents

Angela Dibenedetto and Antonella Colucci

Benjamin Katryniok, François Jérôme, Eric Monflier, Sébastien Paul,
and Franck Dumeignil

Jean-Luc Couturier and Jean-Luc Dubois

Heiko Lange, Elisavet D. Bartzoka, and Claudia Crestini

Raf Roelant, Fabrizio Cavani, Carla S.M. Pereira, and Alírio E. Rodrigues

Michele Aresta

Contributing authors

Gennaro Agrimi
Department of Biosciences, Biotechnologies
and Biopharmaceutics
University of Bari
Bari, Italy
CIRCC – Interuniversity Consortium Chemical
Reactivity and Catalysis
Bari, Italy
e-mail: gennaro.agrimi@uniba.it
Chapter 5

Efthymia Alexopoulou
CRES – Center for Renewable Energy Sources
Biomass Department
Pikermi Attikis, Greece
Chapter 2

Michele Aresta
CIRCC – Interuniversity Consortium
Chemical Reactivity and Catalysis
Bari, Italy
Department of Biomolecular and
Chemical Engineering
NUS
Singapore
e-mail: cheam@nus.edu.sg
Chapters 1 and 13

Elisavet D. Bartzoka
Department of Chemical Sciences
and Technologies
Università degli Studi di Roma 'Tor Vergata'
Rome, Italy
Chapter 11

Antonio Buonerba
Dipartimento di Chimica e Biologia
Università degli Studi di Salerno
Fisciano, Italy
NANOMATES, Research Center for
NANOMAterials and nanoTEchnology
at Salerno University
Fisciano, Italy
Chapter 7

Fabrizio Cavani
CIRCC – Interuniversity Consortium Chemical
Reactivity and Catalysis
Dipartimento di Chimica Industriale
"Toso Montanari"
Università di Bologna
Bologna, Italy
Chapter 12

Myrsini Christou
CRES – Center for Renewable Energy Sources
Biomass Department
Pikermi Attikis, Greece
e-mail: mchrist@cres.gr
Chapter 2

Antonella Colucci
CIRCC – Interuniversity Consortium
Chemical Reactivity and Catalysis
Bari, Italy
Department of Chemistry
University of Bari
Bari, Italy
Chapter 3

Jean-Luc Couturier
Arkema France
Centre de Recherche Rhône-Alpes
Pierre-Bénite, France
e-mail: jean-luc.couturier@arkema.com
Chapter 10

Claudia Crestini
Department of Chemical Sciences
and Technologies
Università degli Studi di Roma 'Tor Vergata'
Rome, Italy
e-mail:crestini@stc.uniroma2.it
Chapter 11

Angela Dibenedetto
CIRCC – Interuniversity Consortium Chemical
Reactivity and Catalysis
Bari, Italy
Department of Chemistry
University of Bari
Bari, Italy
e-mail: angela.dibenedetto@uniba.it
Chapters 1 and 3

Jean-Luc Dubois
Arkema France
420 Rue d'Estienne d'Orves
92705 Colombes, France
e-mail: jean-luc.dubois@arkema.com
Chapter 10

Franck Dumeignil
IUF, FRSC
Université de Lille
UCCS, UMR CNRS 8181
59655 Villeneuve d'Ascq Cedex, France
e-mail: franck.dumeignil@univ-lille1.fr
Chapters 1 and 6

Ioannis Eleftheriadis
CRES – Center for Renewable Energy Sources
Biomass Department
Pikermi Attikis, Greece
Chapter 2

Maria Georgiadou
European Commission DG Reearch-K.3
New & Renewable Energy Sources
CDMA 5/170
1049 Brussels
e-mail: Maria.Georgiadou@ec.europa.eu
Introduction

Alfonso Grassi
Dipartimento di Chimica e Biologia
Università degli Studi di Salerno
Fisciano, Italy
NANOMATES, Research Center for
NANOMAterials and nanoTEchnology at Salerno
University
Fisciano, Italy
e-mail: agrassi@unisa.it
Chapter 7

Christin Groeger
Institute of Bioprocess and Biosystems
Engineering
Hamburg University of Technology
Hamburg, Germany
Chapter 4

Eleni Heracleous
Chemical Process and Energy Resources
Institute (CPERI)
Center for Research and Technology Hellas
(CERTH)
Thessaloniki, Greece
School of Science and Technology
International Hellenic University
Thessaloniki, Greece
Chapter 8

Eleni F. Iliopoulou
Chemical Process and Energy Resources
Institute (CPERI)
Center for Research and Technology Hellas
(CERTH)
Thessaloniki, Greece
Chapter 8

François Jérôme
Institut de Chimie des Milieux et Matériaux de
Poitiers
Université de Poitiers-ENSIP
1, rue Marcel Doré
86022 Poitiers, France
Chapter 6

Benjamin Katryniok
Ecole Centrale de Lille
UCCS, UMR CNRS 8181
59655 Villeneuve d'Ascq Cedex, France
Chapter 6

Michal Krzyzaniak
Department of Plant Breeding
and Seed Production
University of Warmia and Mazury
Olsztyn, Poland
Chapter 2

Heiko Lange
Department of Chemical Sciences
and Technologies
Università degli Studi di Roma 'Tor Vergata'
Rome, Italy
Chapter 11

Angelos A. Lappas
Chemical Process and Energy Resources
Institute (CPERI)
Center for Research and Technology Hellas
(CERTH)
Thessaloniki, Greece
Chapter 8

Angeliki A. Lemonidou
Department of Chemical Engineering
Aristotle University of Thessaloniki
Thessaloniki, Greece
Chemical Process and Energy Resources
Institute
Center for Research and Technology Hellas
Thessaloniki, Greece
e-mail: alemonidou@cheng.auth.gr
Chapter 8

Eric Monflier
Université d'Artois
UCCS, UMR CNRS 8181
Rue Jean Souvraz, SP 18
F-62307 Lens, France
Chapter 6

Luigi Palmieri
Department of Biosciences, Biotechnologies
and Biopharmaceutics
University of Bari
Bari, Italy
CIRCC – Interuniversity Consortium Chemical
Reactivity and Catalysis
Bari, Italy
e-mail: luigi.palmieri@uniba.it
Chapter 5

Ioanna Papamichael
CRES – Center for Renewable Energy Sources
Biomass Department
Pikermi Attikis, Greece
Chapter 2

Sébastien Paul
Ecole Centrale de Lille
UCCS, UMR CNRS 8181
59655 Villeneuve d'Ascq Cedex, France
Chapter 6

Carlo Perego
eni s.p.a., Centro Ricerche per le Energie Non
Convenzionali - Istituto eni Donegani
Novara, Italy
Chapter 9

Carla S.M. Pereira
Laboratory of Separation and Reaction
Engineering
University of Porto
Porto, Portugal
Chapter 12

Isabella Pisano
Department of Biosciences, Biotechnologies
and Biopharmaceutics
University of Bari
Bari, Italy
CIRCC – Interuniversity Consortium Chemical
Reactivity and Catalysis
Bari, Italy
Chapter 5

Marco Ricci
eni s.p.a., Centro Ricerche per le Energie Non
Convenzionali – Istituto eni Donegani
Novara, Italy
versalis s.p.a.,
Green Chemistry
Novara Research Center
Novara, Italy
e-mail: marco.ricci@versalis.eni.com
Chapter 9

Maria Antonietta Ricci
Department of Biosciences, Biotechnologies
and Biopharmaceutics
University of Bari
Bari, Italy
CIRCC – Interuniversity Consortium Chemical
Reactivity and Catalysis
Bari, Italy
Chapter 5

Alírio E. Rodrigues
Laboratory of Separation and Reaction
Engineering
University of Porto
Porto, Portugal
Chapter 12

Raf Roelant
Process Design Center
Breda, The Netherlands
e-mail: roelant@process-design-center.com
Chapter 12

Wael Sabra
Institute of Bioprocess and Biosystems
Engineering
Hamburg University of Technology
Hamburg, Germany
Chapter 4

Sheila Ortega Sanchez
Dipartimento di Chimica e Biologia
Università degli Studi di Salerno
Fisciano, Italy
NANOMATES, Research Center for
NANOMAterials and nanoTEchnology
at Salerno University
Fisciano, Italy
Chapter 7

Mariusz Stolarski
Department of Plant Breeding
and Seed Production
University of Warmia and Mazury
Olsztyn, Poland
Chapter 2

Charles Themistocles
Société Agricole de Befandriana-Sud
& Partners Sarl
Antananarivo 101, Madagascar
Chapter 2

Kostas Tsiotas
CRES – Center for Renewable Energy Sources
Biomass Department
Pikermi Attikis, Greece
Chapter 2

Efterpi S. Vasiliadou
Department of Chemical Engineering
Aristotle University of Thessaloniki
Thessaloniki, Greece
Chapter 8

An-Ping Zeng
Institute of Bioprocess and Biosystems
Engineering
Hamburg University of Technology
Hamburg, Germany
e-mail: aze@tuhh.de
Chapter 4

Abbreviations

bpd	barrel *per* day (1 barrel: 158,98 litres)
BtL	biomass to liquids
CtL	coal to liquids
DBE	1,1-dibutoxyethane
DBM	dibutoxymethane
DMC	dimethylcarbonate
FT	Fischer-Tropsch
GtL	gas to liquids
kt	thousands of metric ton (kt/year, thousands of metric tons *per* year)
MA	maleic anhydride
Mt	millions of metric ton (Mt/year, millions of metric ton *per* year)
MTBE	methyl *tert*-butyl ether
MTO	methanol to olefins
PA	phthalic anhydride
PermSMBR	simulated moving bed membrane reactor
SMBR	simulated moving bed reactor
SWOT	strengths, weaknesses, opportunities, threats
t	metric ton (1000 kg; t/year, ton per year)
TRL	technology readiness level
WGS	Water-gas shift reaction

Introduction to Biorefinery

In a biorefinery, biomass is converted into a wide spectrum of bio-based products including biofuels, biochemicals, biomaterials, feed, heat, and power. Ideally, full-scale, highly efficient, and integrated biorefineries allow the manufacture of bio-based products, which are fully competitive with their conventional equivalents.

Biomass availability in Europe is limited, however, and its sustainable production and cost-efficient use is the key element for the competitiveness of biorefineries. Consequently, diversification of biomass and conversion flexibility are crucial to the biorefinery realization. Versatile biomass feedstocks consist of dedicated lignocellulosic crops including agricultural and forestry, as well as residues including agricultural and forestry, agri-food and urban organic waste, and unmobilized biomass, together with algal and aquatic biomass. A diverse and flexible biomass production sector should be in place to supply biorefineries with biomass continuously and cost-effectively. Conversion technologies that can deal with multiple biomass feedstock streams, ultimately through a combination of several processes in an integrated way, are necessary to provide optimal processing solutions to multiple and valuable marketable products.

The development of biorefineries is nevertheless driven by economic constraints and sustainability criteria. Low-cost operation and efficient use of resources are the critical requirements for their viability. Meanwhile, sustainable production throughout the full product life cycle is a necessary condition for their public acceptance. Standardization, public awareness, and eco-labeling are measures to reassure the consumer about the product sustainability, in other words, that the product is used to generate bioenergy at the end of its life (biofuels, bioheat, biopower) or it is recyclable and biodegradable (biochemicals and biomaterials). Successful biorefineries should build upon both conditions of low cost and sustainability, identifying the golden equilibrium: they should produce environmentally friendly products with small carbon and water footprint, equal or better properties than their fossil equivalents, and a competitive price.

The biorefinery concept faces many challenges along the entire value chain: from the biomass cultivation, logistics, and harvesting to pretreatment and conversion processes, separation techniques, process optimization, life cycle analysis, sustainability, and (new) product specifications. Only through a multidisciplinary integration approach can the problems be understood and the corresponding research needs be identified.

Reliable supply of quality biomass feedstock at reasonable prices will determine the economic viability of biorefineries. Bottlenecks exist as much for the feedstock development as for the logistics and supply chains. For example, improving the yield and the properties of the produced biomass, developing sustainable agricultural and forestry practices for biomass production, assessing the biomass potential for biorefineries, developing logistic systems that provide continuous biomass supply, developing cost-effective preprocessing methods for better storage and transport of feedstock, and improving management, traceability, and automation of biomass production are some of the areas where further research is necessary for a secure and low-cost feedstock supply.

The ability to optimize value extraction from the major lignocellulosic biomass components is another major challenge for efficient biomass processing. Extraction and fractionation methods are currently used for the pretreatment of biomass and may cause significant loss of potential value of one or more biomass components. Hence, major research is needed to develop processes that maximize the value extraction of the three major biomass components, like the organosolv and the ionic liquid extraction type techniques.

Improving conversion processes has a direct impact on the overall performance and efficiency of the biorefinery. Thermochemical, biochemical, and chemical catalysis processes are the three generic categories for biomass conversion, which today still require substantial progress beyond the state of the art for achieving technological breakthroughs with serious advances regarding the cost-competitiveness of the end products. Thermochemical processes include gasification, pyrolysis, torrefaction, and hydrothermal liquefaction; they are confronted with product quality challenges and low overall energy and carbon efficiency, hence underpinning the need for process integration. Biochemical processes suffer from low yields and overall productivity as well as low selectivity, thus making separation and purification of products expensive and energy intensive; limiting material loss during the sequential processing, developing robust biocatalysts (enzymes and yeasts or bacteria) capable of working at high concentrations, as well as improving downstream processing for cost and efficiency will have a large effect on the biorefinery's economic feasibility. The role of systems biology and synthetic biology is central to cost and time optimization of the biocatalyst design. Chemical catalysis is the process used most often in all industrial chemical technology applications. However, developing novel, highly stable, readily available, and low-cost catalysts that can treat oxygen-rich biomass feedstocks and at the same time be multifunctional is the major research challenge in this field. Furthermore, combining biocatalysts and chemical catalysts looks promising for performance improvement.

The success of "tomorrow's" biorefineries depends heavily on integration occuring at various levels: business integration, when new biorefineries are integrated with existing business and cross-sectorial synergies between the value chains are established; technical integration, when the value chain is optimized for all different stakeholders, multiple feedstocks, multiple products, and multiple processing; process integration, when optimal use of mass, energy, and water is obtained within the overall processing while still maximizing production. The main challenges encompass the requirement for new business models, new biorefinery value chains, and new recycling methods, process control systems, and process analysis methods.

Assessing the sustainability of the biorefinery concept spans over the entire value chain, from the feedstock production to the use of the end product, and covers three pillars: environmental, economic, and social. Life cycle assessment is a standardized approach that addresses most of environmental impacts (greenhouse gas emissions, primary energy consumption, water use), while other important issues (direct and indirect emissions from land use, biodiversity) should be also included. Life cycle costing and social life cycle assessment could be adopted to address economic and social impacts (profit aspects, rural development, job creation, re-industrialization, public acceptance). Integration of tools and impacts is crucial for a systemic approach, but valid sustainability criteria for the production of biomass and all types of bio-based products along with standards and universal certification schemes are vital for public appreciation of biorefineries. Although sustainability requirements will initially restrict biomass availability, they will ensure long-term security of supply, minimizing environmental impact.

Europe has the lead in science and technology for most of the processes relevant to the biorefinery concept. However, innovation and market uptake of biorefineries depend on overcoming major challenges related to the bio-based products: biofuels and similarly biochemicals and biomaterials that replace their fossil equivalents are not yet cost-competitive, whereas new bio-based products are difficult to introduce into a market where no demand exists; upscaling of biorefineries requires high investment costs and financing of high-risk technologies is very difficult, restricting further deployment (thus termed the "valley of death"); the European regulatory framework is not stable or holistic; further research and development is necessary for improving the performance and the cost-efficiency of the underlying technologies. Market and socioeconomic research can facilitate assessing the consumer behavior, identifying promising business areas and positioning new products into markets.

Achieving "tomorrow's" biorefineries in Europe requires better integration, more flexibility, and improved sustainability. Their feasibility will be driven by the extent to which they can address societal challenges and needs, like energy security, competitiveness, and growth, as well as climate, environment and resource efficiency. Their commercialization potential is strong, implying significant growth and job creation. However, their full deployment will depend on whether they aim at sustainable products of high quality that can be competitive in European markets.

Maria Georgiadou
Research Programme Officer European Commission[1]

---

[1] All views expressed herein are entirely of the author, do not reflect the position of the European Institutions,"or bodies," and do not, in any way, engage any of them.

Michele Aresta, Angela Dibenedetto, and Franck Dumeignil

1 Catalysis, growth, and society

In developed countries, modern society enjoys unprecedented levels of well-being and comfort enabled by access to a variety of man-made goods in addition to natural goods, efficient medical treatments and prevention, abundance of high-quality food, rapid access to information, comfortable and rapid global travel, and "easy-to-use and transport (batteries)" energy. Chemistry underpins almost all of the technological innovations that make everyday human life as comfortable as possible, being transversal to food-health-energy-mobility-environment-housing. Ninety-five percent or more (in volume) of all the products of the chemical industry are synthesized using at least one catalytic step. Approximately 20% of the world economy depends directly or indirectly on catalysis.

However, the dependence of society on chemistry and catalysis will increase with the growth in population and the demand for higher standard and quality of life.

Catalysis has been underlined as a key enabling technology for the future by several international (e.g., the SusChem Programme) and national (USA, Japan, China, UK, the Netherlands, France, Italy, among others) organizations. Sustainability in chemistry is one of the key targets of the industry of the future, and catalysis is the basis of sustainable chemistry. Several societal programs, while often not openly including the word "catalysis", are nevertheless built around it.

Production processes can be broadly categorized into two large families, namely (i) biotechnological transformations and (ii) chemical conversions, driven or not by catalysis. Variations of the latter are homogeneous and heterogeneous catalysis if the catalyst and the reactants are dissolved in the same medium or if the catalyst is in the solid state while the reactant(s) is(are) in the gas and/or liquid state, respectively. Biotechnological transformations can be carried out using entire microorganisms or enzymes (Figure 1.1b), soluble or heterogeneized. Chemical catalysts (Figure 1.1a,c) can be inorganic salts, synthetic molecular edifices, metal-organic systems, or metals, metal oxides, solid composite materials, organic species, protons, and so on.

The boundaries between the above categories are becoming progressively blurred, and such technologies are often sequentially combined, or more recently, in an even more integrated one-pot approach known as "hybrid catalysis". The different types of catalysis have their respective advantages and disadvantages, as summarized in Table 1.1.

Irrespective of the concerned catalytic species, a catalyzed process involves the faster transformation of molecules under more "facile" conditions than in the absence of a catalyst.

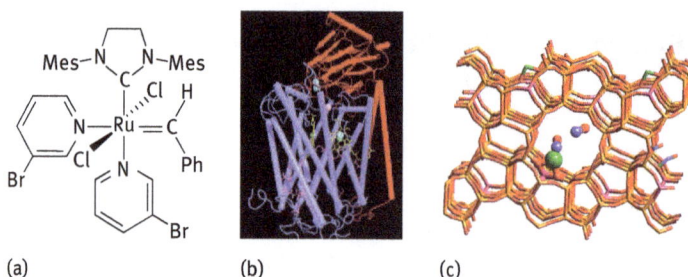

(a) (b) (c)

Fig. 1.1: Representation of a *Homogeneous catalyst* (*used in metathesis,* **a**), an *Enzyme* (*Cytochrome C Oxidase,* **b**) and a *Heterogeneous catalyst* (*zeolite structure,* **c**).

Table 1.1: Comparison of catalysts.

Property	Homogeneous catalysts	Homogeneous or heterogeneous catalysts	Enzymes
Active centers	All metal atoms	In general, the surface atoms/ions	All centers
Used concentration	Low	High	Very low
Selectivity	Good to very high	Poor to good	Very high
Diffusion issues	No	Yes	No
Reaction conditions	Mild	Harsh	Very mild
Application	Restricted	Large	Very restricted
Stability	Poor to good	Good to excellent	Poor
Active center structure	Defined	Not always well defined	Very well defined
Structure stability	Moderate	Very high	Limited
Thermal stability	Moderate	High	Poor
Post-process separation	Difficult	Easy	Easy if supported
Recyclability	Moderate	Poor to very good	Good if supported
Cost	Medium	Low	High

Figure 1.2 shows the global energetic variation of a non-catalyzed (a) and a catalyzed (b) reaction. It is clear that the catalyst modifies the pattern of the energy profile. A "positive" catalyst lowers the energetic barrier that must be overcome for the chemical transformation to occur and ensures that a higher number of reagent molecules (reactant) be converted per unit of time under specific reaction conditions (temperature, pressure, solvent) than under non-catalyzed conditions.

Typically, a catalyst makes a reaction occurring in milder conditions (lower temperature) and increases the rate of the reaction, while the equilibrium position remains unaffected. Nevertheless, there are also catalysts that can hinder a reaction

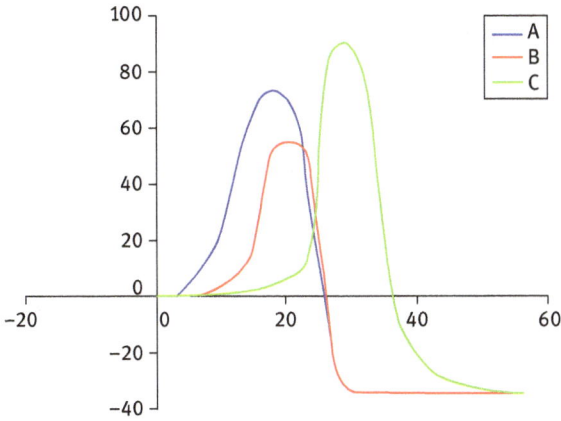

Fig. 1.2: The effect of catalysts on the reaction energy profile.

by increasing the activation energy (Figure 2.2c). Such "negative" catalysts are not rare in bioprocesses and also exist for chemical reactions.

While a basic treatise of catalysis will be a good guide to beginners, we briefly summarize hereafter some basic concepts that may be useful when reading this volume.

The properties of a catalyst are expressed in different ways such as activity, conversion of reagents, selectivity, stability, turnover number (TON), and turnover frequency (TOF). The activity of a catalyst can be represented by the reaction rate expressed as

$$r = \text{mol}_{substrate\ converted}/m_{cat} \cdot t \quad [\text{mol/kg·h}] \tag{1.1}$$

Alternatively, the volume of catalyst can be used, and the expression of the rate becomes

$$r = \text{mol}_{substrate\ converted}/V_{cat} \cdot t \quad [\text{mol/L·h}] \tag{1.2}$$

The kinetic activities are experimental values.

$$dn_p/dt = k \cdot C_p \tag{1.3}$$

where k is the kinetic constant and depends on the Arrhenius equation:

$$k = k_o\, e^{-(Ea/RT)} \tag{1.4}$$

The equations above suggest that the activity of a catalyst can be expressed in three different ways: through the rate of a reaction, the rate constant of the reaction, or the activation energy, *Ea*. The comparison of catalysts can be correctly done using the rate at zero-time, r_o.

Conversion gives the fraction of the jth reagent (percentage or molar fraction) converted. Selectivity represents the fraction of jth reagent converted into a target product. The stability of a catalyst is a measure of its life expressed in units of time (*h*, *d*, *y*). Sometimes, deactivation rate is expressed, which is the parameter that gives information on the loss of activity of the catalyst during a given time.

TOF gives the number of units (or moles) of jth reagent converted per mole of catalyst in the unit of time (*s*, *h*). It is expressed in units of reciprocal time (t^{-1}). TON gives the number of substrate units (moles) converted per catalytic center until it becomes inactive. TON is easily defined for enzymes and for homogeneous catalysts, but it is much lesser for heterogeneous catalysts, because, often, it is not straightforward to identify the number of active sites.

Catalytic processes that use heterogeneous catalysts can be carried out in batch or in flow reactors (see specialized books for their properties). For a flow reactor, the space velocity of the catalyst is defined as the ratio of the specific volume in flow (V/t) per unit mass (or volume) of the catalyst.

$$SR = V/m \cdot t \quad [\text{m}^3 \text{ kg}^{-1} \text{ s}^{-1}] \quad (1.5)$$

In the above equation, if the mass of catalyst *m* is substituted by its volume *V*, the *SR* becomes equal to the reciprocal of the residence time ($SR = t^{-1}$). Another quantity often used to characterize the solid catalysts is the space-time yield (*STY*) for the jth species given by

$$STY = \text{mol}_{j/} V_{cat} \cdot t \quad [\text{mol L}^{-1} \text{ s}^{-1}] \quad (1.6)$$

where mol_j represents the number of moles of the jth species and V_{cat} is the volume of catalyst.

In the case of biochemical (or biotechnological) processes, the catalysts are enzymes, which are characterized by the following properties:
- They are all proteins with complex structures and an active center with a well-defined geometry.
- They are either pure organic species or contain a metal (metal-enzymes) as the active site.
- The way enzymes work is affected by the temperature, pH, solvent, and pressure. They can be denatured (unfolded or destroyed) by excessive heat and other factors.
- All reactions are reversible.
- Enzymes are very specific, that is, they control only one reaction or convert a well-defined substrate into a specific product.

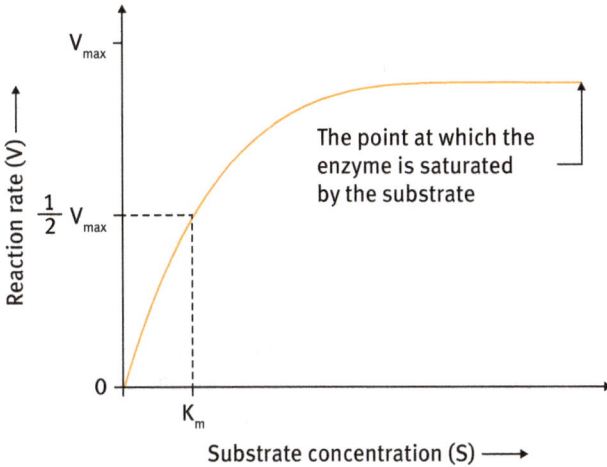

Fig. 1.3: Correlation of the reaction rate v to the substrate concentration [S] in an enzymatic reaction.

The mechanism of enzymatic reactions has long been studied by analyzing the composition of enzymes and their structures. Leonor Michaelis and Maud Menten proposed a method for describing enzyme kinetics based on experimental facts.

In an enzymatic reaction, the initial reaction rate v is increased by increasing the concentration [S] of the substrate S. However, in cases where the amount of enzyme added to the reaction system is fixed, no matter how much [S] is increased, v cannot increase over a certain limit (Figure 1.3). Such extrapolated value is called v_{max}, which means that v is saturated with regard to [S].

The phenomenon of saturation is characteristic of catalytic reactions and is explained as the binding of a substrate to a catalyst. The phenomenon of saturation is observed because even if there is an abundance of substrate molecules, the number of substrate binding sites available over the catalyst is limited. For a generic and simple enzymatic reaction such as Eq. (1.7),

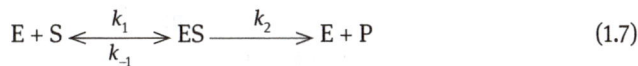

$$E + S \underset{k_{-1}}{\overset{k_1}{\rightleftharpoons}} ES \xrightarrow{k_2} E + P \tag{1.7}$$

a general equation has been established, known under the "Michaelis-Menten equation":

$$v = \frac{v_{max}}{1 + k_m/[S]} \tag{1.8}$$

with

$$k_m = \frac{k_1 + k_2}{k_1} \tag{1.9}$$

Here, v_{max} refers to the maximum initial reaction rate and k_m is the Michaelis-Menten constant. It represents the value of [S] for which the initial reaction rate is equal to $\frac{1}{2} v_{max}$.

However, this equation is essentially a hyperbolic function demonstrating the phenomenon of saturation. It makes some assumptions that do not always hold true for each and all the enzymatic reactions. In practice, the equation holds true for many enzymatic reactions, and k_m is used as an index showing affinity between the enzyme and the substrate (the smaller the k_m, the greater the affinity).

The properties of the catalysts categorized above are compared in Table 1.1, where the essential properties and some inherent characteristics of each class (homogeneous, heterogeneous, enzymatic catalysts) are highlighted. Table 1.1 may be helpful in identifying the best conditions to use in a given class of catalyst or in excluding the application of a catalyst in some other conditions under which the catalyst may not be stable enough.

Catalysis is thus a highly transversal technology that is of utmost interest in many crucial domains, as mentioned above, at the most inner core of current societal demands. In this context, the improvement of existing catalytic (chemical or biochemical) processes is a topical target, and the development of new catalysts with both high performances and innovative synthetic routes is a highly strategic issue. Reaction and process design are of vital importance for the chemical industry as well as for the manufacturing sector in general. It comprises overarching technologies that can be applied to all areas of chemistry and biotechnology. Technology leadership in this area is of growing importance as commercial pressure on Europe (USA and Japan as well) from lower-cost-production regions increases. Product life cycles are becoming shorter. Today, specialty chemicals can rapidly grow in size and become commodity products. However, flexibility in innovative processes is a key issue in order to dominate the market. The time to the market of innovative solutions must be halved from the actual 15–20 years. In this context, process integration and intensification is a key area of excellence.

Over the years, chemical processes have continually improved in terms of their diversified utilization of raw materials, improved safety, and increased productivity, while minimizing wastes production and energy use. In order to reduce CO_2 emission, new quasi-zero emission processes are under study. The substitution of fossil carbon with renewable C (biomass) is one of the key strategies for controlling CO_2 emission, besides CO_2 direct utilization in chemical, fuel, and material production.

Hybrid materials (polymers made of monomers derived partly from fossil and partly from renewable C) are already playing a great role, and their use will grow with the increase in concern of population about sustainable solutions to existing ones that are characterized by a high environmental impact. Integration and exploitation of innovation are leading to highly efficient, inherently safe production technologies that minimize any environmental impact.

It is now clear that integration of biomass utilization as a feedstock into chemical processes has a great value for (i) an efficient CO_2-emission control, (ii) reduction in the extraction of fossil C (preserving resources for next generations), (iii) reduction of wastes at source, and (iv) production of more easily biodegradable or compostable new materials.

The future of humankind demands a step back from fossil C utilization and a long step toward a sustainable bio-economy: the use of biomass for producing chemicals and fuels is of topical interest, and new catalytic processes are one of the main drivers of the innovation. Catalysis offers a response to societal demand for the development of production processes that are more selective and environmentally friendly, with, as a major challenge, the substitution of the current chemistry based on fossil fuels (oil, gas, coal) with a chemistry based on biomass.

If catalysis occupies a central role in the case of a "conventional" production chain based on processing and converting fossil resources, it will play a more important role in the utilization of biomass, which, with respect to coal, oil, and gas, presents a much larger variety of chemical species with individual high structural complexity.

Cellulose, hemicellulose, starch, sugars, vegetal oils, lignin, etc. have different behaviors due to different compositions and require different technologies for their transformation into usable goods. The shift from hydrocarbons (mainly C–C and C–H bonds) to carbohydrates (C–C and C–OH bonds) demands new catalytic processes (elimination of "O") and new catalysts (able to convert multiple functional groups, most likely in more harsh reaction conditions). Furthermore, the presence of water when treating biomass-derived substrates (from the raw materials feeds themselves or generated during the reactions) can strongly alter the properties of catalysts affecting their stability. Therefore, the formulation of catalysts must be adapted to operate optimally in the presence of water.

Another issue is that in such a new frame, the integration of biotechnology and catalysis may raise new issues not common to actual systems. Raw materials can be partially converted into intermediates using biotechnological routes, and the intermediates can be then converted into the target products by implementing catalytic processes. It must be noted that the same intermediate produced by a chemical route from fossil C and by a biotechnological route starting from biomass may present a completely different pattern of low-concentration (even at parts per million level) "contaminants", which may affect downstream catalysts in a quite different way. As an example, we can mention "amino acids", which are common products in biotechnological processes but are practically absent in chemical routes. Such compounds may strongly affect catalysts by binding the catalytic centers.

A new era opens for catalysis

Further, catalysis is such a more and more multidisciplinary science, involving inorganic materials, supramolecular systems, organic chemistry, surface science, chemistry of solids, spectroscopy (operando techniques), engineering, kinetics, that scientists dealing with catalysis must be trained to a multifaceted knowledge, to a variety of techniques and technologies, and have a good command of their research objects. The following chapters give an overview of the basic concepts and objects that a scientist or an engineer must manipulate to efficiently design, understand, and implement catalytic processes.

Because catalysis is a key component of sustainable chemistry, it is therefore important to ask: "are all catalyses as sustainable as possible?" The answer to this question is itself multifaceted and encompasses issues such as the availability of elements that are required to make the catalyst, the catalyst lifetime, durability, and recyclability, as well as perhaps less obvious features such as the environmental footprint associated with catalyst preparation and disposal. In recent years, huge efforts have been done in the direction of sustainability, and catalysis is progressively cleaning the chemical industry.

Myrsini Christou, Efthymia Alexopoulou, Ioannis Eleftheriadis,
Ioanna Papamichael, Kostas Tsiotas, Mariusz Stolarski, Michal
Krzyzaniak, and Charles Themistocles

2 Terrestrial biomass production

Abstract: The aim of this chapter is to review, evaluate, and analyze sustainable biomass chains for several biorefinery options. The analysis comprised feedstock production and supply logistics chain, including storage, and referred to different regional scales (south, central, and northern Europe). This aimed to enable capture of the geographic specificities in terms of ecosystems, climate variation, land-use patterns as well as resource types, crop management, feedstock handling, and associated logistics. The identification of areas in Europe suitable for cultivation of selected crops was based on spatial distribution of parameters influencing conditions for cultivation (e.g., germination, growing, flowering, seed production). For that purpose, available climatic data sets, land-use and land-cover data sets, and elevation data sets were used. In the Eurobioref, project focus is given on specific non-food oil crops and perennial crops based on their favorable oil properties for the various green chemical products dealt in this project. New non-food oil crops were studied in field trials in Greece and Poland. The selected oil crops are grown in Europe only marginally, in small plots or in gardens for ornamental reasons. Average yields over the 3 years of trial in Greece were 2500 kg/ha of seeds for castor, 2380 kg/ha for safflower, and 1300 kg/ha for crambe. Crambe was third in the rank, with seed yields up to 1500 kg/ha in the small plots in Poland and 1000 kg/ha in the 10-ha demonstration field. Lignocellulosic crops recorded yields >15 t/ha of dry matter (apart from cardoon that was unable to survive after the 10 year) and around 9 t/ha of dry matter in average for willow, confirming their potential to efficiently exploit less favorable lands of Europe.

2.1 Introduction

There is a huge potential for agriculture to support the bio-industries as we move toward a bio-based economy. Europe is composed of different environments, which vary with factors like air temperatures, rainfall, and soil quality. No single plant species is optimal for all environments; thus, identifying promising plant species at an EU27 context to enhancing biodiversity will be necessary. The most important driving forces for the selection of plant species to be grown in the EU are currently the demand and supply for certain crops and the rules of the Common Agricultural Policy and Renewable Energy Directives. The recent specific policy targets for biofuels and bioliquids proved to have an important impact on land use.

The aim of this work is to review, evaluate, and analyze sustainable agricultural biomass chains for a biorefinery. The analysis will outline land suitability for growing several feedstocks, biomass production, and logistics including harvesting and storage and refer to different regional scales (south and central Europe). This will enable capture of the geographic specificities in terms of ecosystems, climate variation, land-use patterns as well as resource types, crop management, feedstock handling, and associated logistics. The feedstock will comprise non-food oil crops and lignocellulosic feedstocks.

2.1.1 Land availability

Land availability and sustainability of crop production are recent critical issues of immense importance. The agricultural land use in the EU is already intensive in most regions, and increased production of crops for non-food uses could cause additional pressures on agricultural lands and biodiversity, on soil and water resources, and on the food/feed markets. According to the latest studies [1, 2], the current available land is around 13 Mha, but from this land, only 20% is being used, whereas a total area of 20–30 Mha could be released for growing energy crops from 2020 to 2030, either because marginal lands will be used or because agricultural lands will be released due to agricultural crops' yield improvements. At the same time, it has been reported that with the target of 10% biofuels by 2020, 25 Mha will be needed to be cultivated with crops for biofuel production; 15 Mha will be used for liquid biofuels (biodiesel and bioethanol), 5 Mha for biogas, and 5 Mha for solid biofuels [3]. In all the above scenarios, arable lands are also included.

Taking into account only abandoned and underutilized lands, a recent study [4] suggests that an area of around 1.35 Mha of land (approximately one third of the area cultivated for biofuel feedstock production in 2010) could be dedicated for growing energy crops. This area includes abandoned cropland (~800,000 ha), fallow land in agricultural rotation – most of which is needed for agronomic purposes (~200,000 ha), other underutilized land within the current Used Agricultural Area (UAA) but not permanent grassland (~300,000 ha), and suitable contaminated sites (excluding areas suited only for afforestation). This study gets along with [5] recent position on biofuels, suggesting that 1.5–2 Mha of land (in some form) remains uncultivated since 2009.

2.1.2 The crops

A wide range of crops are available and can substitute the petrochemical feedstocks for energy, biofuels, and bio-based products. The plant database on the IENICA website (2002) [6] lists over 100 plant species with known or potential industrial applications and could be grown in existing farming systems. Some of the species are cultivated solely for one non-food application (e.g., castor); some have a range of non-food applications or their by-/co-products can be used to add further value to the primary use (e.g., hemp); some are grown for food purposes and their by-products have non-food applications (e.g., sunflower, rapeseed, wheat). A wide variety of non-food crops that could be domestically grown in EU27 countries was studied in the European research project Crops2Industry [7]. The crops were allocated in four main categories oil, fiber, carbohydrate, and pharmaceutical and other specialty crops and were evaluated for their suitability for selected industrial applications, namely oils and chemicals, fibers, resins, pharmaceuticals, and other specialty products.

In the Eurobioref project, work focus is given on specific non-food oil crops and perennial crops based on their favorable oil and chemical properties for the various green chemical products dealt in this project.

2.1.2.1 The oil crops

The oil crops under study were castor seed (*Ricinus communis* L., Euphorbiaceae), crambe (*Crambe abysinica* Hochst ex R.E. Fries, Brassicaceae/Cruciferae), cuphea (*Cuphea* sp., Lythraceae), lesquerella (*Lesquerella fendlheri* L., Brassicaceae/Cruciferae), lunaria (*Lunaria annua* L, Brassicaciae/Crusiferae), and safflower (*Carthamus tinctorius* L., Compositae). Their selection has been based on their favorable oil characteristics that will serve the biorefinery concept of this project. These crops do not compete with food crops in terms of agricultural lands as they can grow on less fertile lands, with low inputs (water, nitrogen, pesticides, etc.) In addition, they can be grown in rotation with food crops, having the advantage of being grown in good agricultural lands and at the same allowing a better management of the agricultural lands, machinery, and the human resources as well as assuring internal nutrient recycling, limitations of pests and diseases, avoidance of mono-cultures, etc.

The selected oil crops are grown in Europe only marginally, in small plots or in gardens for ornamental reasons. For most of the crops, yields are reported mainly from the USA. For certain crops like castor seed and safflower, there is an already established market in Europe with imported oils. Castor seed and safflower are the only crops that are commercialized for a number of industrial uses; castor oil is used mainly in industry for technical polymer (polyamide-11), fragrances, coating fabrics, high-grade lubricants, inks, textile dyeing, leather preservation, etc. as well as in medicine. Safflower has been known since ancient times as a source of orange and yellow dyes and food colorings, and more recently, it has been grown for oil, meal, birdseed for the food, and industrial product markets, such as paints and varnishes as well as for the oil food market.

Crambe is closely related to rapeseed and mustard and thus can be cultivated with existing agricultural methods and machinery. Crambe production would not compete directly with domestic seed oils since it would provide a substitute for erucic acid extracted from imported rapeseed. However, there is no broad commercial outlet for crambe seed; therefore, its commercial deployment depends on the market needs.

Cuphea, lesquerella, and lunaria still need experimentation on agronomic methods and plant breeding to improve crop characteristics in order to allow their industrial exploitation. The major constraint to the development of cuphea for industrial uses, apart from its frost sensitiveness, sequential maturation, and release of seeds from seed pots, is the seed shattering, stickiness, and dormancy, which is at present being studied by plant breeders. Thus, the highest priority to assure maximum seed yields is genetic and plant breeding research to obtain determinate flowering and non-shattering cultivars. Lesqerella is still under experimentation, as it is a desert crop not likely to be grown in many parts of the world. At present, lesquerella seed is not sold on any market, and genetic and breeding efforts are focused on the faster growing of the crop – which is perennial but grown as annual in southern USA – and on the improvement of its yielding capacity. Lunaria is also at the development stage. Its mechanical harvesting and cleaning of the seeds is a problem, but the major limitation to progress is the biennial nature of the plant and its high vernalization requirement. The production potential and agronomy of the crop requires further investigation, as the crop often does not thrive in large open fields. Thus, at present, commercial production of lunaria is limited to seed multiplication for ornamentals.

Adaptability and productivity of crops were tested in field trials established in Greece and Poland, representing the Continental and Mediterranean environmental zones. The crops were grown for 3 years in field trials along with rapeseed and sunflower as reference crops. Castor seed, safflower, and crambe have shown a very good establishment and produced relatively high yields in Greece, with castor seed and safflower having the leadership, whereas in the cold climate of Central Europe in Poland, only crambe seem to be appropriate crop.

Following a literature survey [8] and the results of the 3-year field trials, safflower, crambe, and castor were selected for further analysis as candidate crops for a European sustainable agriculture in a short- to medium-term time frame. Value chains starting from the vegetable oils of these crops and producing high-value monomers, short fatty acids, and fuels have been analyzed.

2.1.2.2 The lignocellulosic crops

The lignocellulosic perennial crops were selected for their low input requirements, making maintenance costs low. Such crops have permanent rooting/rhizome systems that reduce risk of erosion, which can be an important benefit in some regions. Grass species are relatively easily harvested and collected, but systems need to be improved. Short-rotation coppice (SRC) provides near-permanent vegetative cover, with only short periods every 3 years or so when the crop has been cut to ground level. With non-intensive chemical inputs and its tall and varied architecture, coppice can provide important habitat for flora and fauna. However, apart for the large fields of willow that have been established in Sweden for heat and power production, the selected energy crops have barely reached beyond the level of R&D. According to Cocchi et al. [9], in Europe, solid biomass energy crops cover about 50–60,000 ha of land, of which reed canary grass occupies around 20,000 ha mainly in Finland, willow around 20,000 ha, half of which is located in Sweden, and miscanthus 2600 ha mainly in the UK and France. Predicting the performance of perennial crops for over 15–20 years, which is the period used in the economic analyses of the crops, is somewhat speculative, unless long-term and reliable data are collected. Although there have been several sizeable pan-Europe initiatives, there is little information on crop development from ~4 years onward to the end of crops life span at 15–25 years. The lignocellulosic crops studied in this project were cardoon (*Cynara cardunculus* L., Compositae), giant reed (*Arundo donax* L., Graminae), miscanthus (*Miscanthus x giganteus* Greef et deu, Poaceae), switch-grass (*Panicum virgatum* L., Graminae), and willow (*Salix* sp, L Salicaceae).

Cardoon has originated from the Mediterranean and then was westerly distributed. It has been investigated in European level in the following projects: AIR CT92 1089 (*Cynara carduculus* L. as a new crop for marginal and set-aside lands), ENK CT2001 00524 (Bioenergy chains), and recently in the BIOCARD project. Being a perennial rain-fed crop that can be grown in marginal lands, its biomass productivity is highly variable, depending on the climate and soil conditions, as well as on the growing period. Thus, cardoon cultivation still needs investigation over a longer period before commercial yields are defined. Harvesting of the crop in order to separate the seeds from the whole plant is still under investigation.

The giant reed is an indigenous species to the Mediterranean basin, but systematic research on the performance of giant reed as an energy crop and its appropriate cultivation techniques has started in 1997. It is a very promising non-food crop for central and southern Europe due to its high biomass yields, low irrigation and agrochemical inputs, high resistance to drought, good biofuel characteristics, and because it can be stored outdoors without major losses. The high biomass yield potential of giant reed has been confirmed in all trials conducted throughout Europe. One of the most critical points of *A. donax* cultivation, which influences productivity and economical viability, is the establishment of the plantation. Giant reed has been investigated in European level in the following projects: FAIR3-CT96-2028 (*A. donax* network), ENK CT2001 00524 (Bioenergy chains), and recently in the EUROBIOREF (www.eurobioref.org), OPTIMA (www.optima-fp7.eu), and BIOLYFE (www.biolyfe.eu) projects. In BIOLYFE project, giant reed was chosen along with fiber sorghum, miscanthus, and switch-grass. Crop cultivation will be at the demonstration level (~25 ha per feedstock) in order to demonstrate the whole supply chain, from feedstock sourcing via fuel production to product utilization. The main result will be the construction of an efficient second-generation industrial demonstration unit with an annual output of about 40,000 tons of lignocellulosic bioethanol.

Miscanthus originates from East Asia, and it was introduced in Europe in the 1930s. Up to now, several fields of miscanthus have been established in many southern and northern European countries. A number of R&D projects have been conducted dealing with cultivation, biomass potential, biofuel characteristics, and other aspects, as mentioned before. The main reasons that miscanthus has gained interest in energy market are its high biomass potential, perennial nature, low inputs, high nutrient and water use efficiency, and good biofuel characteristics (i.e., low moisture content at harvest time in spring). Miscanthus has been investigated in several EU projects such as FAIR CT97 1707, FAIR 1392, AIR1 CT92 0294, ENK CT2001 00524 (Bioenergy chains), and recently in the EUROBIOREF, OPTIMA, and OPTIMISC (http://optimisc.uni-hohenheim.de/92383?L=1) projects.

Switchgrass is native to North America – where it is thoroughly investigated. Switchgrass is more suitable for central and southern Europe and is characterized by the following advantages: easy establishment by seeds; high biomass potential; high competitiveness to weeds once it is well established; high nutrient and water use efficiency; easily harvested with existing equipment; long harvest window expanded from late autumn to early spring; low moisture content at late harvest; good combustion qualities of the biomass; high genetic variability. Switchgrass is a very promising crop for energy production. Management issues such as establishment, time and frequency of harvest, and nitrogen and fertilization practices considerably affected the utilization of switchgrass as a bioenergy crop. Switchgrass has been investigated in European level in the following projects: FAIR 5-CT97-3701 Switchgrass, ENK CT2001 00524 (Bioenergy chains), and recently in the EUROBIOREF and OPTIMA projects.

Willow is characterized as an ideal woody crop because it has several characteristics such as high yields obtained in a few years, ease of vegetative propagation, a board genetic base, a short breeding cycle, and the ability to re-sprout after several harvests. Willow is best suited for the northern part of Europe, and it is grown mainly in Sweden, UK, Finland, Denmark, Ireland, and the Netherlands. Several R&D projects dealing with willow cultivation aspects have been carried out since 1957. Since 1991, willow production has been commercialized.

Cardoon, giant reed, and miscanthus were tested for their productivity in field trials established in Greece. The crops were harvested every year in February when they were in their 9th, 10th, and 11th growing periods, respectively. Willow was studied in Poland in a 3-year field. The crops performed well in Greece and Poland, proving their good adaptability and yielding capacity. The focus of this work was placed on giant reed as candidate crop for a European sustainable agriculture in a short- to medium-term time frame.

2.2 Land suitability

2.2.1 Oil crops

2.2.1.1 Data sets

The identification of areas suitable for the cultivation of safflower, crambe, and castor in Europe was based on the spatial distribution of parameters influencing crop growth (e.g., germination, growing, flowering, seed production). For that purpose, finding available data sets for these parameters was necessary, as well as the classification in classes based on specific references available in literature: climatic data sets, land use-land cover data sets, and elevation data sets.

Climatic data sets are available for downloading at the website of the European Climate Assessment and Dataset project, providing information about changes in weather and climate conditions as well as the daily data set needed to monitor and analyze these conditions. E-OBS is a daily gridded observational data set for precipitation, temperature, and sea-level pressure in Europe, available for download at http://eca.knmi.nl/. After selection, the data included in the downloaded data set reference were daily precipitation and daily temperature (average, minimum, maximum).

Land use and land cover data are derived from reliable databases, like EEA's database and Eurostat. EEA's Corine Land Cover program provides data set about land cover in Europe, classified in specific classes. Corine Land Cover 2000 data, version 16 (04/2012), are available for download at http://www.eea.europa.eu/. Selected attributes from this data set refers to areas where the crops could be established. More specifically, arable lands are classified in two main classes: (a) agricultural areas, arable land, and non-irrigated arable lands and (b) agricultural areas, arable land, permanently irrigated lands.

The Eurostat's LUCAS (Land Use/Cover Area frame Statistical survey) data set provides information about cover/land use as well as agro-environmental and soil data identified through on-site observations of spatially selected geo-referenced points. LUCAS 2009, Land Use/Cover Area frame statistical survey is available for download at http://epp.eurostat.ec.europa.eu/portal/page/portal/eurostat/home.

ETOPO1 is a 1-arc-minute global relief model of the Earth's surface developed by the National Geophysical Data Center of the National Oceanic and Atmospheric Administration (NOAA) in the USA. Land topography and ocean bathymetry are included in this data set. In the further process, only land topography data will be used. 1-Minute Gridded Global Relief Data (ETOPO_1) is available for download at http://www.ngdc.noaa.gov/mgg/global/.

2.2.1.2 Determination of selection criteria

The selection process for the identification of agricultural areas in Europe suitable for the cultivation of the oil crops was based on the determination of specific criteria that influence sowing, germination and growth of the oil crops, and finally, the production of seeds and oil. The selection criteria, which are based on literature, for each crop are presented and detailed in Table 2.1.

Castor seed is a tropical season crop and cannot tolerate temperatures as low as 15°C. It needs a frost-free period of 5–8 months and 450–1000 mm of well-distributed rainfall during the growing season [10]. Previous studies [11, 12] suggested that soil temperature should be the determining factor for planting. Soil temperature for germination should be in the range of 18–23°C, but is also possible in the 12–18°C

Table 2.1: Selection criteria for the oil crops (castor, crambe, and safflower).

Parameters	Castor	Crambe	Safflower
Climatic[a]	1000 mm ≥ R_{total} ≥ 300 mm FFP not ≥ 120 days T_{min} > 15°C T_{max} < 40°C, during period of flowering[b]	800 mm ≥ R_{total} ≥ 380 mm, during the growing season T_{min} ≥ −4°C, during seedling stage and early flowering[c] 25°C ≥ T_{avg} ≥ 15°C, during main vegetative period	800 mm ≥ R_{total} ≥ 380 mm, during the growing season 16°C ≥ T_{avg} ≥ 15°C, during period of seed germination[d] 25°C ≥ T_{avg} ≥ 20°C, during period of flowering and seed formation[e]
Topographic	Alt < 1100 m	Alt < 1100 m	Alt < 1100 m
Soil	SD > 0.5 m Well-drained soils Texture: loam to sandy loam Moderate coarse to fine texture Not tolerant to alkali soils Tolerant to semi-arid soils 5.5 ≤ pH ≤ 8 or more pH < 5 have to be limed	SD > 1 m Well-drained soils Texture: sandy, sandy-loam, clay-loam, loam 5.5 ≤ pH ≤ 8	SD > 1 m Well-drained soils Texture: sandy, sandy-loam, clay-loam, loam 5.5 ≤ pH ≤ 8

[a] Climatic data (precipitation and temperature) only for the year 2012.
[b] Flowering period for castor was determined from day 152 to day 181 (June).
[c] Seedling stage until early flowering for crambe was determined from day 305 to day 336 (December) for autumn sowing and from day 106 to day 136 (15 April to 15 May) for spring sowing.
[d] Seed germination period for safflower was determined from day 274 to day 304 (October).
[e] Flowering and seed formation period for safflower was determined from day 121 to day 151 (May).
T_{avg}, daily average temperature (°C); R_{total}, total annual rainfall (mm); Alt, altitude (m); SD, soil depth (m); FFP, frost-free period; T_{min}, daily minimum temperature (°C); T_{max}, daily maximum temperature (°C); R_{total}, total annual rainfall (mm).

range. Temperature should remain below 40°C during flowering to avoid failure of cross-pollination. The crop requires 120–140 days from planting to maturity.

In this process, May was determined as the germination period.

While castor requires adequate soil moisture during pod set and filling, a subsequent dry period as the plant matures promotes high yields. Temperatures above 35°C and water stress during the flowering and oil formation as well as early harvesting of immature plants can reduce the seed oil content [13].

Well-drained, deep (at least 1.5 m) fertile soils of moderately coarse to fine texture with a pH of 6.0 to 7.0 or slightly higher are best suited for castor production. Silt-laden and clayed soil (from pH 5.5 and up to >8 are tolerated) give good results. The crop will tolerate draught and semi-arid soils, but the yield is lower [14]. Rainfalls of 1000–1400 mm favors the growth of the castor crop, while maturation is favored if the rain is followed by a dry period lasting a few months.

A similar process was followed for the determination of areas having ecological requirements suitable for cultivation of crambe. *Crambe abyssinica* is essentially a cool-season crop but could potentially be a spring crop when grown in the Corn Belt of the USA [15, 16]. As a winter crop, it showed a good potential in areas of central and South Italy. Crambe is susceptible to frost and moderately tolerant to saline soils during germination [17, 18].

Crambe sowing date is a crucial factor that affects seed yields and oil content [15, 17, 19–21]. As reported by Mastebroek et al. [19] in several studies, it was revealed that high temperatures before anthesis accelerated crop development, stimulated early flowering, and reduced the number of flowers and seed yields. In the same study, it was reported that low temperatures until July extended the period before anthesis, and consequently, a high number of branches was formed. The effect of environmental conditions on seed yield and oil content may hamper varietal selection; however, this effect does not apply to the content of erucic acid and glucosinolates. Crambe should be sown in late April to May when the frost risk has passed in the colder climates of North Dakota/USA [18]. In the low desserts of the southern USA, the highest yields of crambe and rape were planted in November, which, however, had caused plant lodging, which was partly due to the extended period of growth [17]. Morrison and Stewart [21] reported that flowering of rape is inhibited above 27°C; high temperatures in mid-May lead to early flowering, thus reducing plant vegetative growth. However, high temperatures between the end of May to the beginning of June may lead to incomplete seed filling [15]. Therefore, the sowing date is important to avoid high temperatures at the end of the growing season. The crop requires an average of 54 days from sowing to flowering (ranges from 42 to 100 days) [16]. On the whole, sowing time depends on location and climatic conditions, but as a general rule, advanced sowing favors higher yields [15]. In temperate climates, it could be sown from September to November like rape, whereas in colder climates, it is advised to be sown in late April or May, as a spring crop. Under favorable conditions, two crambe crops could be harvested in the same year if the first crop is sown in early spring and the second about mid-July [15].

The result of the land suitability analyses is the calculation of spatial distribution of average temperature suitable for seed germination of crambe for two different planting dates, April and May.

Safflower is frost-tolerant in the seedling stage, withstanding temperatures of −7°C; however, it does best in areas with warm temperatures and sunny, dry conditions during the flowering and seed-filling periods [22]. Early spring sowing is done in April/May in areas that have at least 120 days of frost-free periods and hot summers. Planting prior to April 10 usually shows no advantage since cool soil temperatures (below 4°C) prevent germination and encourage seedling blight. Planting after May 20 increases the risk of fall frost injury and diseases that reduce seed yield and quality. The crop may not mature if planted after mid-May [22]. In the temperate regions of the Mediterranean basin (Greece, Turkey, and Lebanon), safflower can be sown either in October-December as a winter crop or in March-April as a spring crop [23–27].

This crop does best in areas with warm temperatures and sunny, dry conditions during the flowering and seed-filling periods. Yields are lower under humid or rainy conditions since the seed set is reduced and the occurrence of leaf spot and head rot diseases increases. Consequently, this crop is adapted to semi-arid regions. Deep, fertile, well-drained soils that have a high water-holding capacity are those ideal for safflower. This crop is also productive on coarse-textured soils with low water-holding capacity when adequate rainfall or moisture distribution is present. Soils that crust easily can prevent good stand establishment. High levels of soil salinity can decrease the frequency of seed germination and lower seed yield and oil content. Safflower has approximately the same tolerance to soil salinity as barley [22].

Early planting allows the crop to take full advantage of the entire growing season [28]. Further to that, under water-scarce regions such as the Mediterranean, spring-sown safflower is more sensitive to water than winter-sown safflower. In addition, winter sowing is more preferable to spring sowing in order to meet vegetable oil requirements [27].

For that purpose, it was necessary to use data set of spatial distribution of daily average temperature in order to calculate the average temperature for three different planting dates across Europe. More specifically, it was calculated for potential planting in October, in April, and in early May. The result of the analyses is the calculation of spatial distribution of average temperature suitable for seed germination of safflower for three different planting dates: October, April, and early May.

The first step of the selection process was to exclude all areas not having climatic data for all days of the year 2012, like Sicily in Italy and Peloponnese in Greece. This was decided in order to avoid wrong estimations or calculations for variables like the total annual precipitation. All geo-data were plotted using the European Terrestrial Reference System 1989 (ETRS89). In order to meet the requirements of climatic criteria concerning the total precipitation during the growing season and the average temperatures during seed germination, flowering, and seed formation for the crops, it was necessary to calculate the total precipitation only for the growing season and not for the whole year, summarizing daily data only for the days of the growing season. Thereafter, topographic data from global elevation data set were selected and plotted after "masking" on boundaries of EU27+ areas with elevation lower than 1100 m.

Intersection between arable lands and areas matching the climatic criteria was also used for selection of arable irrigated and non-irrigated lands with precipitation during growing season suitable for cultivation of safflower, crambe, and castor. The next step of this work was to include temperature requirements in the selection process. To do so, a crucial parameter had to be taken into account: the sowing date of the crops, which is crop-specific.

On the final phase of this work, previously produced geo-data were analyzed and the identification of agricultural areas suitable for the cropping of specific plants was the expected result. Data used in the process were (i) the spatial distribution of arable lands in EU27+, where the precipitation level is suitable for cropping of three specific crops (safflower, crambe, and castor) and (ii) the spatial distribution of average temperature suitable for seed germination of all three crops, taking into account case-by-case scenarios about their sowing periods. In that way, the combination of selection parameters was activated, providing the opportunity for more detailed determination of lands for cultivation.

Figures 2.1, 2.2, and 2.3 present the arable agricultural lands in EU27+ suitable for castor seed, crambe, and safflower cropping, taking into account climatic and topographic parameters, as well as possible scenarios about their germination period as presented in Table 2.1.

In conclusion, safflower is an annual oilseed crop grown throughout the semi-arid region of the temperate climates in many areas of the world. Based on its climate requirements, it can be grown mostly in Bulgaria, Romania, France, Italy, Greece, and Spain.

Crambe is an annual cold-season oilseed crop that belongs to the same family as rapeseed; thus, it can be grown in areas where rapeseed grows. Based on its climatic requirements, crambe can be grown in mostly in Poland, Germany, France, Romania, Spain, Italy, Bulgaria, Czech Republic, Hungary, UK, Denmark, Greece, Sweden, Lithuania, Slovakia, Turkey, Finland, Austria, Latvia, and Portugal.

Castor is an annual oilseed crop of tropic origin. Based on its climatic requirements, castor could be grown in mostly in Romania, Spain, France, Hungary, Italy, Bulgaria, Turkey, Greece, and Portugal.

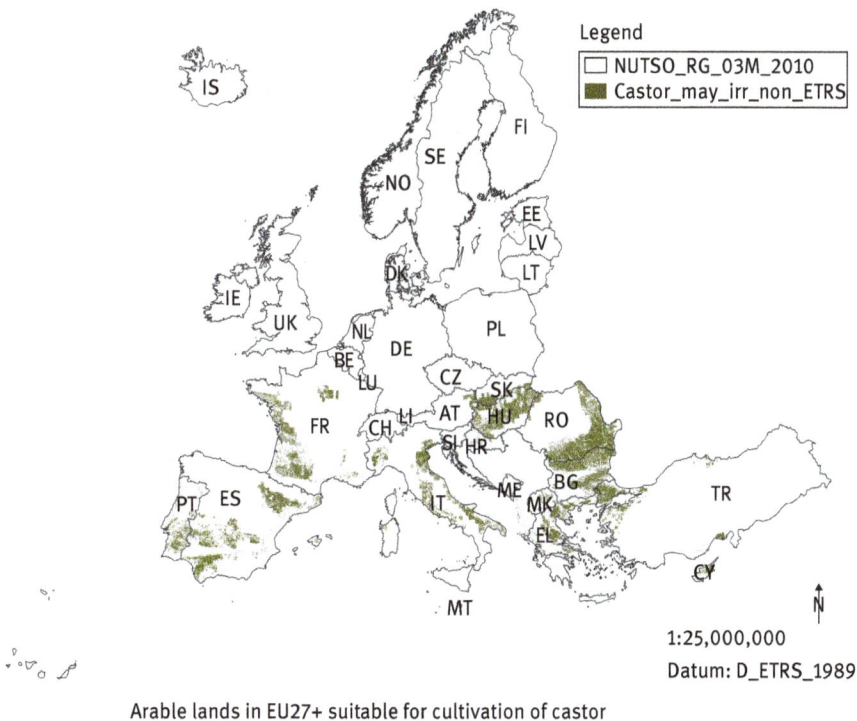

Arable lands in EU27+ suitable for cultivation of castor

Fig. 2.1: Arable agricultural lands in EU27+ suitable for cultivation of castor.

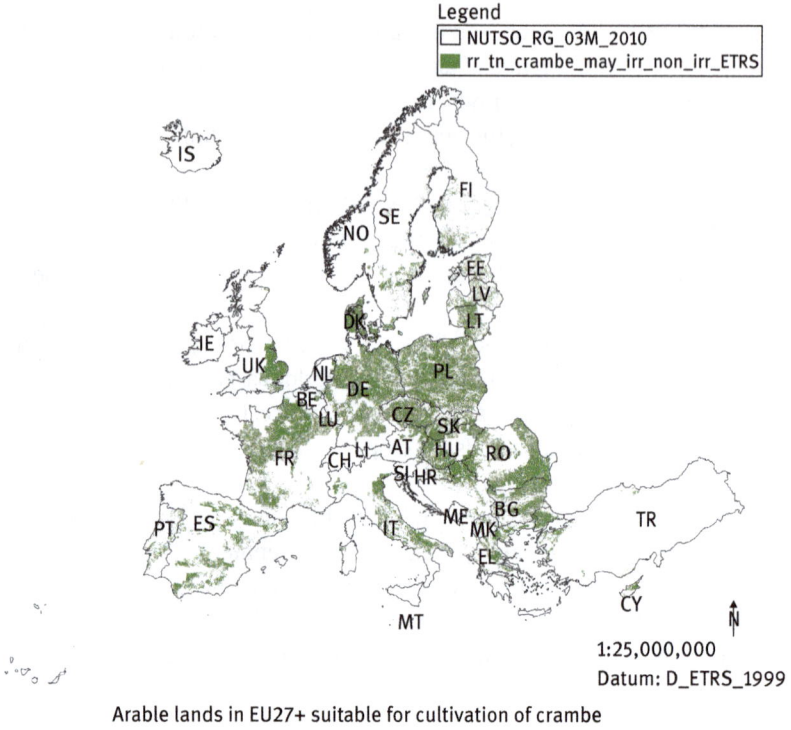

Legend
☐ NUTSO_RG_03M_2010
■ rr_tn_crambe_may_irr_non_irr_ETRS

1:25,000,000
Datum: D_ETRS_1999

Arable lands in EU27+ suitable for cultivation of crambe

Fig. 2.2: Arable agricultural lands in EU27+ suitable for cultivation of crambe.

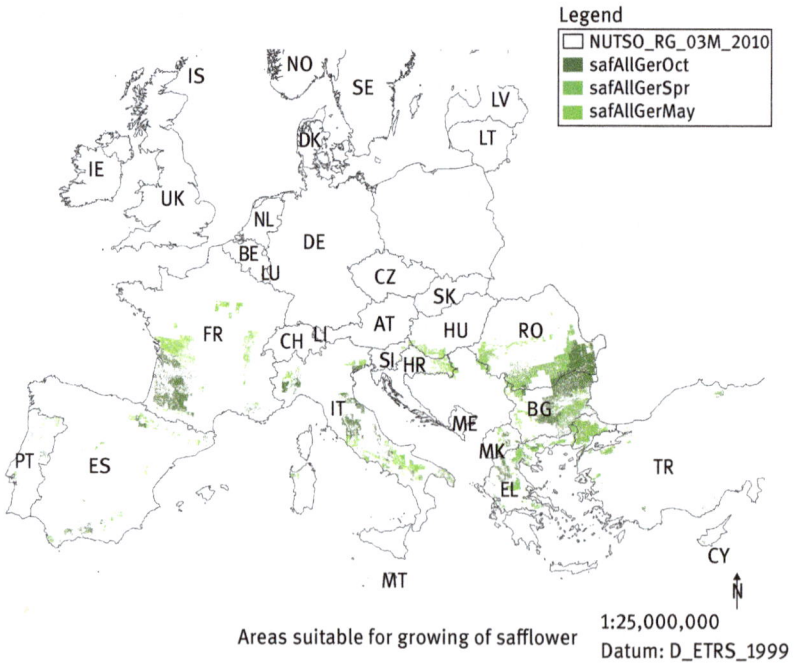

Legend
☐ NUTSO_RG_03M_2010
■ safAllGerOct
■ safAllGerSpr
■ safAllGerMay

1:25,000,000
Areas suitable for growing of safflower   Datum: D_ETRS_1999

Fig. 2.3: Arable agricultural lands in EU27+, suitable for growing of safflower.

It has to be noted here, however, that although crops may appear suitable to be cultivated in a range of environments, for instance, castor as north as Romania, there is no evidence of potential seed and oil yields in these environments, which is a decisive parameter for the cultivation of the crop in larger scale. Consequently, the maps are indicative of the potential cultivation of the crops in several European environments, which has to be supported by appropriate research in the future.

2.2.2 Lignocellulosic crops

Giant reed is a warm-temperate or subtropical species, but it is able to survive frost. When frosts occur after the initiation of spring growth, it is subject to serious damage [29]. It tolerates a wide variety of ecological conditions. Estimations on the land suitability for giant reed were based on their main climate requirements and existing literature [30]. Giant reed can be grown in the Mediterranean north, Mediterranean south, Lusitanian, and Atlantic central climatic zones (Figure 2.4). Earlier trials for growing

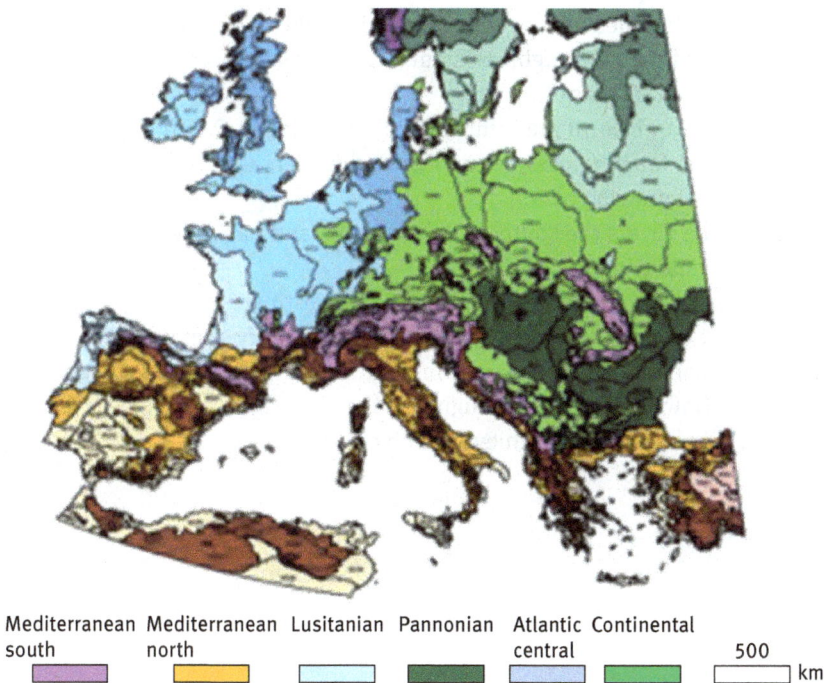

Mediterranean Mediterranean Lusitanian Pannonian Atlantic Continental
south   north   central   500 km

Fig. 2.4: Crop suitability for the environmental zones of Europe according to Metzger et al. [31]. The Continental zone covers most part of Europe and is characterized by high temperatures in summer and very low in winter, followed by relatively high precipitation.

giant reed in Germany and UK showed that adaptation of *A. donax* L. in Germany and UK had not encountered any problems due to low temperatures over the first winter. This might be an effect of the moderate air temperatures prevailing in January and February in these regions. However, the crop did not appear to naturally dry over the winter or to flower. This may be a potential problem for growing *A. donax* L. in North European conditions, as there could be a lack of sufficient nutrients sequestered to the rhizomes for the following year's growth. Besides, none of the populations grown in the UK and Germany show as good growth and yield characteristics as in Mediterranean climatic conditions. Undoubtedly, it was mainly owed to the prevailing climate, which was rather unfavorable for the newly inserted giant reed populations [33].

Giant reed prefers well-drained soils with abundant soil moisture. It can withstand a wide variety of climatic conditions and soils from heavy clays to loose sands and gravelly soils and can tolerates soils of low quality such as saline ones. Giant reed is classified as a mesophyte or almost a hydrophyte or xerophyte. These classifications were given because it can survive not only under very wet conditions, but also under very dry conditions for longer periods. Commonly, it is referred to as a drought-resistant species because of its ability to tolerate extended periods of severe drought accompanied by low atmospheric humidity. This ability is attributed to the development of coarse drought-resistant rhizomes and deeply penetrating roots that reach deep-seated sources of water.

Atlantic Central and North are influenced by the Atlantic Ocean and the North Sea, with rather low temperatures in summer and winter, abundant rainfalls, and satisfactory length of growing period.

Lusitanian zone covers the southern Atlantic area and has rather high summer temperatures and mild winters.

Mediterranean North and South cover the southwest part of Europe and North Africa. It has short precipitation periods and long hot and dry summers. The length of the growing season is long, and the air temperature is favorable for growing a wide number of crops. However, summer drought is a limiting factor that imposes the use of irrigation for crop survival and achieving high crop yields.

2.3 Crop setting up

2.3.1 Oil crops

2.3.1.1 Soil preparation
Castor. It is essential for castor to have a well-prepared seedbed with adequate moisture; thus, deep ploughing or sub-soiling is recommended in light soils, whereas in heavy soils, ploughing at 30–50 cm give better results [10]. Soil preparation consists of

Fig. 2.5: Crambe soil preparation.

deep ploughing, integration of organic fertilization (if available), ploughing (tillage) with a disk plough or chisel, and earth clump spraying with a cover crop.

Crambe. Crambe crops are managed in the same way as rapeseed crops. Machinery used for tillage, planting, spaying, and harvesting crambe is similar to that used for small grains. Soil preparation includes winter ploughing, cultivator, basic fertilization, and harrowing.

In spring 2013, a commercial field of crambe (*C. abyssinica*) was established in the northeast of Poland, at the Didactic and Research Station in Łężany (53°35′ N, 20°36′ E), which is a unit of the University of Warmia and Mazury in Olsztyn (Figure 2.5). The soil was prepared according to good agricultural practice. The following procedures were conducted before sowing in the spring 2013: ploughing to a 20-cm depth, harrowing, and cultivating (shortly before sowing). Crambe was sown on a total area of 10 ha.

Safflower. Soil preparation is carried out in September/October for autumn sowing and February for spring sowing. It is done when the soil is at an optimum stage, to ensure optimal ventilation and moisture conditions for seed germination and the best possible soil fragmentation. The favorable conditions of humidity and ventilation along with the favorable temperature (20°C) and good crushing of soil are the basic requirements for quick and uniform germination of seedlings and a normal development of the crop. This crop is also productive on coarse-textured soils with low water-holding capacity when adequate rainfall or moisture distribution is present. High levels of soil salinity can decrease the frequency of seed germination and lower seed yield and oil content. Safflower has approximately the same tolerance to soil salinity as barley [22].

Soil preparation should be similar to wheat and other small grains cultivation using the same machines and consists of winter ploughing, soil cultivation, harrowing with a riper, disk harrowing for better soil fragmentation, and soil leveling.

2.3.1.2 Sowing

Castor. Castor is sown in March-April (as spring crop) and harvested in October.

The most suitable equipment is the conventional corn sower with a disk diameter properly modified to fit the castor oil plant seeds. The most appropriate equipment is the standard corn drill with 6 mm plates and integrated fertilization.

Seeds should be planted 5–8 cm deep, but can range from 0.6 cm in humid regions and up to 2.54 cm in drier areas, depending on the texture and condition of the soil. Castor can be planted with a seeding rate of 11 to 16 kg/ha and spacings of 100 × 20, 100 × 25, 100 × 90, and 100 × 50 cm. Seed densities can vary from 20,000 to 30,000 plants kg/ha. Special care must be taken to prevent crushing the fragile seed in the planter box [11]. In the castor large field established within the Eurobioref project in Madagascar, a plant density of 60 × 80 cm was applied, resulting in 18,000 plants kg/ha, at a soil depth of 5 cm.

Castor seeds are large and slow to germinate; emergence of the seedlings may take 7 to 14 days.

Crambe. Sowing time depends on location and climatic conditions, bus as a general rule advanced sowing favors higher yields [15]. In temperate climates, it could be sown from September to November like rape, whereas in colder climates, it is advised to be sown as spring crop sown in late April or May. Under favorable conditions, two crambe crops could be harvested in the same year, if the first crop is sown in early spring and the second about mid-July [15].

According to Enders and Schatz [18], the recommended seeding rate is 17–22 kg/ha, which results to 2500000 plants/ha, because crambe plants are more competitive with weeds and mature uniformly. Lower seeding rates resulted in lower plant densities but better yields due to increased plant branching and extended flowering period. Sowing could be done with mechanical broadcasting or in rows 15–35 cm apart in irrigated areas where weeds are not a problem and 45–60 cm apart in drier areas. In Italy, sowing density of 50 seeds/m² with an inter-row spacing of 40 cm showed good results, whereas inter-row spacing wider than 75 cm could result to plant lodging, which makes harvest difficult [15].

Planting depth is a critical factor in obtaining good crambe yields. Seed should be planted 0.6 cm deep in humid regions and up to 2.54 cm deep in drier areas.

In a large field of crambe in Poland, seeds at 15 kg/ha were sown by drill and the inter-rows were 12.5 cm (Figure 2.6).

Safflower. In the temperate regions of the Mediterranean basin (Greece, Turkey, and Lebanon), safflower can be sown either in October-December as a winter crop

(a)   (b)

Fig. 2.6: Crambe sowing (left) and in the early growth stage (right).

or in March-April as a spring crop [23–27]. Early planting allows the crop to take full advantage of the entire growing season [28].

Early spring sowing is done in April/May in areas that have at least 120 days of frost-free periods and hot summers. The crop is frost-tolerant in the seedling stage, withstanding temperatures of −7°C and is typically grown at less than 1100 m altitude. Planting prior to April 10 usually shows no advantage since cool soil temperatures (below 4°C) prevent germination and encourage seedling blight. Planting after May 20 increases the risk of fall frost injury and diseases that reduce seed yield and quality. The crop may not mature if planted after mid-May [22].

It is reported that summer sown safflower is more sensitive to water than winter sown safflower; thus, under water scarce regions as in the Mediterranean region, winter sowing is more preferable than summer sowing in order to meet vegetable oil requirements [27].

Sowing is done with the sowing machinery for small grains. Safflower seedlings are not vigorous. Soil crusting can be a major deterrent to adequate stand establishment. Planting depths of 3–4 cm are optimum. Recommended seeding rates are from 7 to 14 kg/ha of pure live seed. Several densities can be applied, with rows 24 cm apart (9–10 kg/ha of seeds), 48 cm apart (5–6 kg/ha of seeds), rows 75 cm away and 4 cm apart within the row, with pneumatic sowing machine (3–4 kg/ha of seeds). Usually, it is planted in 15 to 25 cm row spacing. Row spacing greater than 35.56 cm increase air movement and penetration of sunlight into the crop canopy. This may reduce leaf disease incidence but can favor weed competition, delay maturity, and decrease branching and seed oil content. Narrow rows are best for competing with weeds and usually result in more uniform stands that mature earlier [22]. In Greece, 250,000 plants/ha have been recommended [24, 25].

In the large field established in Greece (Figure 2.7), several sowing dates were tested: October, November, and March.

(a)　　　　　　　　(b)

Fig. 2.7: Safflower large field in early growth.

Sowing was done with the sowing machinery for small grains.
Several densities were applied.

a. at rows 24 cm apart (9–10 kg of seeds/ha)
b. at rows 48 cm apart (5–6 kg of seeds/ha)
c. at rows 75 cm away and 4 cm apart within the row, with pneumatic sowing machine (3–4 kg/ha of seeds)

2.3.1.3 Irrigation

Castor. Irrigation is necessary for castor. Depending on the intensity of atmospheric evaporation and the water-retaining capacity of the soil, 600 mm of water during the growing period of 4.5 months is required. Under normal conditions, 12 to 14 days between irrigations should keep plants from stressing for moisture, but high temperatures and high winds during the peak growing and fruiting periods may cause the plants to need more frequent irrigation. Castor requires 20–25 cm/ha of water annually to produce high yields. The time of last irrigation is usually from 1 to 10 September [11].

Safflower. Safflower was significantly affected by water stress during the sensitive late vegetative stage [27]. The highest seed yields of 5220 kg/ha were obtained in non-stressed conditions, which, according to the author, included the following irrigation schedule: the first irrigation at the vegetative stage, when after 40–50 days from sowing/elongation and branching stage, which is the end of May; the second irrigation is at the late vegetative stage, after 70–80 days from sowing/heading stage, which is in the middle of June; the third irrigation is at the flowering stage, approximately 50% level, which is the first half of July; and the fourth irrigation is at the yield formation stage, seed filling, which is the last week of July.

In a study [34], it is reported that moderate stress induced an increase in total lipid content in all lipid classes. However, severe water deficit induced a sharp decrease in the total lipid content. Concerning the fatty acid composition, water deficit induced a decrease in their degree of unsaturation, expressed by a reduction in the proportions of linolenic (18:3) and linoleic (18:2) acids and most lipid classes. Overall, tolerance of safflower toward drought is expressed by structural modifications, which allow the plants to adjust their membranous fluidity by not only an appropriate rearrangement of their glycerolipids but also by an adjustment of their unsaturated fatty acid composition.

2.3.1.4 Fertilization

Castor. Fertilization requirements vary according to location. In general, castor responds well to phosphate application; thus, 30–60 kg/ha is usually applied before or at sowing. If the soil is deficient in nitrogen, then 90–135 kg/ha of nitrogen is needed for maximum yields. Depending on soil fertility, 30–60 kg/ha of nitrogen is recommended in two applications; half applied at sowing and the rest just before the flowering period [11]. When castor follows alfalfa or a heavily fertilized previous crop, fertilization may not be needed, as it has a strong rooting system that extracts the nutrients from the soil. In the project trial, 150 kg/ha NPK fertilizer was used.

Crambe. Fertilizer requirements are similar to other spring oilseed crops; best results may be achieved with around 150 kg/ha of nitrogen applied to seedbeds [6]. Reports of fertilizer trials on Crambe are rare, other than to establish whether or not fertilizers are necessary, and application levels, types, and timing must be locally determined. Crambe can be deep-rooted and thus draw on soil nutrients at depth [13]. P_2O_5 (50 kg/ha) and K_2O (90 kg/ha) are recommended for basic fertilization. Crambe also responds to nitrogen fertilizer with approximately 90 to 112 kg/ha of the actual N recommended [16].

In the large field of crambe established in Poland, before the experiments were set up, phosphorus as triple superphosphate was applied at 40 kg/ha P_2O_5, potassium as potassium chloride at 60 kg/ha K_2O, and nitrogen as ammonium nitrate at 40 kg/ha. Subsequently, another dose of the same nitrogen fertilizer was applied as top dressing at 60 kg/ha.

2.3.1.5 Weeding

Castor. Weeding is required just after sprouting, as the castor oil plant sprouts are sensitive. It is suggested to apply trifluarine pre-sowing at 3 L/ha and a few days before sprouting, Round Up (glyphosate) or Basta F1 at 5 L/ha. Weeding should be done for 2 months, January and February. Mechanical or manual harrowing should be done to avoid weed development. Trophee/Harness (5 L/ha) or Adengo (2 L/ha) was used as post-sowing weeding, whereas during cultivation, Fusillade Max (2 L/ha) was applied for the Graminae weeds.

Crambe. Crambe has been found to be susceptible to *Alternaria* and *Sclerotinia*, and a well-timed fungicide application at the mid-flowering stage has had a yield response (up to 1 t/ha) and may also improve oil content. Fungicide-dressed seeds may also beneficial. Plants are susceptible to the same range, pests, and diseases as those of oilseed rape including beet cyst nematode (*Heterodera schachtii*) [6]. In the large field of the project, spraying with glyphospate at 5 L/ha was applied before the crambe plantation was set up. Subsequently, immediately after sowing, the soil-applied herbicide Butizsan Star 416 SC was applied at 2.7 L/ha.

Safflower. N levels affected the crop phonological stages from full bloom to maturity as well as the grain filling period. The application of 100 and 200 kg/ha of N decreased the number of days required for safflower to reach full bloom compared with no nitrogen application. In contrast, maturity was delayed with N application, which caused an increase in the grain filling period by an average of 20%. N fertilization increased seed yield by an average of 19%, the seed weight per plant by 60%, the number of heads per plant by 32%, and the number of seeds per plant by 41% compared to no fertilization, under rain-feed conditions [24, 25].

In another study [35], it was demonstrated that safflower has the ability to use residual soil nitrogen efficiently and also to compensate for low plant densities in low-input farming conditions in temperate climates. The ability of safflower to remove N from the soil was also reported by [36].

Basic fertilization applied in the field trial was NPK fertilizer, 200 kg/ha (16-20-0). The quantity was determined after a soil analysis at several parts of the field.

2.3.1.6 Crop rotations

Castor. Best maize yields are obtained after castor in crop rotation, which confirms the synergy between a non-edible crop and an edible crop on food production cropping systems [10].

Crambe. Crambe is advised to be grown in 4-year crop rotation schemes to keep insect, disease, and weed pressure to a minimum [15, 18]. Crambe should follow small grains, corn, grain legumes, or fallow, while it can be sown as companion crop for alfalfa and other biennial or perennial forage-type legume establishment [18]. It should never follow crambe or other akin crops, such as colza, rape, or wild mustard [15]. Small grains should perform well following crambe. The stubbles of crambe provide cover for trapping snow, controlling erosion, and establishing winter crops in a no-till production system. In the latter case, care must be taken to minimize stubble disturbance because they break easily [18].

Safflower. Safflower most often is grown on fallow or in rotation with small grains and annual legumes. Safflower should not follow safflower in rotation or be grown in close rotation with other crops susceptible to the disease sclerotinia (white mold). These crops include dry bean, field peas, sunflower, mustard, crambe, and canola/rapeseed [22]. A crop following safflower should be grown only if there has been a significant recharge of soil moisture. Very little crop residue remains after harvesting safflower. Therefore, reduced tillage or chemical fallow after safflower may help reduce wind and water erosion of the soil. The production practices and equipment needed for safflower are similar to those used for small grains [22].

2.3.2 Lignocellulosic crops

2.3.2.1 Soil type requirement

Arundo donax tolerates a wide variety of ecological conditions. It prefers well-drained soils with abundant soil moisture. It can withstand a wide variety of climatic conditions and soils from heavy clays to loose sands and gravelly soils and tolerates soils of low quality such as saline ones too [29].

Fig. 2.8: The furrow opener.

(a) (b)

Fig. 2.9: Rhizome planting and plants sprouted from rhizomes.

2.3.2.2 Soil preparation

Arundo donax has no special soil preparation requirements. Prior to the establishment, the field is ploughed, sub-soil tilled, milled, and fertilized with basic fertilization. In the large field of giant reed, the applied fertilizer was an 11-15-15 one and the application dose was 500 kg/ha. The furrows for the rhizome planting were opened with a carried two-strips furrow opener at distances 1.6 m (Figures 2.8 and 2.9).

2.3.2.3 Planting

The establishment is the most critical point of *A. donax* cultivation and has strong influences on productivity and economical viability. The two main factors determining establishment success and costs are the propagation material and the planting density. Because of seed sterility, only vegetative propagation is foreseen for the commercial production of *A. donax*.

The large thick-woody rhizomes have to be divided. Each rhizome section should have one to three viable and well-developed buds. Rhizomes should be placed in rows 1.5 m apart and 70 cm within the row. Then rhizomes have to be covered with soil in a depth of about 10–15 cm. Care should be taken to ensure that they do not dry out, especially during the first few weeks.

(a)   (b)

Fig. 2.10: Plants sprouted from whole stem planted on furrows.

Planting of rhizomes, whole stems, and stem cuttings have been tested (Fig. 2.10) but appropriate machinery for these operations is not yet available [37, 38]. In the tests done so far, rhizome establishment turned out most promising. The planting of large rhizome pieces with well-developed buds directly into the field early in spring in southern European areas had nearly 100% success [39]. However, this is a very costly labor-intensive method, as this includes digging the rhizomes, transporting them to the site, keeping them wet for a certain period, cutting them in smaller pieces, and then planting them in the new field.

The in vitro response of *A. donax*, the evaluation of micropropagation as a commercial propagation technique, and the ex vivo acclimatization of macropropagated plantlets were studied in the frame of the "Giant Reed Network". The study of the suitability of in vitro culture showed that shoots can be propagated satisfactorily from axillary buds of mature giant reed plants. In addition, the rooting and acclimatization results were extremely satisfactory, entailing that micropropagation is an accessible and efficient method for giant reed mass production.

2.3.2.4 Irrigation

Giant reed is reported to be grown without irrigation under the semi-arid southern EU conditions. Laboratory experiments conducted by Rezk and Ebany [40] confirmed that *A. donax* can endure a wide range of water table levels. However, plants should be well watered during the first year to ensure a successful establishment. In general, application of irrigation had a considerable effect on growth and biomass production since the plant used effectively any possible amount of water. The irrigated plants formed denser stands and higher yield [41]. Water use efficiency (WUE) of the giant reed depended on the irrigation rate applied. Highly irrigated plants tended to use water less effectively, whereas in non-irrigated treatments, WUE was improved. It could be partly attributed to the relative stability of the photosynthesis in a certain range of rate of transpiration and stomatal conductance [33].

2.3.2.5 Fertilization

Before establishing the plantation, a sufficient amount of K and P should be incorporated when the nutrient status of the soil is poor. Because *A. donax* is a perennial high-yielding crop, 200 kg/ha of phosphorus is required, especially in fields poor in phosphorus. As regards potassium, most fields in the semi-arid Mediterranean conditions are rich in potassium. However, in soils poor in potassium, potassium fertilization will be needed, since large quantities of biomass are removed every year.

Nitrogen application, in general, in the first year is not required, as it will promote the development of weeds. However, in poor soils, nitrogen applications of up to 100 kg/ha may be needed in the first year, early in spring, before new vegetation emerge. In fertile fields, nitrogen application in the first year has to be omitted.

According to our experience, there are no significant differences in yields between high (120 kg/ha) and low (60 kg/ha) nitrogen fertilization rates. Annual applications of 50 kg N/ha (maximum 100 kg/ha) early in the spring, before new vegetation emerge, are considered as adequate.

2.3.2.6 Weeding

Due to huge leaf mass and high growth rates, *A. donax* does not face significant weed competition from the second year on. *Arundo donax* can rapidly invade the area from a few planted individuals. When established, it has a strong ability to out-compete and completely suppress native weeds. For safe establishment, however, herbicide application is recommended for the first year. A pre-planting application of herbicides for broad-leafed weeds could be used.

2.4 Yields

2.4.1 Oil crops

Five oil crops have been tested: castor seed (*Ricinus communis* L.), crambe (*C. abyssinica* L.), cuphea (*Cuphea* spp. L), lunaria (*Lunaria annua* L.), and safflower (*Carthamus tinctorius* L.). Along with the selected crops, rapeseed, sunflower, and camelina were also grown as reference crops.

2.4.1.1 Experimental design

Two series of trials were set to study the growth and yields of the selected oil crops, in Greece and Poland. Greece represents the Mediterranean environmental zone, while Poland, the Continental environmental zone (Figure 2.4), covering thus a wide European territory.

The crops have been cultivated in a farm in Central Greece for three subsequent growing periods. The experimental field covered a total area of 2480 m^2 and consisted of 24 plots of 100 m^2 (10 × 10 m). A randomized complete block experimental design was applied in three blocks. In each plot, seeds were sown in rows 0.5 m apart. The field was fallow, and before sowing, it was ploughed, sub-tilled, and tilled. No basic fertilization was applied. Irrigation was provided to enable a good establishment of the seeds. Lunaria, safflower, and rapeseed were sown as winter crops in October and early November. Only in the second year, safflower seedlings failed to survive the low temperatures of the winter, and thus, safflower was sown again at the end of March. Castor, crambe, cuphea, and sunflower were sown as spring crops from the end of March to mid-May. Different varieties of castor have been used each year, coming from France and Israel. Because of lack of seeds in the 2010/2011 growing season, cuphea was cultivated only in the 2011/2012 and 2012/2013. Crops were harvested from the end of June to beginning of October. Important days are depicted in Tables 2.2 and 2.3.

Plant height and seed yields were measured from plant samples taken from the central area of each plot at the final harvest.

In Poland, three oil plants species, crambe, lunaria, and safflower, have been cultivated. Spring rape and camelina were also sown as reference crops. They were sown as spring crops on total area of 3 ha, together with recurrences for agricultural machines. A fore crop for all plants was spring rape. Before sowing, ploughing on 20 cm depth, harrowing, and soil cultivation was conducted. During the growth period, the crops were mechanically weeded and fertilized with 90 kg/ha ammonium nitrate. Camelina was harvested in first half of August, Crambe in the end of August, rapeseed in mid-September and safflower end September.

In spring 2012, a second experiment with two factors in three replications was established in the Didactic and Research Station in Łężany. The first factor was crop species: crambe, camelina, safflower, and spring rapeseed. The second factor was fore crop: fallow land, safflower, and winter triticale.

2.4.1.2 Results

In Greece plant heights ranged from as low as 25 cm for lunaria to as high as 190 cm for safflower (Figure 2.11). Safflower and castor reached an average of 160 and 137 cm height, averaged over the 3 years of the experiment, and crambe followed with 85 cm, in line with the 150 cm reported for castor and safflower and the 100 cm reported for crambe. Seed yields ranged from as low as 167 kg/ha for lunaria to as high as 3180 kg/ha for castor and 3015 kg/ha for safflower (Figure 2.12). Lunaria is the lowest yielding crop,

Table 2.2: Important dates.

	2010/2011				2011/2012				2012/2013			
	Sowing	Emergence	Flowering	Harvesting	Sowing	Emergence	Flowering	Harvesting	Sowing	Emergence	Flowering	Harvesting
Winter crops												
Lunaria	17/10/2010	30/10/2010	27/4/2011	10/7/2011	3/11/2011	9/11/2011	24/4/2012	10/7/2012	16/10/2012	30/11/2012	25/4/2013	12/7/2013
Safflower	17/10/2010	28/10/2010	27/5/2011	31/8/2011					17/10/2012	2/11/2012	27/5/2013	7/8/2013
Rapeseed	23/10/2010	30/10/2010	27/3/2011	26/6/2011	25/11/2011	30/11/2011	30/4/2012	10/7/2012	19/10/2012	21/10/2012	23/3/2013	1/7/2013
Spring crops												
Castor	26/4/2011	7/5/2011	15/6/2011	4/10/2011	19/4/2012	24/4/2012	29/6/2012	20/9/2012	31/3/2013	11/4/2013	13/6/2013	20/9/2013
Crambe	15/5/2011	20/5/2011	2/6/2011	9/8/2011	26/4/2012	2/5/2012	3/6/2012	17/7/2012	26/4/2013	30/4/2013	3/6/2013	11/7/2013
Cuphea					3/5/2012	7/5/2012	20/7/2012	12/9/2012	22/4/2013	4/5/2013	15/6/2013	6/9/2013
Safflower					23/3/2012	4/4/2012	20/6/2012	11/8/2012				
Sunflower	26/4/2011	2/5/2011	23/6/2011	8/9/2011	14/4/2012	20/4/2012	15/6/2012	25/8/2012	12/4/2013	22/4/2013	17/6/2013	23/8/2013

Table 2.3: Length of the growing period.

	2010/2011			2011/2012			2012/2013		
	Emergence	Flowering	Harvesting	Emergence	Flowering	Harvesting	Emergence	Flowering	Harvesting
Winter crops									
Lunaria	13	192	266	6	173	250	45	191	269
Safflower	11	222	318				16	222	294
Rapeseed	7	155	246	5	157	228	2	155	255
Spring crops									
Castor	11	50	161	5	71	154	11	74	173
Crambe	5	18	86	6	38	82	4	38	76
Cuphea				4	78	132	12	54	137
Safflower				12	89	141			
Sunflower	6	58	135	6	62	133	10	66	133

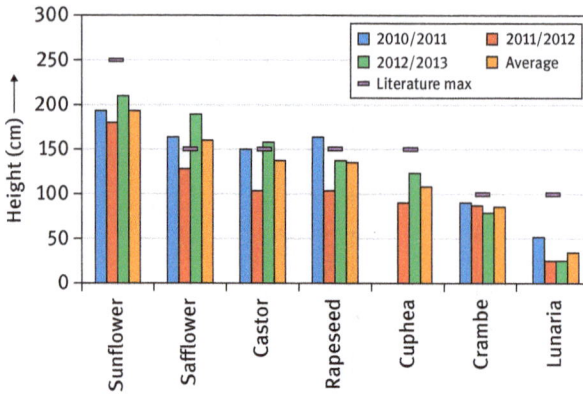

Fig. 2.11: Plant height in three subsequent growing periods.

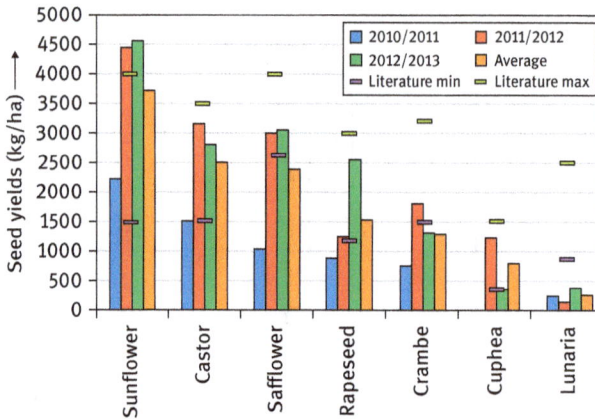

Fig. 2.12: Seed yields of crops in three subsequent growing periods.

whereas crambe produced yields similar to rapeseed. Averaged yields over the 3 years of the trial were 2500 kg/ha of seeds for castor, 2380 kg/ha for safflower, and 1300 kg/ha for crambe.

Yields of castor were very close to the highest yields reported in literature (3500 kg/ha), whereas crambe and safflower yields were closer to the minimum reported values. More specifically, new genotypes of castor are reported with yields varying from 1600 to 2620 kg/ha or even 5140 kg of seeds per hectare [42–44]. Crambe plants are reported to achieve seed yields of 2300–3200 kg/ha in Italy at 32–37% oil content [15], while in the USA, yields range from 1500 to 2250 kg/ha at 27–35% oil content [16]. Finally, for

Table 2.4: Oil content of crops from two growing periods.

	Oil content (% on seed yields) in field trials	Oil content (% on seed yields) from literature
Castor	41–52	40–55
Crambe	26–27.1	27–32
Cuphea	19.8	27–35
Lunaria	22–33	30–40
Safflower	23–26.5	35–45
Sunflower	47.2	25–35
Rapeseed	45.6	30–40

safflower seeds, yields of 500–2000 kg/ha are reported in the USA, and they could be as high as 3000 kg/ha in irrigated regimes [22]. In Greece, seed yield varied greatly among genotypes and ranged from 923 to 3391 kg/ha [23], whereas safflower yields for winter sowing are within a range of 2100–4000 kg/ha, and for summer sowing in watered conditions, 1310–3740 kg/ha. FAO presents that good rain-fed yields are in the range of 1000–2500 and 2000–4000 kg/ha under irrigation [27].

As shown in Table 2.4, the oil content of castor seeds was by far the highest of all crops.

In Poland, lunaria showed a very poor germination and growth, low competitiveness to weeds that completely stopped its growth, and very slow and late sprouting. As a result, the plant had very limited plant survival. Safflower showed a fair germination and growth but seeds fail to mature because the cold and wet summer retarded plant maturation. In addition, they were eaten by wild animals (roe-deer, deer). No yield was acquired.

In contrast to the poor performance of the two previous plants, crambe showed very good germination and growth and the plants successfully matured and produced seeds, although irregular maturation occurred, which resulted in seed spill, probably to the high precipitation during the vegetation period. Consequently, crambe could be considered as the best among the studied plants for the Polish climate conditions.

Yield of crambe in 2013 in the large plantation was lower than in 2012, mainly due to late date of sowing, which resulted in shorter vegetation of the crop. The main reason for that was the long winter of 2012/2013; thus, field works started at the beginning of May 2013. Seed yield of 944 kg should be considered as low in comparison to rapeseed. However, crambe cultivation enables the utilization of land unsuitable for rape cultivation. Oil in crambe fruits (husks and seeds) amounted to 26.09% DM.

In the second growing period, experiments included crambe and safflower, as well as spring rape and camelina for comparison. Safflower proved once again to be unsuitable for cultivation in Polish climate conditions, and only few plants grew.

Table 2.5: Yield of the analyzed species cultivated in crop rotation in Poland (kg/ha).

Fore crop	Spring rape	Safflower	Crambe	Camelina (spring variety)
Fallow land	1340	0.00	1170	1080
Safflower	1330	0.00	1130	1120
Winter Triticale	1920	0.00	1320	1290
Average	1530	0.00	1210	1160

Moreover, the plants did not bloom or produce seeds. Thus, this species will not be further considered for cultivation.

Crambe, camelina, and rape had successfully grew, bloomed, and produced seeds. However, in the blooming period, rape flowers were partially eaten by wild boars. It is worth to emphasize that camelina and crambe were not damaged by wild animals. Among the three species, the highest yield was produced from spring rapeseed (1530 kg/ha on average), while camelina and crambe yielded almost the same level. The best yields of seeds acquired from plots were winter triticale sown previously (Table 2.5).

According to the results of the 3-year field trials, it can be stated that castor and safflower are the best suited plants to be grown in the Mediterranean agro-climatic zone, compared to the rest of the crops studied in these field trials. They grew satisfactorily and produced considerably high seed yields. On the contrary, safflower proved to be unsuitable for cultivation in Polish climate conditions. Only few plants survived, which fail to bloom and produce seeds. Among crambe, camelina, and rape, the highest yield were produced from spring rapeseed, while camelina and crambe yielded almost the same level. The best yields of seeds acquired from plots were winter triticale sown previously.

After the harvest, a considerable amount of field residues was collected from safflower and castor, which amounted to an average of 7600 and 3347 kg of dry matter, respectively, for each crop (Table 2.6). The moisture content of the castor field residues was the highest of all crops (75% on average, due to the high percentage of leafy material of the crop). Samples from all crops were subjected to fuel characterization analysis (Table 2.7).

2.4.2 Lignocellulosics

2.4.2.1 Giant reed

Giant reed was grown for nine consecutive growing periods in a marginal land in Greece.

Experimental design. An experimental trial using the split plot design was carried out with giant reed. The factors were studied with three irrigation and three fertilization levels in three replications. The three irrigation (I) and nitrogen fertili-

Table 2.6: Field residues from the oil crops.

	Amount of field residues (kg of dry matter)	Moisture content (%)
Castor		
Mean	3347	75
Min	2500	74
Max	4600	76
Crambe		
Mean	2078	13
Min	1300	13
Max	2800	14
Safflower		
Mean	7600	26
Min	6400	21
Max	9000	31
Cuphea		
Mean	3748	9
Min	880	6
Max	6000	12.6
Lunaria		
Mean	1583	9
Min	230	9
Max	3900	
Sunflower		
Mean	8608	45
Min	4200	16
Max	5800	45
Rapeseed		
Mean	5541	25
Min	4300	24
Max	7300	26

zation (N) levels were the following: I_0, no irrigated; I_1, 50% of ET_0; I_2, 100% of ET_0 (potential evapotranspiration); N_0, 0 kg N/ha; N_1, 40 kg N/ha; N_2, 120 kg N/ha. Irrigation and fertilization rates were uniform during the first growing period in order to assure well establishment of plants. They differentiated from the second growing period onward. The experimental field of Arundo consisted of 27 plots of 100.8 m^2 (12 × 8.4 m) each. Rhizomes were planted with distances of 1.5 m between rows and 0.7 m along the row.

The irrigation needs for each crop were determined from soil water content and reference evapotranspiration (ET_0) according to modified FAO Penman-Monteith method. Soil water content was measured by means of time domain reflectometry sensors placed at three depths (15, 35, and 75 cm) for continuous recording and plastic probes up to 180 cm below ground level for periodic measuring. For the soil analysis, samples from all treatments were collected, and electrical conductivity, pH, total calcium, organic matter, total nitrogen, phosphorus, potassium, sodium, and chlorine content, and some important characteristics were measured at the laboratory. Values of each year were compared in order to evaluate the effect of giant reed growth on soil. In general, a slight increase to electrical conductivity, pH, and calcium concentration was noted, and there was a considerable increase in phosphorus and potassium because of fertilizer application. On the contrary, a slight decrease in organic matter and sodium was recorded. Chlorines were measured because it is an important factor for the conversion process since a high chlorine content of feedstock causes enormous problems in conversion. However, the soil analysis indicated that chlorine content is low in all treatments.

Climatic data were taken by an automatic meteorological station established in the field to provide minimum, maximum, and mean air temperature, relative humidity, wind speed at 2 and 6 m above ground, precipitation, photosynthetically active radiation, and evaporation from a pan A class. A mast was established with two sensors, and tube solarimeters recorded incoming, reflected, and net short wave and total solar radiation.

Growth data were taken at monthly intervals during the first 3 years of the plantation and thereafter only once at the final harvest. Yielding data were collected at harvest. Statistical differences were detected by performing ANOVAs at $p = 0.05$ level. Samples from each harvest were laboratory analyzed for feedstock characterization (gross and net calorific values, hydrogen, carbon, and nitrogen content, volatile, ash, and fixed carbon content).

Results. The height of the plants as it progressed during the growing period was influenced by irrigation and nitrogen fertilization (Figure 2.13). As anticipated, the plants receiving the higher irrigation amount were taller than the non-watered ones, and this superiority was statistically significant at $p = 0.05$ level. The same applied to plants receiving medium and high irrigation rates from June onward, but at the end of the growing period, the difference in height was not confirmed statistically.

Nitrogen fertilization affected the height of the plants as well. Strong differences (at $p = 0.05$ level) were detected between N_0 and the two fertilized treatments from May until the end of growing period.

Results from the 9-year field trial indicated that growth and yields were still significantly affected by irrigation, whereas the effect of nitrogen was not pronounced. The final plant height of giant reed grown without irrigation was around 4 m (3.7–4.5 m) from the 4th until the 9th growing period. For the medium and high irrigation rates, it ranged from 5 to 6 m from the 5th growing period onward (Figure 2.14).

The evolution of the dry matter yields along the growing period was largely influenced by both irrigation and nitrogen fertilization as early as the first sampling harvest, and it became more pronounced after irrigation and nitrogen fertilization were differentiated (Figure 2.15). All differences between the non-irrigation and high irrigation level (I_0–I_2) were statistically significant at $p = 0.05$ level during the third growing period. Also, statistically significant differences between the non-irrigation and the medium irrigation as well between the two irrigation levels were depicted but only in some sampling harvests. The lack of fertilization in most cases resulted

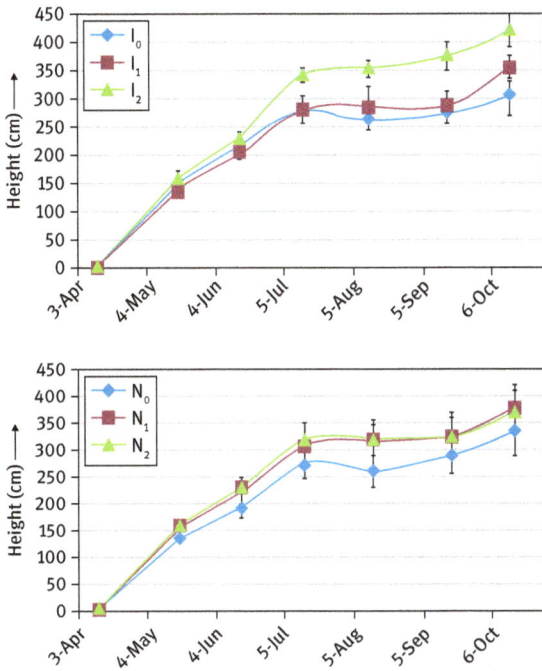

Fig. 2.13: Evolution of plant height of giant reed grown in marginal lands during the third growing period for three irrigation rates and three nitrogen fertilization rates.

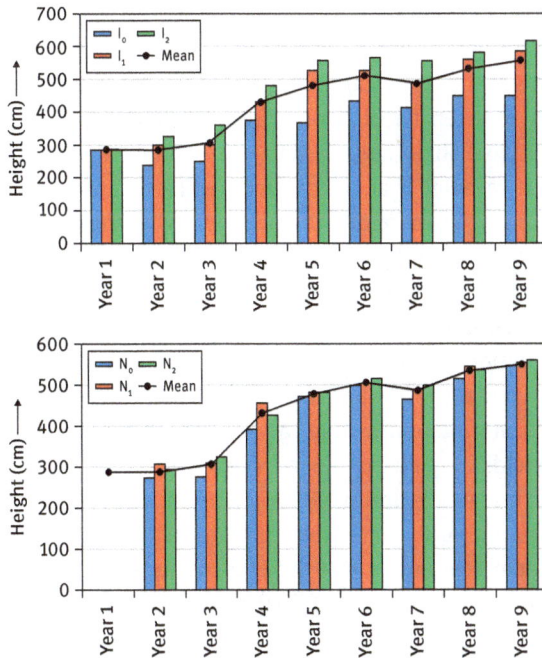

Fig. 2.14: Plant height of giant reed grown in marginal lands during nine growing periods for three irrigation rates and nitrogen fertilization rates.

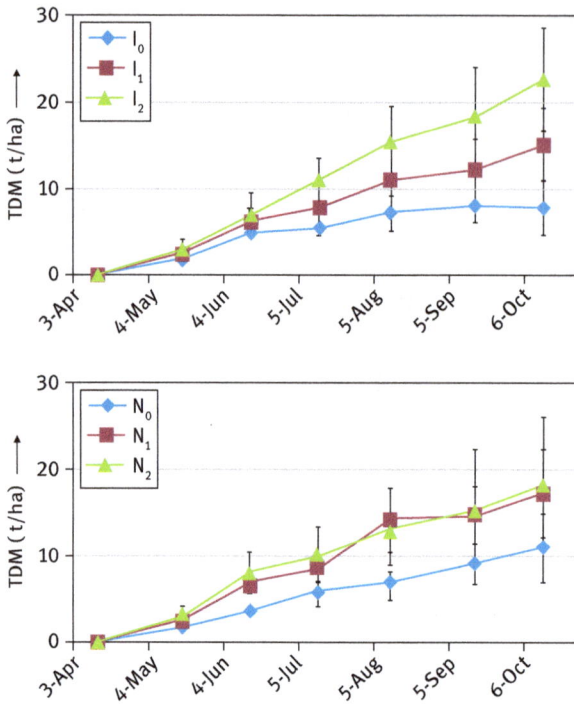

Fig. 2.15: Evolution of the total dry matter yields of giant reed grown in marginal lands during the third growing period for three irrigation rates and nitrogen fertilization rates.

in lower biomass production (statistically significant at $p = 0.05$ level) compared to the yields of plants that received nitrogen fertilization. Yield differences between the two nitrogen fertilization rates were slight and without statistical confirmation. Comparisons among the treatments revealed that all irrigated treatments were more productive than the non-irrigated ones. Furthermore, the plots receiving the highest irrigation and fertilization rates (I_2N_2 treatment) proved to be more productive compared to any other treatment, followed by the plots receiving the highest irrigation and the medium fertilization rates (I_2N_1 treatment).

When grown at low-input conditions, giant reed reaches full maturity after the 3rd year and thereafter yields remained at the level of around 7 t/ha for non-irrigated plants (Figure 2.16). Yields of medium irrigation were at the level of 15 t/ha, whereas the high irrigation rate exhibited yields ranging from 16 to 18 t/ha.

Stem fraction of the harvested biomass is around 90% at harvest. Following the growing season, the stems/leaves ratio sharply increased from 65% on average in May to 87% in July, followed by a slighter increase up to 90% until the harvest (Figure 2.17). A further increase in stem fraction was also recorded if final harvest is delayed from

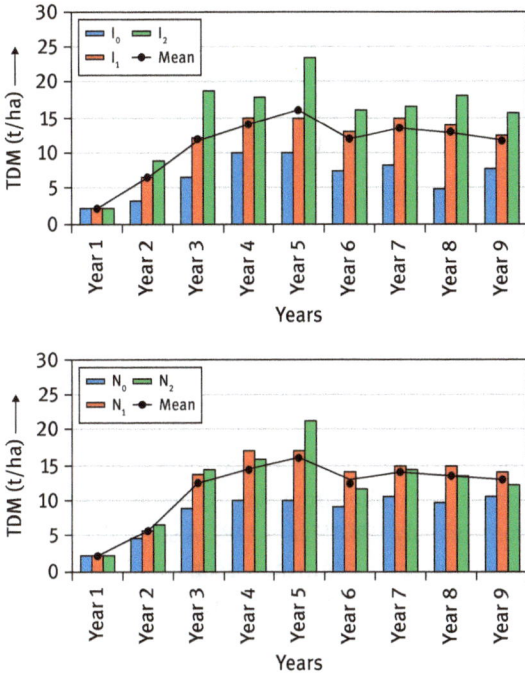

Fig. 2.16: Dry matter yields of giant reed grown in marginal lands during nine growing periods for three irrigation rates.

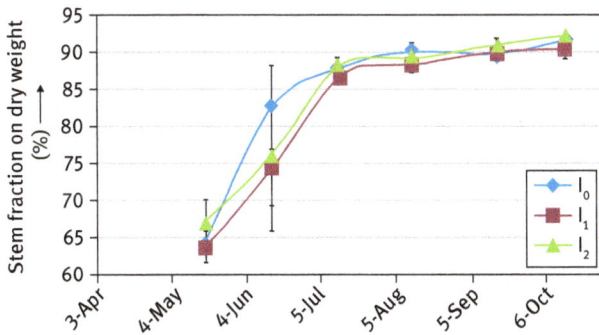

Fig. 2.17: Evolution of stem fraction on total dry matter during the third growing period.

February to March because of leaves falling. In such a delayed harvest, leaf fraction amounted to less than 2% of total harvested biomass on average. Irrigation resulted in a higher stem fraction on total fresh and dry matter in both harvests, February and March.

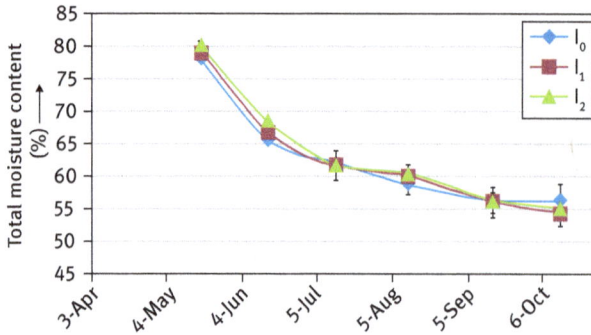

Fig. 2.18: Evolution of moisture content in the harvested material.

The moisture content of giant reed is around 50% at the final harvest of the plant. A progressive decrease in moisture content is recorded from May to October when moisture falls from more than 80% to approximately 55% (Figure 2.18). A slight decrease is also noted, reaching 45% on average with harvest delays in the season mainly because of the loss of leaves. Neither irrigation nor nitrogen fertilization affected moisture content.

2.5 Crop harvesting

2.5.1 Oil crops

2.5.1.1 Castor

In dry regions, it is best to begin harvesting when all fruits are mature. Harvesting in Europe starts in October, cutting the spikes off and stripping off the capsules into a wagon or sled or into containers strapped on the farmers. Unless the capsules are dry, they must be spread out to dry quickly. In the tropics, fruits are collected and spread in piles to dry under the sun until they blacken. In Europe, North America, and Australia, drying may be accomplished by frost or by the use of chemical defoliants. In some mechanized countries, harvesting is done with a modified wheat headers, but in the USA, more expensive harvesters, which shake capsules from plants, are used. For mechanical harvesters, a relative humidity of 45% or less is necessary for efficient operation. Seed capsules shatter easily in most cvs. Some indehiscent varieties are threshed by ordinary grain thresher. After harvesting, seeds must be removed from the capsules or hulls, usually with hulling machines if capsules are dry. The percentage of seed to hull averages 65–75, depending upon the maturity of the seed at harvest [11]. Castor oil is manufactured by running clean seeds through the decorticating machines to remove the seed coat from the kernel. Castor seeds cannot be ground or tempered unlike flaxseed or soybeans. Preheating may make heavy viscous oil more mobile. Seeds are put in a "cage" press, and number 1 oil is obtained, which needs little refining and bleaching. Oil remaining in the press-cake is extracted by solvent methods and is called number 2 oil, which contains impurities and cannot be effectively refined. Castor bean oil can be stored 3–4 years without deterioration [45].

2.5.1.2 Crambe

After flowering, crambe matures rapidly in 1 to 2 weeks. Timely harvest is important to avoid high shattering losses. During warm dry weather, the crop should be frequently monitored (daily or every other day) to determine correct harvest stage. Crambe is physiologically mature when 50% of the seeds have turned brown. According to Weiss [13], crambe is ready to be harvested when the majority of leaves have fallen, the upper part of the stem is yellow, and at least 75% of the capsules have turned yellow. This is usually some 90–100 days after planting. Extensive branching is considered to be a disadvantage for mechanical harvesting, and although branching may increase individual plant yield, it may also increase the number of harvested immature seeds.

Crambe can be swathed to dry in the field, but most growers prefer to harvest crambe directly with combined headers commonly used with wheat.

In the large field of crambe in Poland, desiccation of the plants was performed before the crambe seeds were harvested with Klinik 360 SL at 4 L/ha to ensure uniform ripening of plants. The harvest was performed with a combined harvester in the fourth week of August 2013 (Figures 2.19 and 2.20). The straw after harvest was left in the field to enrich soil with organic carbon.

(a)   (b)

Fig. 2.19: Crambe at harvest.

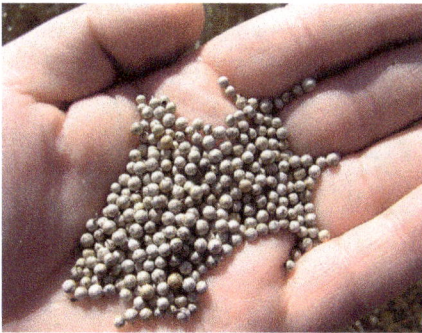

Fig. 2.20: Crambe seeds.

2.5.1.3 Safflower

Safflower is ready to harvest when most of the leaves turn brown and very little green remains on the bracts of the latest flowering heads. The stems should be dry, but not brittle, and the seeds should be white and easy to hand thresh. This crop should be harvested as soon as it matures in order to avoid the seed discoloration or sprouting in the head that can occur with fall rains [22].

Safflower is ready to harvest when most of the leaves turn brown and very little green remains on the bracts of the latest flowering heads. The stems should be dry, but not brittle, and the seeds should be white and easy to hand thresh. This crop should be harvested as soon as it matures in order to avoid seed discoloration or sprouting in the head that can occur with fall rains [22]. Safflower is an excellent crop for direct combining since it stands well and does not shatter easily. Direct combining may require artificial drying or waiting until green weeds are killed by frosts. The crop can be windrowed to dry green weeds when the moisture content of the seed is as high as 25%. The time for harvesting safflower in Europe can vary from early to late September due to the environmental conditions during the growing season. The combined cylinder speed should be set low at 550 RPM to avoid seed cracking. The reel speed should be about 25% faster than the ground speed. To prevent clogging of the machine from plant residue, the shaker speed must be greater than speeds used for small grains. Air speed should be sufficient to remove most unfilled seeds, straw, and bulls. The combined radiator and air intake should be checked regularly to avoid blockages from the white fuzz of seed heads. Accumulations of this white fuzz can be a fire hazard [22].

Safflower harvesting from the large field in Greece is shown in Figure 2.21.

2.5.2 Giant reed

Giant reed forms dense plantations similar to maize and can be harvested when the plant reaches maturity and the moisture content of the stems is the lowest possible (45–50%). That is from late winter to early spring, before new growth starts.

(a)   (b)

Fig. 2.21: Safflower large field at harvest.

In order to identify the most appropriate agronomic practices for successful introduction of giant reed into EU agricultural system, the suitability of equipment used for forage crop harvesting was tested.

The utilization of commonly harvesting equipment could offer the possibility to increase the working window of machines, and therefore a reduction of their unit cost, while, on the contrary, the utilization of equipment still available in the farms could increase the suitability of introduction of giant reed in agriculture.

In order to determine the suitability of common equipment for giant reed harvesting, a three-row mower-fodder-loader combining machine (HESTON 7650, 250HP) generally used for harvesting maize was tested in the frame of the European project FAIR CT96 2028 Giant Reed Network (Figure 2.22). During harvesting, the machine cuts some giant reed stems obliquely. These slanting stumps represent a risk of tire puncture. Therefore, it would be interesting to utilize a harvest machine with a larger cutter, in order to cut all the stems horizontally. Instead, it could be possible to assemble chains on the tires to prevent the risk of puncture.

A three-row CLAAS Jaquar 690cl mower-fodder-loader combining machine was also used for harvesting giant reed and maize silage (Figure 2.23). The efficiency of this machine was quite high, indicating that it should be regarded as suitable for giant reed harvesting. However, in dense plantations, harvesters are operated in a much slower speed and the cutting knives are fixed at a height that is quite high (about 0.5 m) in order to avoid tire puncture.

Fig. 2.22: The three-row mower-fodder-loader combining machine – HESTON 7650 (source: CETA).

Fig. 2.23: The three-row CLAAS Jaquar 690cl mower-fodder-loader combining machine.

After harvesting, the biomass was conveyed to the farm, where a storage site was prepared with used pallets on the floor and walls in order to allow pile ventilation (Figure 2.24). The pile was covered with PVC film. Samples of the harvested biomass were taken at regular intervals, and their weight and water moisture content were measured.

The storage of the giant reed biomass in piles covered with PVC film for 1 month strongly decreased the biomass moisture content from 48% (at starting up) to 23% (Figure 2.25). The average of the water content during the storage time was about 19%. At the same time, due to microbial degradation of the biomass, a loss of dry matter was observed and determined. The degradation of giant reed biomass was in the range of 9%, after the first month of storage, to 18%, at the last sampling in November. The yearly average of biomass loss was about 15%.

Fig. 2.24: Storage trials of gient reed (source: CETA).

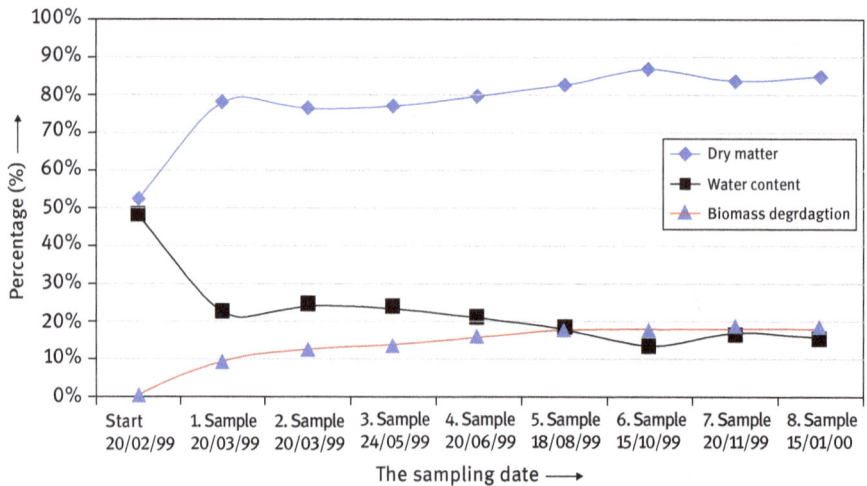

Fig. 2.25: Influence of storage on the giant reed biomass characteristics (source: CETA).

2.6 Fuel characterization

Laboratory analysis of samples from all final harvests of each crop (from small and large fields) was conducted in the CRES and UWM laboratories for feedstock characterization. The following characteristics were investigated:
- Proximate analysis was carried out to determine moisture, volatile, ash, and fixed carbon content, using a thermogravimetric analyzer.
- Gross calorific value was determined using an oxygen bomb calorimeter.
- Hydrogen content, together with carbon and nitrogen content, was determined by a microprocessor-controlled elemental analyzer.
- The dry samples were ground using a Fritch Pulverizette 15′ mill to pass a 60-mesh (0.25-mm) screen.

Proximate analysis (moisture, volatiles, ash, and fixed carbon content determination) was performed using a fully automatic thermogravimetric analyzer LECO TGA-501. The procedure followed is outlined in ASTM D3174-93 and ASTM D3175-89a (modified procedure for sparking fuels). Ash content was determined at 550°C. Gross calorific value measurement was carried out using an oxygen bomb calorimeter Parr-1261 according to the procedure outlined in ASTM D3286. Elemental analysis (C, H, and N content determination) was carried out using a Perkin Elmer elemental analyzer according to classic organic elemental analysis techniques.

The results of laboratory analyses that have been carried out for the harvested biomass from all the aforementioned crops are listed in Tables 2.7 for the Greek trials and 2.8 for the Polish trials. In addition, data from SRC species are presented in order to compare SRCs to the studied crops and crop residues.

As it is shown, cardoon, along with the field residues of sunflower, cuphea, and crambe, exhibited the highest ash content in comparison to the rest of the crops, indicating that their energy conversion may face problems. Among the oil crops, only safflower produces field residues with ash content similar to the perennial crops studied in this project. Also, cardoon and sunflower produce the least energy since its energy content, measured as gross calorific value, is much lower than the measured one for the rest of the crops and crop residues. Last, the nitrogen content for cardoon is also higher than the other crops, which may also have negative effect due to the higher possibility for NO_x formation during energy conversion.

When comparing the four energy crops and the field residues of the oil crops to SRC, it is obvious that most of the energy characteristics of the three energy grasses are within the ranges for SRC, indicating their high value as energy crops. SRC is better than grasses if ash and carbon content is considered.

Table 2.7: Comparison of laboratory analysis results for the four studied non-food crops in Greece.

	Proximate			Gross calorific value (kcal/kg dm)	Elemental		
	Volatile matter (%)	Ash (%)	Fixed carbon (%)		C (%)	H (%)	N (%)
Castor							
Mean	78.22	7.81	13.98	4144	28.12	3.68	0.38
Min	76.90	5.72	12.53	3974	40.97	5.50	0.41
Max	80.09	10.58	15.22	4240	43.38	5.53	0.74
Crambe							
Mean	76.24	10.53	13.24	4220	41.99	5.45	0.76
Min	75.70	9.42	12.66	4060	41.63	5.29	0.75
Max	76.77	11.64	13.82	4380	42.35	5.61	0.78
Safflower							
Mean	77.03	5.77	17.20	4324	44.94	5.72	0.59
Min	75.56	4.95	16.67	4175	44.79	5.64	0.29
Max	78.07	7.11	17.60	4454	45.10	5.79	0.89
Cuphea							
Mean	75.35	9.79	14.86	4115	41.48	5.69	0.91
Min	73.98	8.62	14.66	4106	40.72	5.53	0.83
Max	76.73	10.96	15.07	4124	42.25	5.84	0.99
Lunaria							
Mean	76.94	8.54	14.52	4201	42.62	5.42	1.33
Min	76.23	6.64	13.32	4089	42.40	5.33	1.29
Max	77.64	10.45	15.72	4312	42.85	5.52	1.37
Sunflower							
Mean	76.01	11.47	12.53	3917	40.46	5.36	0.26
Min	74.07	8.45	11.44	3746	39.06	5.23	0.2
Max	78.06	14.50	13.24	4024	41.73	5.54	0.45
Rapeseed							
Mean	78.09	6.77	15.15	4359	43.75	5.82	0.27
Min	78.04	5.48	13.89	4238	43.53	5.77	0.11
Max	78.13	8.07	16.40	4480	43.96	5.87	0.44
Cardoon							
Mean	73.62	15.63	10.75	3923	41.76	5.50	1.40
Min	72.82	15.05	10.40	3799	39.51	5.13	0.74
Max	74.33	16.79	11.23	4102	43.64	6.01	2.05
Miscanthus							
Mean	81.47	3.05	15.48	4330	45.76	6.29	0.12
Min	81.09	2.63	15.43	4217	45.47	6.26	0.10
Max	81.85	3.48	15.54	4451	46.05	6.32	0.14
Arundo							
Mean	77.18	4.75	18.06	4267	45.07	5.96	0.64
Min	74.81	3.75	16.21	4106	42.44	5.58	0.19
Max	78.70	8.85	19.07	4405	46.47	6.25	1.78

Continued

Table 2.7: *Continued*

	Proximate			Gross calorific value (kcal/kg dm)	Elemental		
	Volatile matter (%)	Ash (%)	Fixed carbon (%)		C (%)	H (%)	N (%)
Switchgrass							
Mean	78.66	4.93	16.41	4179	44.83	6.08	0.33
Min	76.61	2.56	15.42	4024	42.90	5.86	0.11
Max	81.09	6.35	17.35	4272	45.91	6.47	0.85
SRC[a]							
Min	80.94	0.52	16.35	4596	48.18	5.71	0.15
Max	82.55	1.33	18.26	4711	50.73	5.92	0.57

[a] *An Atlas of Thermal Data for Biomass and Other Fuels*, National Renewable Energy Laboratory, US Department of Energy, June 1995.

Table 2.8: Characteristics of crambe straw in the large field in Poland.

Item	Crambe (chemical weed control)	Crambe (without chemical control)	Average	Crambe cake
Moisture content (%)	14.24	17.56	15.90	7.16
Oil content (% dm)	25.66	26.46	26.06	20.30
Higher heating value (MJ/kg dm)	18.65	18.66	18.65	23.83
Lower heating value (MJ/kg)	15.64	14.96	15.30	21.95
Ash content (% DM)	5.93	5.42	5.68	6.41
Volatile matter (% DM)	73.35	73.36	73.36	75.39
Fixed carbon (% DM)	17.48	17.90	17.69	16.51
N (% DM)	6.53	6.99	6.76	3.82
C (% DM)	48.24	47.20	47.72	53.85
H (% DM)	4.99	4.83	4.91	7.21
S (% DM)	0.203	0.173	0.188	0.918

Source: [46]

Bibliography

[1] Krasuska, E., Cadórniga, C., Tenorio, J. L., Testa, G., Scordia, D., Potential land availability for energy crops production in Europe, Biofuels Bioprod Bioref 4 (2010) 658–673.

[2] Elbersen, B., Startisky, I., Hengeveld, G., Schelhaas, M.-J., Naeff, H., Böttcher, H., Atlas of EU Biomass Potentials, 2012, available from: http://www.biomassfutures.eu/public_docs/final_deliverables/WP3/D3.3_Atlas_of_technical_and_economic_biomass_potential.pdf.

[3] Jossart, J. M., Overview of energy crops and their uses in Europe, Presentation in Pulawy, Poland, 2009, http://www.encrop.net.

[4] Allen, B., Kretschmer, B., Baldock, D., Menadue, H., Nanni, S., Tucker, G., Space for energy crops – assessing the potential contribution to Europe's energy future, Report produced for BirdLife Europe, European Environmental Bureau and Transport & Environment, IEEP, London, 2014.

[5] Copa-Cogeca, Copa-Cogeca's position on the EU's biofuels policy, position paper, Copa-Cogeca, Brussels, Belgium, available from: www.copa-cogeca.be/Download.ashx?ID=1028131.

[6] IENICA Crops Database 2002 Crambe. Agronomy Guide, Generic Guidelines on the Agronomy of Selected Industrial Crops, August 2004, available from: http://ienica.net/cropsdatabase.htm/.

[7] Christou, M., Fritsche, U., Papadopoulou, E., Monti, A., Nissen, L., Schurr, U., Schmid, E., Alatsidis, I., Stefanidou, R., Panoutsou, C., Heller, K., Baraniecki, P., Milioni, D., Margaritopoulou, T., Zucchini, V., Grigore, A., Pages, X., Alfos, C., Reiders, M., Non-food crops to industry schemes for a European bio-based industry and sustainable agriculture, 21st European Biomass Conference and Exhibition, 3–7 June 2013, Copenhagen, Denmark.

[8] Christou, M., Alexopoulou, E., The terrestrial biomass: formation and properties (crops and residual biomass), in Aresta, M., Dibenedetto, A., Dumeignil, F., editors, Biorefinery: From Biomass to Chemicals and Fuels, De Gruyter, Berlin, Boston, 2012, chapter 3.

[9] Cocchi, M., Grassi, A., Capaccioli, S., Laitinen, T., Lehtomäki, A., Rechberger, P., Lötjönen, K., Pahkala, T., Xiong, S., Finell, M., Salve, M., Gabauer, W., Dörrie, D., Köttner, M., Opportunities and barriers of energy crops at European level, success stories and strategies for promoting the production and utilization of energy crops in different EU regions, 18th European Biomass Conference and Exhibition, 3–7 May 2010, Lyon, France.

[10] Arkema, Castor (Rimicus communis L) Production Technology Guide, Arkema, Paris.

[11] Oplinger, E. S., Oelke, E. A., Kaminski, A. R., Combs, S. M., Doll, J. D., Schuler, R. T., Castorbeans. Alternative Field Crops Manual, 1990, available from: http://www.hort.purdue.edu/newcrop/afcm/castor.html.

[12] Baldwin, B. S., Cossar, R. D., Castor yield in response to planting date at four locations in the south-central United States, Ind Crops Prod **29** (2009) 316–319.

[13] Weiss, E. A., Oilseed Crops, Longman, New York, 1983, 660.

[14] Babita, M., Maheswari, M., Rao, L. M., Shanker, A. K., Rao, D. G., Osmotic adjustment, drought tolerance and yield of castor (Ricinus communis L.) hybrids, Environ Exp Bot **69** (2010) 243–249.

[15] Laghetti, G., Piergiovanni, A. R., Perrino, P., Yield and oil quality in selected lines of Crambe abyssinica Hochst. ex R. E. Fries and C. hispanica L. grown in Italy, Ind Crop Prod **4** (1995) 203–212.

[16] Oplinger, E. S., Oelke, E. A., Kaminski, A. R., Putnam, D. H., Teynor, T. M., Doll, J. D., Kelling, K. A., Durgan, B. R., Noetzel, D. M., Crambe, in Alternative Field Crops Manual, 1991, available from: http://www.hort.purdue.edu/newcrop/AFCM/crambe.html.

[17] Adamsen, F. J., Coffelt, T. A., Planting date effects on flowering, seed yield, and oil content of rape and crambe cultivars, Ind Crops Products **21** (2005) 293–307.

[18] Enders, G., Schatz, B., Crambe production, A-1010 (revised), NDSU, November 1993, available from: http://ag.ndsu.edu/pubs/plantsci/crops/a1010w.htm, accessed 4 May 2010.

[19] Mastebroek, H. D., Wallenburg, S. C., van Soest, L. J. M., Variation for agronomic characteristics in crambe (Crambe abyssinica Hochst. ex Fries), Ind Crop Prod **2** (1994) 129–136.

[20] Johnson, B. L., McKay, K. R., Schneiter, A. A., Hanson, B. K., Schatz, B. G., Influence of planting date on canola and crambe production, J Prod Agric **8** (1995) 594–599.

[21] Morrison, M. J., Stewart, D. W., Heat stress during flowering in summer brassica, Crop Sci **42** (2002) 797–803.

[22] Oelke, E. A., Oplinger, E. S., Teynor, T. M., Putnam, D. H., Kelling, K. A., Durgan, B. R., Noetzel, D. M., Safflower, 1992, available from: http://www.hort.purdue.edu/NEWCROP/AFCM/safflower.html.

[23] Koutroubas, S. D., Papakosta, D. K., Doitsinis, A., Nitrogen utilization efficiency of safflower hybrids and open-pollinated varieties under Mediterranean conditions, Field Crop Res **107** (2008) 56–61.

[24] Dordas, C., Sioulas, C., Safflower yield, chlorophyll content, photosynthesis, and water use efficiency response to nitrogen fertilisation under rainfed conditions, Ind Crop Prod **27** (2008) 75–85.

[25] Dordas, C., Sioulas, C., Dry matter and nitrogen accumulation, partitioning and translocation in safflower (Carthamus tinctorious L.) as affected by nitrogen fertilization, Field Crop Res **110** (2009) 35–43.

[26] Istanbulluoglu, A., Effects of irrigation regimes on yield and water productivity of safflower (Carthamus tinctorius L.) under Mediterranean climatic conditions, Agric Water Manage **96** (2009) 1792–1798.

[27] Istanbulluoglu, A., Gocmen, E., Gezer, E., Pasa, C., Konukcu, F., Effects of water stress at different development stages on yield and water productivity of winter and summer safflower (Carthamus tinctorius L.), Agric Water Manage **96** (2009) 1429–1434, available from: www.elsevier.com/locate/agwat.

[28] Yau, S. K., Winter versus spring sowing of rainfed safflower in a semi-arid high elevation Mediterranean environment, Eur J Agron **26** (2007) 249–256.

[29] Perdue, R. E., Arundo donax – source of musical reeds and industrial cellulose, Econ Bot **12** (1958) 368–404.

[30] Zegada-Lizarazu, W., Elbersen, W., Cosentino, S. L., Zatta, A., Alexopoulou, E., Monti, A., Agronomic aspects of future energy crops in Europe, Biofuels Bioprod Bioref **4** (2010) 674–691.

[31] Metzger, M. J., Bunce, R. G. H., Jongman, R. H. G., Mücher, C. A., Watkins, J. W., A climatic stratification of the environment of Europe, Global Ecol Biogeogr **14** (2005) 549–563.

[32] Rexen, F., Blicher-Mathiesen, U., IENICA, report from the State of Denmark, 1998.

[33] Christou, M., Mardikis, M., Alexopoulou, E., Kyritsis, S., Cosentino, S., Vecchiet, M., Bullard, M., Nixon, P., Gosse, G., Fernandez, J., El, Bassam, N., Arundo donax productivity in the EU. Results from the Giant Reed (Arundo donax L.) Network (1997–2001), in Palz, W., Spitzer, J., Maniatis, K., Kwant, K., Helm, P., Grassi, A., editors, Biomass for Energy, Industry and Climate Change. Volume I. Proceeding of the 12th European Biomass Conference, Amsterdam, the Netherlands, 17–21 June 2002, ETA-Florence and WIP-Munich, 2002, 127–130.

[34] Hamrouni, I., Hammmadi, B. S., Marzouk, B., Effects of water deficit on lipids of safflower aerial parts. Phytochemistry **58** (2001) 277–280.

[35] Elfadl, E., Reinbrecht, C., Frick, C., Claupein, W., Optimization of nitrogen rate and seed density for safflower (Carthamus tinctorius L.) production under low-input farming conditions in temperate climate. Field Crops Research **114** (2009) 2–13.

[36] Yau, S. K., Ryan, J., Response of rainfed safflower to nitrogen fertilisation under Mediterannean conditions. Industrial Crops and Products **32** (2010) 318–323.

[37] Pari, L., First trials on Arundo donax and miscanthus rhizomes harvesting, in Chartier, P., Ferrero, G. L., Henius, U. M., Hultberg, S., Sachau, J., Wiinblad, M., Chartier, P., editors, Biomass for Energy and the Environment: Proceedings of the 9th European Bioenergy Conference, Copenhagen, Denmark, 24–27 June 1996, Pergamon, New York, 1996, 889–894.

[38] Vecchiet, M., Jodice, R., Pari, L., Schenone, G., Techniques and costs in the production of giant reed (Arundo donax L.) rhizomes, in Chartier, P., Ferrero, G. L., Henius, U. M., Hultberg, S., Sachau, J., Wiinblad, M., Chartier, P., editors, Biomass for Energy and the Environment: Proceedings of the 9th European Bioenergy Conference, Copenhagen, Denmark, 24–27 June 1996, Pergamon, New York, 654–659, 1996.

[39] Christou, M., Mardikis, M., Alexopoulou, E., Propagation material and plant density effects on the Arundo donax yields, in Kyritsis, S., Beenackers, A. A. C. M., Helm, P., Grassi, A., Chiaramonti, D., editors, Biomass for Energy and Industry, Proceeding of the 1st World Conference, Seville, Spain, 5–9 June 2000, James & James (Science Publishers) London, 2001, 1622–1628.

[40] Rezk, M. R., Edany, T. Y., Comparative responses of two reed species to water table levels, Egypt J Bot **22** (1979) 157–172.

[41] Christou, M., Mardikis, M., Alexopoulou, E., Research on the effect of irrigation and nitrogen upon growth and yields of Arundo donax L. in Greece. Aspects Appl Biol **65** (2001) 47–55.

[42] Laureti, D., Marras, G., Irrigation of castor (Ricinus communis L.) in Italy, Eur J Agron **4** (1995) 229–235.

[43] Koutroubas, S. D., Papakosta, D. K., Doitsinis, A., Adaptation and yielding ability of castor plant (Ricinus communis L.) genotypes in a Mediterranean climate, Eur J Agron **11** (1999) 227–237.

[44] Soratto, R. P., Souza-Schlick, G. D., Fernandes, A. M., Zanotto, M. D., Crusciol, C. A. C., Narrow row spacing and high plant population to short height castor genotypes in two cropping seasons. Industrial Crops and Products **35** (2012) 244–249.

[45] Domingo, W. E., The development of domestic Castor bean production. Econ Bot **7**(1) (1953) 65–75.

[46] Stolarski, M., Krzyżaniak, M., Śnieg, M., Christou, M., Alexopoulou, E., Production costs and residues evaluation of Crambe abyssinica as an energy feedstock. Environmental Biotechnology **9**(2) (2013) 59–64.

Angela Dibenedetto and Antonella Colucci

3 Production and uses of aquatic biomass

Abstract: This chapter covers the production and uses of aquatic biomass. Algae are considered to be able to contribute to the fast fixation of CO_2 either under natural or enhanced conditions. The conditions for growing algae rich in lipids are presented, the extraction techniques and the conversion of glycerides are discussed.

3.1 Introduction

The increase in world energy consumption is raising worries due to the effects of greenhouse gases generated in the combustion of fossil fuels. Biofuels, produced with quasi-zero CO_2 emission, are considered an alternative to fossil fuels, at least in the transport sector. Particular attention is paid to bioethanol, biodiesel, and fatty acid methyl esters (FAMEs). The latter can have properties similar to diesel. They are produced by transesterification of lipids, Eq. (3.1), with an alcohol, such as methanol, ethanol, or butanol in the presence of a catalyst [1].

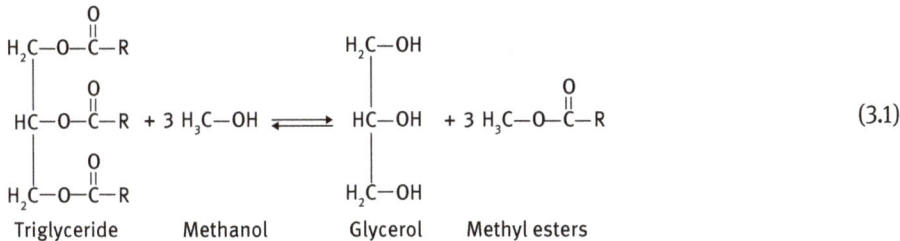

$$\tag{3.1}$$

| Triglyceride | Methanol | Glycerol | Methyl esters |

First-generation biofuels, which have now attained economic levels of production, have been mainly extracted from food crops, including cereals and rapeseeds, sugarcane, sugar beets, as well as from vegetable oils and animal fats using conventional technologies [2, 3]. The use of first-generation biofuels has generated a lot of controversy, mainly due to their impact on global food markets and on food security, especially with regard to the most vulnerable regions of the world economy. This has raised pertinent questions on their potential to replace fossil fuels and the sustainability of their production [4]. Currently, about 1% (14 million hectares) of the world's available arable land is used for the production of biofuels, providing 1% of global transport fuels. Clearly, increasing such share to anywhere close to 100% is impractical, owing to the severe impact on the world's food supply and the large areas of production land required [5].

The advent of second-generation biofuels is intended to produce fuels from the whole plant matter of dedicated energy crops or agricultural residues, forest harvesting residues, or wood processing waste, rather than from food crops.

Aquatic biomass is currently considered as an ideal second- or third-generation biodiesel (or biofuel in general) feedstock. They do not compete with food and feed crops, do not require arable land for their cultivation, and can grow under enhanced CO_2 concentration in process soft water or saltwater.

Biodiesel can be obtained from vegetable oils and animal fats [6], and it can be used in diesel engines blended with standard gasoil or alone. From an environmental point of view, biodiesel includes several benefits such as the reduction of sulfur dioxide, carbon monoxide (50%), and carbon dioxide (78%) emissions [7]; it is non-toxic and biodegradable; and its use reduces fossil fuels consumption. The interest in the production and use of liquid biofuels from biomass is not limited to FAMEs. However, based on current knowledge and technology projections, third-generation biofuels specifically derived from aquatic biomass are considered a technically viable alternative energy source that is devoid of the major drawbacks associated with first- and second-generation biofuels. The key issue is their cost, considering the extremely low actual price of oil and methane.

Macroalgae and microalgae are photosynthetic organisms, with simple growing requirements (light, usable organic carbon sugars, CO_2, N, P, and K) that can produce lipids, proteins, and carbohydrates in large amounts over short periods. These products can be processed into both biofuels and valuable co-products.

3.2 Classification of aquatic biomass

3.2.1 Macroalgae

Macroalgae (Figure 3.1) are classified as *Phaeophyta* (or brown algae), *Rhodophyta* (or red algae), and *Chlorophyta* (or green algae) based on the composition of photosynthetic pigments. The green macroalgae have evolutionary and biochemical affinity with higher plants. The life cycles of macroalgae are complex and diverse, with different species displaying variations of annual and perennial life histories, combinations of sexual and asexual reproductive strategies, and alternation of generations.

3.2.2 Microalgae

Microalgae are microscopic organisms and are currently cultivated commercially as feed for fish around the world in several dozen small- to medium-scale production systems, producing from a few tens to several hundred tons of biomass annually. The main algae genera currently cultivated photosynthetically (e.g., with light) for

Fig. 3.1: Different types of macroalgae.

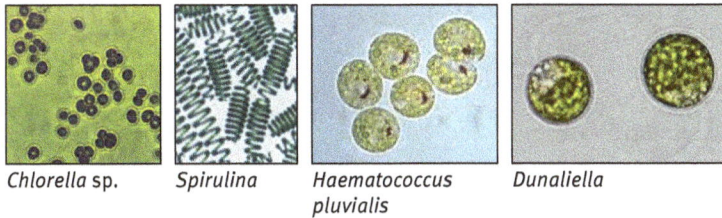

Fig. 3.2: Different type of microalgae.

various nutritional products are *Spirulina*, *Chlorella*, *Dunaliella*, and *Haematococcus* (Figure 3.2).

3.2.3 Plants

Aquatic plants (Figure 3.3), also known as hydrophytes, grow in all ponds, shallow lakes, marshes, ditches, reservoirs, swamps, canals, and sewage lagoons. Less frequently, they also live in flowing water, in streams, rivers, and springs. Such macroscopic aquatic flora includes the aquatic angiosperms (flowering plants), pteridophytes (ferns), and bryophytes. They can be divided into four categories according to the habitat of growth: floating unattached, floating attached, submersed, and emergent. Macrophytes play a key role in nutrient cycling to and from the sediments and help stabilize river and stream banks. Plants are often used for water phytodepuration, as they efficiently use N and P compounds present in wastewater; some species can also concentrate heavy metals [8].

Such macrophytes can be spontaneous, but in several countries, they are grown for various purposes, from water treatment to nutrition (human and animal) and to

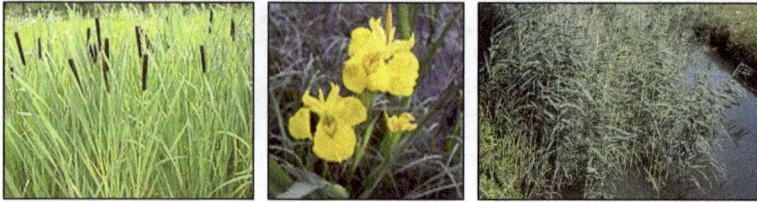

Typha latifolia   Iris pseudacorus   Phragmites australis

Fig. 3.3: Different type of plants.

the production of materials used in several fields, including building. With respect to microalgae and macroalgae, hydrophytes may contain more cellulosic materials and require different technologies for their treatment and for energetic purposes.

3.3 Cultivation of aquatic biomass

In this section, only macroalgae (seaweeds) and microalgae will be considered, as plants do not contribute to biofuels production, but have other utilization.

3.3.1 Macroalgae

The distribution of macroalgae is worldwide. They are abundant in coastal environments, primarily in near-shore coastal waters with suitable substrate for attachment. Macroalgae also occur as floating forms in the open ocean, and floating seaweeds are considered one of the most important components of natural materials on the sea surface [9]. The use of macroalgae for energy production has received attention only very recently [10]. The great advantage of macroalgae with respect to terrestrial biomass is their high biomass productivity (faster growth in dry weight/ha/year than for most terrestrial crops). The productivity of natural basins is in the range of 1–20 kg/m^2/year dry weight (10–150 tons dry weight ha^{-1} year $^{-1}$) for a 7- to 8-month culture. Interestingly, macroalgae are very effective in nutrients (N and P) uptake from sewage and industrial wastewater. The estimated recovery capacity is 16 kg/ha/day [11]. To this end, macroalgae have been used for cleaning municipal wastewater [12] (essentially in Europe), for recycling nutrients, and for the treatment of fishery effluents [13, 14] (either in Europe or in Japan). The latter use has an economic value, as macroalgae can reduce the concentration of nitrogen derivatives like urea, amines, ammonia, nitrite, or nitrate to a level that is not toxic for fishes, allowing the reuse of water and reducing both the cost of their growth and the water use. The capacity of macroalgae as biofilters or nutrient uptake has been tested in the northwestern Mediterranean

Sea, along French coasts [15] using *Ulva lactuca* or *Enteromorpha intestinalis* that adapted to non-natural basins.

In Europe, macroalgae are grown in experimental fields and natural basins. They can be grown on nets or lines and can be seeded onto thin lightweight lines suspended over a larger horizontal rope [16]. Also, in a colder climate, macroalgae grow at an interesting rate. For example, in Denmark, the Odjense Fiord produces ca. 10 kt per day of dry weight biomass equivalent to ca. 10 t per year per ha.

Although macroalgae can grow in both hemispheres, the climatic factors may affect the productivity by reducing either the rate of growth or the growing season. The Mediterranean Sea has ideal climatic conditions for a long growing season, with good solar irradiation intensity and duration, and with a correct temperature. Moreover, along the coasts of several EU countries (Italy, Spain, France, and Greece) fishponds exist, which may be the ideal localization for algae ponds.

Very interesting is the fact that the photosynthesis of macroalgae is saturated at different levels of carbon dioxide, ranging from 500 to 2000 ppm [17, 18], which means that with carbon dioxide concentration up to five times the atmospheric concentration and under the correct light conditions and nutrient supply, macroalgae may grow with the same or better performance than they show in natural environments [19]. Experimentally, it has been demonstrated [20, 21] that concentrations of CO_2 in the gas phase of up to 5% are acceptable for growing macroalgae such as *Gracilaria bursapastoris*, *Chaetomorpha linum*, and *Pterocladiella capillacea*.

In general, macroalgae does not require very sophisticated techniques for growing: coastal farms are the most used techniques for macroalgae.

The world market of seaweeds is remarkable. The aquaculture production is around 11.3 million wet tons. China is the main producer (92% of the world seaweeds supply) [22]. Brown seaweeds represent 63.8% of the production, while red seaweeds represent 36.0% and the green seaweeds 0.2%. Approximately 1 million tons of wet seaweeds are harvested and treated to produce about 55,000 tons of hydrocolloids, valued at almost US$ 600 million [23].

The adaptation of macroalgae from wild conditions to pond culture is not straightforward. Thalli can be cut and used for starting a new culture. In principle, it is more suitable to cultivate macroalgae using the natural climatic conditions, as the adaptation to different climates may not be easy. The knowledge of physiological conditions is essential for the definition of the best operative parameters (optimized growing conditions) [10].

Very interesting is the use of drift macroalgae (when the production is higher than the capacity of the ecosystem), which may represent a way to convert a waste into an energy source. Actually, macroalgae have been used to produce algae-paper and as soil additives in agriculture. Figure 3.4 shows examples of overproduction of macroalgae that can be harvested and used as energy source when the economic conditions exist and the CO_2 emission into the atmosphere will be reduced with respect to the use of fossil fuels.

(a) (b) (c)

Fig. 3.4: (a) *Ulva clathrata* (Aonori Aquafarms), (b) *Himanthalia elongata, Laminaria digitata*, and *Fucus serratus* (in Tralee Bay), and (c) red drift algae.

3.3.2 Microalgae

Microalgae can be grown in open ponds or in photobioreactors (PBRs). The culture in open ponds is more economically favorable with respect to PBRs [24]. It was reported that the production cost ranged from $8 to $15/kg dry algae biomass in open ponds, whereas it was as large as $50/kg algae biomass in closed PBRs due to large operating costs [25].

However, PBRs provide yields that are three to five times higher than for open ponds. The latter may raise the issue of land cost and water availability, appropriate climatic conditions, nutrients cost, and production. Moreover, in the open-pond option, other cultivation aspects should be taken into consideration such as the maintenance of long-term growth of the desired algae strain without interference by competitors, grazers, or pathogens. These results also indicate how strong the overall production cost depends on the reactor (open or closed) cost [26].

Using open-pond systems, the nutrients can be provided through runoff water from nearby living areas or by channeling the water from wastewater treatment plants. CO_2 from power plants or industries (if the content of SO_2 and NO_x is below 150 ppm, see below) could be efficiently bubbled into the ponds and captured by the algae. The water is moved by paddle wheels or rotating structures (raceway systems), and some mixing can be accomplished by appropriately designed guides. Typically, microalgae are cultivated in open ponds (horizontal or circular) as shown in Figure 3.5.

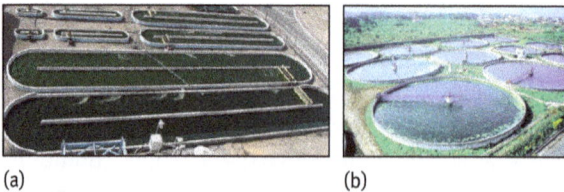

(a) (b)

Fig. 3.5: (a) Raceway pond and (b) circular open pond.

The production of microalgae in open ponds depends on the climatic conditions. The solar irradiation and temperature are the most important factors affecting the farming process and its productivity. These two parameters drive the growing period and thus the economics of the process.

The availability of land and water are the key factors for developing open-pond cultures. So far, semi-desert flat lands unsuitable for tourism, industry, agriculture, or municipal development were also selected if biomass cultivation in such areas is strongly affected by the supply of CO_2 and water. In fact, either CO_2 or water becomes a limiting factor.

In a open-pond system, the loss of water is greater than in closed tubular cultivation or bag cultivation methods. The water can be ground saline water, local industrial water, or water drained from agricultural areas and recycled after harvesting algae. Carbon dioxide for algae growth can be distributed using pipelines, which transport purified CO_2 or directly flue gases from power plants or any other gas rich in carbon dioxide.

Nutrients (N and P compounds, micronutrients) represent one of the major costs for algal growth. The use of wastewater (sewage, fisheries, some industrial waters) rich in N and P nutrients is an economic option with a double benefit represented by the recovery and utilization of useful inorganic compounds, and the production of clean water that can finally be reused or discharged into natural basins. Should nutrients be added to water, the biomass would not produce a "zero-emission fuel", as the production of nutrients are associated with a large emission of CO_2. Therefore, the use of wastewater rich in N and P compounds is a must when growing algae in pounds or PBR. The direct use of flue gases as CO_2 providers requires that algae should be resistant to the pollutants that are usually present in the flue-gas stream, namely nitrogen and sulfur oxides. Studies have shown that 150 ppm of NO_2 and 200 ppm of SO_2 do not affect the growth of some algal species [27].

Anyway, it must be noted that the resistance to NO_x and SO_y is not a common feature of all algal species, and this may represent a limitation to the direct use of flue gases. Another point that demands clarification is the optimal concentration of carbon dioxide in the culture, as CO_2 addition lowers the pH of the medium. Although the response to the pH change generated by the concentration of carbon dioxide may be different for the various algal species, operating at pH close to 6 may in general strongly affect the algal growth. However, one of the key points in culturing microalgae, or algae in general, is to generate the optimal concentration of CO_2 in the gas and liquid phase. CO_2 can be supplied into the algal suspension in the form of fine bubbles. A drawback of this methodology is the residence time in the pond: it must be sufficient to allow CO_2 to be uptaken [28]. In general, in this way, a lot of CO_2 is lost to the atmosphere and only 13–20% of CO_2 is usually used. A different method to supply CO_2 is the gas exchanger, which consists of a plastic frame that is covered by transparent sheeting and immersed in the suspension. CO_2 is fed into the unit and the exchanger floated on the surface. CO_2 needs to be in a concentrated form and

25–60% of it is distributed and used [28]. Also, if it is the most effective method, it presents as drawback the need to use very concentrated and pure CO_2, which is trapped under the transparent plastic frame with very little amount of CO_2 lost into the atmosphere. The growth rate of microalgae is dependent on the temperature and the season (high growth rate in summer and low growth rate in winter). Tropical or semi-tropical areas are the most practical locations for algal culture systems [29]. Before starting to build a culture system, it is necessary to consider several aspects such as the evaporation rate, which may represent a problem in dry tropical areas. Here, the evaporation rate is higher than the precipitation rate: a high evaporation rate increases the salt concentration and pumping costs due to water loss [30]. High precipitation rate can cause dilution and a loss of nutrients and algal biomass. With low relative humidity, high rates of evaporation occur, which can have a cooling effect on the medium [31], while with high relative humidity and no wind, an increase in the temperature of the medium may occur (even up to 40°C). Finally, a location must be chosen where there is a constant and abundant supply of water for the mass culture pond systems.

Methods to cultivate algae have been developed over the years. Recent developments in algae growth technology include vertical PBRs [32] and bag (vertical) reactors [33] made of polythene mounted on metal frames, reducing the land required for cultivation.

Using such bioreactors, microalgae can grow under light irradiation and temperature-controlled conditions, with an enhanced fixation of carbon dioxide that is bubbled through the culture medium. Algae receive sunlight either directly through the transparent container walls or via light fibers or tubes that channel the light from sunlight collectors. A number of systems with horizontal and vertical tubes, bags, or plates are made of either glass or transparent plastic exposed to the sun either in the free air or in greenhouses (Figure 3.6).

Using these kind of reactors, several microalgae productions have been set up as shown in Figure 3.7.

Microalgae may easily adapt to the culture conditions [30, 34]; also, the different parameters that influence the rate of growth and cell composition of microorganisms must be kept under strict control in order to guarantee a constant quality of the biomass, a parameter particularly important for the biomass exploitation.

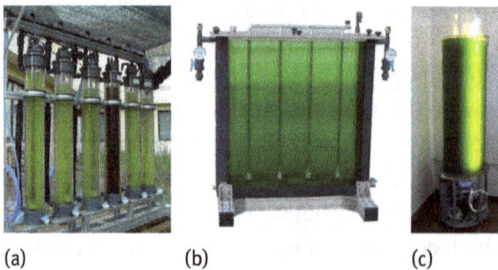

(a)   (b)   (c)

Fig. 3.6: PBRs (CAISIAL, Portici (Na), Italy).

Fig. 3.7: (a) *Chlorella* production in Germany, (b) *Spirulina platensis* production in Hawaii, (c) Arizona State University Polytechnic Laboratory for Algae Research and Biotechnology, LARB, (d) *Haematococcus* production in Negev desert, Israel, (e) pilot plant at Coyote Gulch outside Durango, CO, for biofuel production, and (f) flat-plate "acrylic" PBRs, AzCATI.

Another factor that influences the growth of microalgae is irradiation. Both in ponds and in bioreactors, light availability is of paramount importance. Shadow or short light cycles may slow down the growth rate; conversely, intense light (as may occur in desertic areas or bioreactors) does not guarantee fast growth, as it may modify the cell functions [35–37].

3.4 Harvesting of aquatic biomass

3.4.1 Macroalgae

The harvesting of macroalgae and plants requires more immediate and not very sophisticated technologies. The technique depends on whether the biomass is grown floating-unattached or attached to a hard substrate. In the former case, the biomass can be easily collected using a net (as in fishing); in the latter case, it must be cut from the substrate. Automated or manual devices can be used for the collection [38].

Harvesting of macroalgae is carried out in different ways. The manual harvesting (Figure 3.8a) is common for both natural and cultivated seaweeds. Mechanized harvesting methods (Figure 3.8b,c), which can involve mowing with rotating blades, suction, or dredging with cutters, have been developed.

(a)　　　　　　　　(b)　　　　　　　　(c)

Fig. 3.8: Harvesting of macroalgae.

3.4.2 Microalgae

Differently from macroalgae, microalgae, due to their size and, sometimes, fragility, demand for sophisticated equipment and handling operations.

The choice of harvesting methods depends on a few factors such as
- type of algae that has to be harvested (filamentous, unicellular, etc.)
- type of harvesting (continuously or discontinuously)
- energy demand per cubic meter of algal suspension
- investment costs [30, 36, 37, 39].

The mainly used technologies with microalgae are centrifugation, sedimentation, filtration, screening and straining, and flocculation.

Various flocculants have been used, covering a large variety of chemical structures such as metal compounds [40], cationic polymers [41], and natural polymers such as chitin [42]. They have been employed not only at the laboratory scale, but also at the industrial scale. Such "induced flocculation" may be accompanied by a "spontaneous or auto-flocculation" that can be caused by pH variation of the culture medium upon CO_2 consumption. For example, an increase in pH may cause the precipitation of phosphates (essentially Ca phosphate), which causes flocculation of algae. Aggregation of algae, produced by organic secreted substances [43] or aggregation with bacteria [44, 45] (Figure 3.9), may also occur, which facilitates their sedimentation.

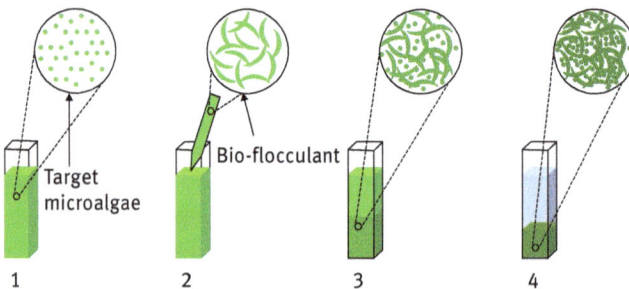

Fig. 3.9: Schematic view of bio-flocculation [45].

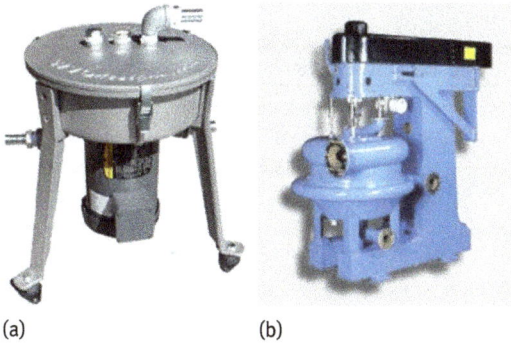

(a)   (b)

Fig. 3.10: (a) An example of centrifuge (CAISIAL) and (b) Alfa Laval CH-36B GOF separator centrifuge.

Centrifugation is a very popular technique today, but still it presents some drawbacks such as the rate of separation and it generally is considered expensive and electricity consuming. It is, however, the best-known method of concentrating small unicellular algae [46]. Centrifuges such as that illustrated in Fig. 3.10a are used; Benemann recommends in Sazdanoff's report [47] to use centrifugation after pond settling, with a specific centrifuge (Figure 3.10b) that has an acceptable energy consumption.

Recently, new technologies have been developed that lower the energy consumption [48]. Most advanced technologies are based on the use of membranes (tubular, capillary, or hollow-fiber membranes) that are becoming more and more popular [49]. The size of the pore decreases in the order from tubular (5–15 mm) to capillary (1 mm) to hollow fiber (0.1 μm), and the risk of plugging increases with the decrease in the pore diameter.

3.5 Composition of aquatic biomass

Aquatic biomass contains several pools of chemicals at different concentration depending on the physical stresses or genetic manipulation induced on the organism. Tables 3.1 and 3.2 show the categories of many products produced by macroalgae and microalgae, respectively.

In general, microalgae and macroalgae can be used in different sectors:
- energy (hydrocarbons, hydrogen, methane, methanol, ethanol, fames, biodiesel, etc.)
- food and chemicals (proteins, oils and fats, sterols, carbohydrates, sugars, alcohols, etc.)
- other chemicals (dyes, perfumes, vitamins/supplements, etc.).

Table 3.1: Products from macroalgae.

Class of products	Chemicals	Extraction technology	Commercial use
Proteins			Pharmacology
Amino acids		Phenol-acetic acid-water	Food industry
Lipids		Sc-CO$_2$, organic solvent, liquefaction, pyrolysis	Biofuels, food, and pharmaceutical industry
Essential oils	Geraniol, geranyl formate or acetate, citronellol, nonanol, eucalyptol	Distillation	
Alkaloids		Solvent extraction	
Sterols	Cholesterol		
Pigments: chlorophylls, carotenoids, xantophylls	Isoprenoids	Solvent extraction	
Amines	Methylamines, ethylamines, propylamine, isobutylamine		Pharmaceutical industry
Inorganic compounds	Iodides, bromides, sulfates, nitrates, etc.		Pharmaceutical industry

Table 3.2: Products from microalgae.

Class of products	Chemicals	Applications
Coloring substances and antioxidants	Xantophylls (astaxantines and canthaxanthin, lutein, β-carotene, vitamins C and E)	Health, food, functional food, feed additive, aquaculture, soil conditioner
FAs	Arachidonic acid, eicosapentenoic acid, docosahaexenoic acid, glinolenic acid, linoleic acid	Food and feed additives, cosmetics
Enzymes	Superoxide dismutase, phosphoglycerate kinase, luciferase and luciferin, restriction enzymes	Health food, research, medicine
Polymers	Polysaccharides, starch, poly-β-hydroxybutyric acid	Food additive, cosmetics, medicine
Special products	Peptides, toxins, isotopes, amino acids (proline, arginine, aspartic acid), sterol	Research, medicine

Aquatic biomass can be used as a raw, unprocessed food, as they are rich in caro-
tenoids, chlorophyll, phycocyanin, amino acids, minerals, and bioactive compounds.

Besides their nutritional value, these compounds have application in pharmaceutical fields as immune-stimulating, metabolism-increasing, cholesterol-reducing,
anti-inflammatory, and antioxidant agents [50]. Also, they are rich in omega-3 fatty

acids (FAs), which can be used in the treatment of heart diseases because of its anti-inflammatory property.

Due to the high product distribution entropy, the extraction of a single low-cost product may have an economic benefit only when the product represents at least 10% of the global dry mass. If it is present at the level of only a few percentage or lower, then it should have a high market value to meet the economic criteria. As mentioned above, the ability of algal organisms to concentrate a type of resource (proteins, starch, lipids) upon stress may help to reduce the entropy and to increase the concentration of a given product in the biomass. This issue is particularly relevant when the use of aquatic biomass for energy purposes is considered. Due to the cost of cultivation, in case of application as energy source, producing biomass with a high content of energy products is a must.

3.5.1 Bio-oil content of aquatic biomass

Microalgae are much considered today for the production of biodiesel, although this is not the only producible fuel: biogas can also be produced as well as bioethanol or bio-hydrogen or hydrocarbons. The quality and composition of the biomass will suggest the best option for the biofuel to be produced. A biomass rich in lipids will be suitable for the production of bio-oil and biodiesel, while a biomass rich in sugars will be better suited for the production of bioethanol. Anaerobic fermentation of sugars, proteins, and organic acids will produce biogas.

Several species of microalgae are very rich in lipids (up to 70–80% dry weight, with a good average standard of 30–40%, Table 3.3), and this makes a given species-strain more or less suitable for bio-oil production. The highest values are relevant to particular growing conditions. In a commercial culture, what is of interest is the productivity of a pond, i.e., the production per unit time.

Table 3.3: Lipid content of same microalgae [51].

Microalgae	Oil content (% dry weight)
Botryococcus braunii	25–75
Chlorella sp.	28–32
Crypthecodinium cohnii	20
Cylindrotheca sp.	16–37
Dunaliella primolecta	23
Isochrysis sp.	25–33
Monallanthus salina	>20
Nannochloropsis sp.	31–68
Neochloris oleoabundans	35–54
Nitzschia sp.	45–47
Phaeodactylum tricornutum	20–30
Schizochytrium sp.	50–77
Tetraselmis sueica	15–23

Table 3.4: Yields (L/ha/year) of fuels for various types of biomass.

Biomass	Yield (L/ha/year)
Corn	170
Soybeans	455–475
Safflower	785
Sunflower	965
Rapeseed	1200
Jatropha	1890
Coconut	2840
Palm	6000
Microalgae	47,250–142,000

Table 3.4 compares the amount (L) of oil per hectare per year produced by different types of biomass including microalgae [52, 53].

Macroalgae, in general, present a lower content of lipids (Table 3.5) than microalgae and a larger variability [21]. The lipid content largely depends on the cultivation technique and on the period of the year macroalgae are collected [54, 55]. These are thus key issues to be taken into consideration in the development of a commercial exploitation of such biomass.

Comparing microalgae and macroalgae, it must be considered that macroalgae are produced at lower costs than microalgae. The energy value of an alga cannot be stated only on the basis of the specific amount of oil it produces. Other very important parameters much be considered, such as the quality of the extracted oil, the possibility to produce other forms of energy from the residue after lipid extraction,

Table 3.5: Lipid content of same macroalgae.

Macroalgae	Oil content (% dry weight)
Chaetomorpha linum	14–24
Chlorophytes	1.5–7.5
Cladophora	12–22
Codium duthiae	12.2–20.7
Codium harveyi	8.8–12.1
Codium fragile	21.1
Fucus serratus	2.1
Palmaria palmate	0.3
Phaophyta	1.7–7.8
Rhodophyta	2–8
Spyridia filamentosa	1–3
Stypopodium schimperi	11–13
Ulva lactuca	12–15
Ulva lobata	2–4

and even the production of chemicals. Today, microalgae are not economically viable as an only source of fuels. Therefore, a biorefinery approach that may afford chemicals and fuels may be the winning option to add value to the feedstock.

3.5.2 The quality of bio-oil

Although the algae biomass can be thermally processed (pyrolysis) [56] to afford an oily product, the acidity and composition of the liquid are such that its direct use is not suited and complex processing is needed before its use. The extraction of lipids will be discussed in the following paragraphs. Lipids are a mixture containing more than a single type of FA, most frequently, the lipid fraction of algae (both microalgae and macroalgae) contains a large variety of FAs, with different number of unsaturation, as shown in Table 3.6. This is an important issue for assessing the energetic value of a biomass. The number of unsaturation in an FA is important, as it determines the usability of the compound as a fuel. In fact, the optimal conditions for having FAMEs or biodiesel with good combustion properties is the presence of only one unsaturation in the C chains [57]. Therefore, the higher is the number of unsaturations, the lower is the quality of the fuel produced.

An increase in CO_2 concentration up to 10% in the gas phase can influence the number of unsaturation and can almost double the total concentration of FAs (from 29.1% to 55.5%) and in particular that of FAs 16:0, 18:1, 20:4, and 20:5 in *C. linum* [20]. In general, it has been found that the number of unsaturation may increase with the concentration of CO_2 [20, 58, 59].

Bio-oil, such as extracted, can be directly used in thermal processes or in combustion but cannot be used in diesel engines, as it presents a low enthalpy value (LHV) (8–12 MJ/kg) and a high viscosity and unsaturation. It can be converted into biodiesel through a transesterification reaction, followed by partial hydrogenation in order to reduce the number of unsaturation to 1. The LHV must be increased as close as possible to >30.

From the environmental point of view, biodiesel introduces several benefits, such as reduction of carbon monoxide (50%) and carbon dioxide (78%) emissions [60], elimination of SO_2 emission, as biodiesel does not contain sulfur, and reduction of particulate. As biodiesel is non-toxic and biodegradable, its use and production has rapidly increased, especially in Europe, USA, and Asia. A growing number of fuel stations are making biodiesel available to consumers, and a growing number of large transport fleets use a fuel that contains biodiesel in variable percentage. Table 3.7 reports some fuel properties of different types of bio-oil. Actually, the low cost of oil and methane is compressing the biofuel market.

Recently, a new area of application became of interest, i.e., the production of avio-fuels. These may include biodiesel and other molecules derived from different fractions of aquatic biomass.

Table 3.6: Distribution of FAs in lipids present in some macroalgae.

FA compound	N of C atoms/unsaturated bonds	Species and percentage of a given compound in the species					
		Fucus sp.	Nereocystis luetkeana	Ulva lactuca	Enteromorpha compressa	Padiva pavonica	Laurencia obtuse
Saturated	$C_{12} \rightarrow C_{20}$	15.6	27.03	15.0	19.6	23.4	30.15
Monounsaturated	$C_{14} \rightarrow C_{20}$	28.55	15.84	18.7	12.3	25.8	9
Polyunsaturated	$C_{16/2} \rightarrow C_{16/4}*$ $C_{18/2} \rightarrow C_{18/4}*$ $C_{20/2}$	55.86	57.11	66.3	68.1	50.8	60.9

Table 3.7: Fuel characteristics of different bio-oils.

	Density (kg/L)	Ash content (%)	Flash point (°C)	Pour point (°C)	Cetane number	Calorific value (MJ/kg)	Reference
Algae	0.801	0.21	98	–14	52	40	Vijayaraghavan and Hemanathan [61]
Peanuts	–	–	271	–6.7	41.8	–	Knothe et al. [62]
Soya bean	0.885	–	178	–7	45	33.5	
Sunflower	0.860	–	183	–	49	49	
Diesel	0.855	–	76	–16	50	43.8	
Biodiesel from marine fish oil	–	–	103	–	50.9	41.4	Lin and Ri [63]

3.6 Technologies for algal oil and chemicals extraction

Oil and chemicals can be extracted from the biomass using a fractionation approach, i.e., a variety of technologies of different intensity (destructive, semi-destructive, and non-destructive) [21, 64]. There is a relation between the softness-hardness of the technology used and the complexity of the structure of the chemicals extracted. Softer technologies will less affect complex molecular structures that will be recovered unchanged. An increase in hard character of technologies will progressively destroy complex networks and complex molecules.

Biomass is suitable for the production of different products, such as bio-oil, biodiesel, bioalcohol, biohydrogen, biogas, all related to the production of energy.

The extraction of chemicals from microalgae and macroalgae requires the breaking of the cell membrane of the algae, which has quite a large variability of mechanical resistance. Oil from microalgae cannot be extracted using the most conventional method used in oil seed processing. Algal lipids are stored inside the cell as storage droplets or in the cell membrane. The small size of the algal cell and the thickness of the cell wall prevent simple expelling to release the oil. Depending on the species-strain, cell membrane can be very hard or elastic, so that crushing of the membrane is recommended prior to the extraction. Such crushing is quite effective if performed at low temperature, typically the liquid nitrogen temperature (183 K). This will obviously increase the cost of the extracted oil and lower the net energetic value of the biomass.

Among the technologies used to produce chemicals from biomass, solvent extraction with conventional organic solvents (with and without in situ transesterification), supercritical fluids, mechanical extraction, and biological extractions are the most used.

3.6.1 Fractionation of algal biomass

Algae biomass have captured the interest of researchers because of the good application potential of the concept of a biorefinery in the production of chemicals and energy, improving biomass utilization in its entirety (Figure 3.11). Lipids are considered the most valuable components of algal biomass in the context of a biofuels process, but other biomass components such as proteins and carbohydrates represent a large fraction of the biomass to convert [65]. The composition of algal biomass is similar to that of traditional plant crops, but the lack of the structural component lignin facilitates the separation of more valuable carbohydrates from less valuable lignin, which is often complicated [66].

Recently, new integrated processes that convert all biomass components into biofuels and chemicals are investigated. The National Renewable Energy Laboratory process assumes high solubilization/recovery of both carbohydrates and lipids and their conversion to ethanol and renewable diesel blendstock (RDB) (paraffins targeted in the C13–C20 range), respectively. The process consists of a dilute acid pretreatment

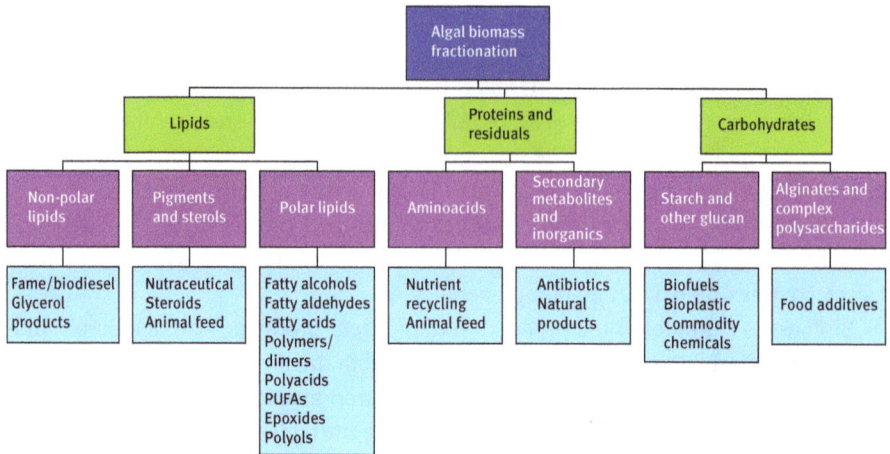

Fig. 3.11: Algal biomass fractionation and co-product generation (**adapted from** [66]).

of algal biomass delivered after upstream dewatering to 20 wt% solids, followed by whole-slurry fermentation of the monomeric sugars to ethanol, followed by distillation and solvent extraction of the stillage to recover lipids [67].

Laurens et al. [68] describe a new route to valorizing algal biomass components with a integrated technology based on moderate temperatures and low pH to convert the carbohydrates in wet microalgal biomass (*Chlorella* and *Scenedesmus*) to soluble sugars for fermentation, while making lipids more accessible for downstream extraction and leaving a protein fraction behind. Such method may offer more co-product flexibility than, for example, a hydrothermal liquefaction, which converts the whole biomass without fractionates to selective components.

Czartoski et al. [69] propose a new way to extract target classes and products and to recover by-products and recycle critical nutrients and water. The invention describes a method of fractionating biomass, including several steps in which two main fractions (one polar and one non-polar) are obtained. With the non-polar fraction, triglycerides, free FAs, and hydrocarbons are collected, while with the polar fraction, all the polar-soluble components (fiber, nutrients, soluble proteins, carbohydrates, residual solid algae cell structural particles, etc.) are collected.

3.6.2 Extraction using chemico-physical methods

Chemico-physical treatments, such as ultrasonication (disruption with high-frequency sound waves) and homogenization (carried out by rapid pressure drops), may be used to disrupt cell walls and lead to enhanced oil recovery. For example, Pursuit Dynamics Ltd [70] manufactures a device based on steam injection and supersonic disruption and claim homogenization of plant material with very low energy input. Systems based on sonication process and centrifugation may provide economic solutions for algal lipid recovery. Very interesting is the application of the reactive extraction using ultrasonication or microwaves in order to have a direct one-pot conversion of biomass into FAMEs.

3.6.3 Conventional solvent extraction

Solvents, such as hexane, have been used to extract and purify soybean seed oils and high-value FAs. These types of solvent-based processes are most effective with dried feedstock or with those with minimal amounts of free water. Of course, when aquatic biomass (which has water content ranging from 70% to 90%) has to be treated, the cost of drying significantly adds to the overall production cost and requires significant energy demand or the use of ovens, each raising specific issues of land requirement or energy cost. A limited number of solvents have been evaluated for large-scale extraction of algal biomass with some success, but at that time, no effort has been made to determine the process economics or material and energy balances of such processes [71]. The drying of algae wet pastes for the large-scale organic solvent extraction may not be economically feasible or sound in terms of energy for biofuels production.

An alternative to the organic solvent-based processes process is the extraction by in situ transesterification [72]. In this approach, in particular using heterogeneous catalysts and methanol as solvent, the bound lipids are released from the biomass directly as methyl esters. Additionally, the catalyst can be easily recovered.

3.6.4 Supercritical fluid extraction (SFE)

Supercritical CO_2 ($scCO_2$) has both liquid and gas properties, allowing the fluid to penetrate the biomass and act as an organic solvent, without the challenges and expense of separating the organic solvent from the final product. Literature describes successful extraction of algal lipids with $scCO_2$ [21, 73]) and the resulting conversion into biodiesel. The ability of SFE to operate at low temperatures preserves both the algal lipid quality during the extraction process and the residue composition, virtually eliminates the degradation of the product extracted, and minimizes the need for additional solvent processing (sometimes methanol can be added as co-solvent in order to increase the extraction yield). In addition, the ability to significantly vary the CO_2 solvation power by changes in pressure and/or temperature adds operating flexibility to the $scCO_2$ extraction process that no other extraction method, including solvent extraction, can claim [74]. Also, for the $scCO_2$ extraction, the biomass should be dried, then the cellular wall has to be broken in order to increase the extraction yield (it is possible to use liquid nitrogen or a different method) [75].

Bench-scale scCO$_2$ experiments on microalgae have been performed on *Botryococcus*, *Chlorella*, *Dunaliella*, and *Arthrospira*, from which different types of valuable products have been extracted as hydrocarbons (up to 85% mass of cell from *Botryococcus*), paraffinic and natural waxes from *Botryococcus* and *Chlorella*, strong antioxidants (astaxanthin, β-carotene) from *Chlorella* and *Dunaliella*, and linolenic acid from *Arthrospira*.

scCO$_2$ may substitute as the organic solvent, as it has some unique advantages and is considered a good candidate for algae treatment because it is a non-toxic and fully "green" solvent [50]. Despite the advantages, using scCO$_2$ to extract valuable compounds from microalgae is not the prevailing technology in use today even though production costs are of the same order of magnitude as those related to classic processes. In fact, for such a technique, anhydrous materials are recommended (water content below 5%); thus, energy should be consumed to dry the biomass.

However, the capital and operating costs for a high-pressure SFE operations currently limits its potential for biofuel production. Over time, SFE applications have targeted added value products, but not yet commodity chemicals. Technology development (e.g., gas antisolvent and subcritical fluid extractions) and further reductions in costs may lead to processes applicable to biofuel production.

3.6.5 Biological extraction

Biological methods used to capture and extract lipids offer low-technology and low-cost methods of harvesting and lipid extraction. Demonstrations in large open ponds of brine shrimp feeding on microalgae to concentrate the algae, followed by harvesting, crushing, and homogenizing the larger brine shrimp to recover oil have been successful [76]. Using crustaceans to capture and concentrate microalgae would appear to be a promising solution for algae oil recovery. The use of enzymes to degrade algal cell walls and reduce the energy needed for mechanical disruption has also been investigated.

3.7 Conclusions

Wild type of microalgae and macroalgae very often are not suitable to produce energy, as they have a chemical composition that may vary according with the growing conditions also within the same strain. For this reason, to produce energy, it is better to use a selected cultivated strain in order to have an optimal energetic yield. Moreover, aquatic biomass has to be considered as a source of several compounds that can be used as chemicals or to produce energy. The co-production of chemicals and fuels from aquatic biomass can be of great importance in order to make the economic balance positive. In fact, if biomass is used only to produce energy, the cost of fuels derived from it is not competitive with that of fossil fuels. The correct application of the concept of biorefinery may lower the cost of producing fuels if high-value chemicals are co-produced. In the near future, aquatic biomass might contribute to the production of transport fuels in a significant volume, supposing that the right conditions for its growth, collection, and processing are developed. In any case, it seems that the co-production of chemicals and fuels is necessary for a profitable exploitation of aquatic biomass.

Bibliography

[1] Meher, L. C., Vidya Sagar, D., Naik, S. N., Technical aspects of biodiesel production by transesterification – a review, Renew Sustain Energy Rev **10** (2006) 248–268.

[2] FAO, Sustainable Bioenergy: A Framework for Decision Makers, UN Energy, 2007, www.fao.org.

[3] FAO, The State of Food and Agriculture, New York: Food and Agriculture Organization, 2008.

[4] Moore, A., Biofuels are dead: long live biofuels(?) – part one, New Biotechnol **25** (2008) 6–12.

[5] IEA, World Energy Outlook 2006, International Energy Agency, Paris, 2006.

[6] Marchetti, J. M., Miguel, V. U., Errazu, A. F., Possible methods for biodiesel production, Renew Sustain Energy Rev **11** (2007) 1300–1311.

[7] Sheehan, J., Dunabay, T., Benemann, J., Roessler, P., A look back at the US Department of Energy aquatic species program: biodiesel from algae, Nat Renew Energy Lab (1998) 1–328.

[8] Dhote, S., Dixit, S., Water quality improvement through macrophytes – a review, Environ Monit Asses **152** (2009) 149–153.

[9] Vandendriessche, S., Vincx, M., Degraer, S., Floating seaweed in the neustonic environment: a case study from Belgian coastal waters, J Sea Res **55** (2006) 103–112.

[10] Aresta, M., Dibenedetto, A., Tommasi, I., Cecere, E., Narracci, M., Petrocelli, A., Perrone, C., The use of marine biomass as renewable energy source for reducing CO_2 emissions, Special Issue Dedicated to GHGT-6, Elsevier, Kyoto, 2002.

[11] Ryther, J. H., DeBoer, J. A., Lapointe, B. E., Cultivation of seaweeds for hydrocolloids, waste treatment and biomass for energy conversion, Proc Int Seaweed Symp **9** (1979) 1–16.

[12] Schramm, W., Seaweed for waste water treatment and recycling of nutrients, in Guiry, M. D., Blunden, G., editors, Seaweed Resources in Europe: Uses and Potential, John Wiley & Sons, Chichester, UK John Wiley & Sons, 1991, 149–168.

[13] Cohen, I., Neori, A., Ulva lactuca biofilters for marine fishpond effluents I, ammonia uptake kinetics and nitrogen content, Bot Mar **34** (1991) 977–984.

[14] Hirata, H., Xu, B., Effects of feed addictive *Ulva* produced in feedback culture system on the growth and color of red sea bream, Pagure Major Suisanzoshoku **38** (1990) 177–182.

[15] Sauze, F., Increasing the productivity of macro-algae by the action of a variety of factors, in Strub, A., Chartier, P., Schleser, G., editors, Energy from Biomass, Elsevier Applied Science, London, 1983, 324–328.

[16] Adams, J. M., Gallagher, J. A., Donnison, I. S., Fermentation study on *Saccarina latissima* for bioethanol production considering variable pre-treatments, J Appl Phycol **21** (2009) 569–574.

[17] Brown, D. L., Tregunna, E. B., Inhibition of respiration during photosynthesis by some algae, Can J Bot **45** (1967) 1135–1143.

[18] Smith, R. G., Bidwell, R. G. S., Carbonic anhydrase-dependent inorganic carbon uptake by the red macroalga *Chondru crispus*, Plant Physiol **83** (1987) 735–738.

[19] Gao, K., Aruga, Y., Asada, K., Kiyohara, M., Influence of enhanced CO_2 on grand photosynthesis of the red algae *Gracilaria* sp. and *G. chilensis*. J Appl Phycol **5** (1993) 563–571.

[20] Aresta, M., Alabiso, G., Cecere, E., Carone, M., Dibenedetto, A., Petrocelli, A., VIII Conference on Carbon Dioxide Utilization, Oslo, Book of Abstracts **56** (2005) 20–3.

[21] Aresta, M., Dibenedetto, A., Carone, M., Colonna, T., Fragale, C., Production of biodiesel from macro-algae by supercritical CO_2 extraction and thermochemical liquefaction, Environ Chem Lett **3** (2005) 136–139.

[22] FAO, The State of World Aquaculture, Electronic Publishing Policy and Support Branch, Rome, Italy, 2006.

[23] McHugh, D. J., A Guide to the Seaweed Industry. FAO Fisheries Technical Paper No. 441, FAO, Rome, 2003.

[24] Oilgae, Comprehensive Oilgae Report, Oilgae, Tamilnadu, India, 2010.

[25] Ozkan, A., Kinney, K., Katz, L., Berberoglu, H., Reduction of water and energy requirement of algae cultivation using an algae biofilm photobioreactor, Bioresource Technol **114** (2012) 542–528.

[26] Darzins, A., Pienkos, P., Edye, L., Current status and potential for algal biofuels production, A report to IEA Bioenergy Task 39, Report T39-T2, 2010.

[27] Zeiler, K. G., Heacox, D. A., Toon, S. T., Kadam, K. L., Brown, L. M., The use of micro-algae for assimilation and utilization of carbon dioxide from fossil fuel-fired power plant flue gas, Energy Convers Manage **36** (1995) 707–712.

[28] Becker, E. W., Micro-algae: Biotechnology and Microbiology, Cambridge University Press, New York, 1994.

[29] Borowitzka, M. A., Borowitzka, L. J., Micro-algae Biotechnology, Cambridge University Press, Cambridge, 1988.

[30] Collins, S., Sueltemeyer, D., Bell, G., Changes in C uptake in populations *Chlamydomonas reinhardtii* selected at high CO_2, Plant Cell Environ **29** (2006) 1812–1819.

[31] Richmond, A., Handbook of Microalgal Mass Culture, CRC Press, Boca Raton, FL, 1986.

[32] Hitchings, M. A., Algae: the next generation of biofuels, fuel, fourth quarter, Hart Energy Publishing, Houston, TX, 2007.

[33] Bourne, Jr., J. K., Green dreams, National Geographic (October 2007), 3–5.

[34] Cecere, E., Aresta, M., Alabiso, G., Carone, M., Dibenedetto, A., Petrocelli, A., International Conference on Applied Physiology, Kunming, China, 24–28 July 2006.

[35] Dibenedetto, A., Tommasi, I., Biological utilization of carbon dioxide: the marine biomass option, in M. Aresta, editor, Carbon Dioxide Recovery and Utilisation, Kluwer Publisher, the Netherlands, 2003, 315–314.

[36] Ono, E., Cuello, J. L., Design parameters of solar concentrating systems for CO_2– mitigating algal photobioreactors, Energy **29** (2004) 1651–1657.

[37] Ono, E., Cuello, J. L., Carbon dioxide mitigation using thermophilic cyanobacteria, Biosyst Eng **96** (2007) 129–134.

[38] Morineau-Thomas, O., Legentilhomme, P., Jaouen, P., Lepine, B., Rince, Y., Influence of a swirl motion on the interaction between microalgal cells and environmental medium during ultrafiltration of a culture of *Tetraselmis suecica*, Biotechnol Lett **23** (2004) 1539–1545.

[39] Pulz, O., Gross, W., Valuable products from biotechnology of micro-algae, Appl Microbiol Biotechnol **65** (2004) 635–648.

[40] Bare, W. F. R., Jones, N. B., Middlebrooks, A. J., Algae removal using dissolved air flotation, J. Water Poll Control Fed **47** (1975) 153–159.

[41] Tenny, M. W., Echelberger, Jr., W. F., Scnessler, R. G., Pavoni, J. L., Algal flocculation with synthetic organic polyelectrolytes, Appl Microbiol **18** (1969) 965–971.

[42] Venkataraman, L. V., Algae as food/feed, Proceedings of Algae Systems, India Society of Biotechnology, HT India, 83, 1980.

[43] Benemann, J., Koopman, B. C., Weissman, J. R., Eisenberg, D. M., Goebel, R. P., Development of micro-algae harvesting and high rate pond technologies in California, in Shelef, G., Soeder, C. J., editors, Algal Biomass, Elsevier/North Holland Biomedical Press, Amsterdam, 1980; 457–495.

[44] Kogure, K., Simidu, U., Taga, N., Bacterial attachment to phytoplankton in sea water, J Exp Mar Biol Ecol **5** (1981) 197–204.

[45] Salim, S., Bosma, R., Vermuë, M. H., Wijffels, R. H., Harvesting of micro-algae by bio-flocculation, J Appl Phycol **23** (2011) 849–855.

[46] Grima, E. M., Belarbi, E. H., Fernandez, F. G. A., Medina, A. R., Chisti, Y., Recovery of microalgal biomass and metabolites: process options and economics, Biotechnol Adv **20** (2003) 491–515.

[47] Sazdanoff, N., Modeling and Simulation of the Algae to Biodiesel Fuel Cycle, Department of Mechanical Engineering, the Ohio State University, 2006.

[48] Boele, H., Broken, M., Evodos SPT proven algae harvesting technology, in Workshop on algae: Technology Status and Prospects for Deployment, EU BC&E Conference, 2011, ISBN-10 8889407557, ISBN-13 978-8889407554.

[49] Mohn, H. F., Experiences and strategies in the recovery of biomass from mass cultures of micro-algae, in Shelef, G., Soeder, C. J., editors, Algae Biomass: Production and Use, Elsevier/North Holland Biomedical Press, Amsterdam, 1980, 547–571.

[50] Singh, S., Kate, B. N., Banerjee, U. C., Bioactive compounds from cyanobacteria and micro-algae: an overview, Crit Rev Biotechnol **25** (2005) 73–95.

[51] Chisti, Y., Biodiesel from microalgae, Biotechnol Adv **25** (2007) 294–306.

[52] Briggs, M., Widescale Biodiesel Production from Algae, University of New Hampshire Biodiesel Group, 2004, available from: http://www.unh.edu/p2/biodiesel/article_alge.html.

[53] Riesing, T. F., Cultivating algae for liquid fuel production, 2006, available from: http://oakhavenpc.org/cultivating_algae.htm.

[54] Khotimchenko, S. V., Fatty acids of species in the genus *Codium*, Bot Mar **46** (2003) 455–460.

[55] Al-Hasan, R. H., Hantash, F. M., Radwan, S. S., Enriching marine macro-algae with eicosatetraenoic (arachidonic) and eicosapentaenoic acids by chilling, Appl Microbiol Biot **35** (1991) 530–305.

[56] DalmasNeto, C. J., Sydney, E. B., Assmann, R., Neto, D., Soccol, C. R., Production of biofuels from algal biomass by fast pyrolysis, in Biofuels from Algae, 2013, chapter 7, 143–153.

[57] Renaud, S. M., Luong-Van, J. T., Seasonal variation in the chemical composition of tropical Australian marine macro-algae, J Appl Phycol **18** (2006) 381–387.

[58] Fu, F. X., Warner, M. E., Zhang, Y., Feng, Y., Hutchins, D. A., Effects of increased temperature and CO_2 on photosynthesis, growth, and elemental ratios in marine Synechococcus and Prochlorococcus (Cyanobacteria), J Phycol **43** (2007) 485–496.

[59] Andersen, T., Andersen, F. Ø., Effects of CO_2 concentration on growth of filamentous algae and *Littorella uniflora* in a Danish softwater lake, Aquat Bot **84** (2006) 267–271.

[60] Ben-Amotz, A., Polle, J. E. W., Subba Rao, D. V., The Alga *Dunaliella*: biodiversity, Physiology, Genomics and Biotechnology, Science Publ., 2008.

[61] Vijayaraghavan, K., Hemanathan, K., Biodiesel production from freshwater algae, Energy Fuel **23** (2009) 5448–5453.

[62] Knothe, G., Dunn, R. O., Bagby, M. O., Biodiesel: the use of vegetable oils and their derivatives as alternative diesel fuels in fuels and chemicals from biomass, in ACS Symposium Series, Vol. 666, 1997, chapter 10, 172–208, ISBN 13:9780841235083.

[63] Lin, C.-Y., Li, R.-J., Fuel properties of biodiesel produced from the crude fish oil from the soapstock of marine fish, Fuel Process Technol **90** (2009) 130–136.

[64] Dibenedetto, A., Angelini, A., Colucci, A., di Bitonto, L., Pastore, C., Aresta, B. M., Giannini, C., Comparelli, R., Tunable mixed oxides: efficient agents for the simultaneous trans-esterification of lipids and esterification of free fatty acids from bio-oils for the effective production of FAMEs, Int J Renew Energy Biofuels, 2014.

[65] Templeton, D. W., Quinn, M., Van Wychen, S., Hyman, D., Laurens, L. M., Separation and quantification of microalgal carbohydrates, J Chromatogr A **1270** (2012) 225–234.

[66] Foley, P., Beach, E., Zimmerman, J., Algae as a source of renewable chemicals: opportunities and challenges. Green Chem **13** (2011) 1399–1405.

[67] Davis, R., Kinchin, C., Markham, J., Tan, E. C. D., Laurens, L. M. L., Sexton, D., Knorr, D., Schoen, P., Lukas, J., Process design and economics for the conversion of algal biomass to biofuels: algal biomass fractionation to lipid and carbohydrate-derived fuel products. Technical Report, 2014 NREL/TP-5100-62368.

[68] Laurens, L. M. L., Nagle, N., Davis, R., Sweeney, N., Van Wychen, S., Lowell, A., Pienkos, P. T., Acid-catalyzed algal biomass pretreatment for integrated lipid and carbohydrate-based biofuels production, Green Chem **17** (2015) 1145–1158.

[69] Czartoski, J. T., Perkins, R., Villanueva, J. L., Richards, G., Algae biomass fractionation, US Patent Application Publication US2011/0086386 A1, 2011.

[70] European Patent, EP20080737243, 09/28/2011, 2011, http://pursuitdynamics.com.

[71] Nagle, N., Lemke, P., Production of methyl ester fuel from micro-algae, Appl Biochem Biotechnol **24/25** (1990) 355–361.

[72] Dibenedetto, A., Aresta M., Ricci, M., ENI Patent Appl MI2010A001867, 2010.

[73] Couto, R. M., Simões, P. C., Reis, A., Da Silva, T. L., Martins, V. H., Sánchez-Vicente, Y., Supercritical fluid extraction of lipids from the heterotrophic microalga *Crypthecodinium cohnii*, Eng Life Sci **10** (2010) 158–164.

[74] Mendes, R. L., Nobre, B. P., Cardoso, M. T., Pereira, A. P., Palavra, A. F., Supercritical carbon dioxide extraction of compounds with pharmaceutical importance from micro-algae, Inorg Chim Acta **356** (2003) 328–334.

[75] Gaspar, F., Leeke, G., Comparison between compressed CO_2 extracts and hydrodistilled essential oil, J Essent Oil Res **16** (2004) 64–68.

[76] Brune, D. E., Beecher, L. E., Proceedings of the 29th Annual Symposium on Biotechnology for Fuels and Chemicals, 29 April to 2 May 2007, Denver, CO, 2007.

Christin Groeger, Wael Sabra, and An-Ping Zeng

4 Introduction to bioconversion and downstream processing: principles and process examples

Abstract: For the development of biorefinery process a combination of many different process steps has to be considered, like biomass generation, upstream processing, bio-reaction engineering, downstream processing, and also transport and logistics. The heart of a biorefinery is the bioconversion itself, where substrates are transferred into different products by microorganism or enzymes. The next important step is the separation of the product(s), which has a main impact on the overall production costs. This chapter gives an introduction into bioconversion and downstream processing, showing their strong dependence to each other.

4.1 Introduction

A fundamental challenge in the 21st century is the conversion of an industry based on fossil resources and a consumption-oriented society into a sustainable industry and society based on realistic needs and natural resources. From a sustainability viewpoint, the renewable carbon inherent in biomass is vast, making biomass an attractive candidate among alternative and conventional carbon sources.

On average, herbal biomass consists of 75% carbohydrates, 20% lignin, and 5% other compounds like lipids and proteins [1]. Separation of plant biomass generally provides three process streams (Fig. 4.1): (1) carbohydrates, in the form of starch, cellulose, hemicellulose, and monomeric sugars; (2) aromatics, in the form of lignin; and (3) hydrocarbons (lipids or microbial oils), in the form of plant triglycerides and fatty acids. Compared to plants, algal biomass contains no lignin, but about 10% carbohydrates and up to 75% fats, which offers great potential, e.g., for lipids-based chemical conversion routes [2]. Proteins might also be part of the biomass feedstock, like in algae or protein-rich grains, but the percentage is far too low to create an economical biorefinery process. Protein-rich residues are usually converted via fermentation into biogas (energy integration) or are directly used as animal feed. In general, the cost of biomass increases in the following order: cellulosic<starch/sugar<triglyceride base biomass, whereas the cost of conversion technology is vice versa [3]. Unlike the petrochemical refinery, where impressive arrays of selective, high-yield structural transformations were developed for the conversion of crude oil or natural gas to an initial set of simpler building blocks and eventually to thousands of chemical products used by consumers, biomass usage as a raw material suffers from a much narrower range of discrete building blocks and limited methods for conversion. This technology gap is not the result of any inherently higher level of difficulty in processing of biomass. Instead, it is the result of chemical production research and technology to date being focused almost

exclusively on highly reduced, oil-based hydrocarbons, rather than highly oxygenated carbohydrate based materials. The increase in research interest in renewable sources in recent years is an effort to narrow this technology gap and to develop methodology for renewable carbon as efficient as that available for non-renewable carbon.

One of the major burdens of today's biorefineries is the energy-consuming degradation of highly complex molecules into intermediates and the subsequent formation of monomers or new intermediates and finally the desired products with various structures. This multiple consumption of process energy might be overcome by proper choice and engineering of microorganisms, which metabolize complex substrates and create molecules (e.g., biofuels, chemicals) efficiently with steps of only small alterations of the substrate and intermediates [4]. Hence, instead of just replacing fossil oil with biomass-based fuels, new bioconversion processes are needed to save energy, convert biomass efficiently, and ensure a sustainable usage of natural resources. Being almost the last step in biorefinery, bioconversion is important in determining the overall efficiency of a biorefinery process. Whether biofuels and bulk chemicals can be produced biotechnologically from renewable resources in the future will be determined mainly by the availability of the substrates and the production costs.

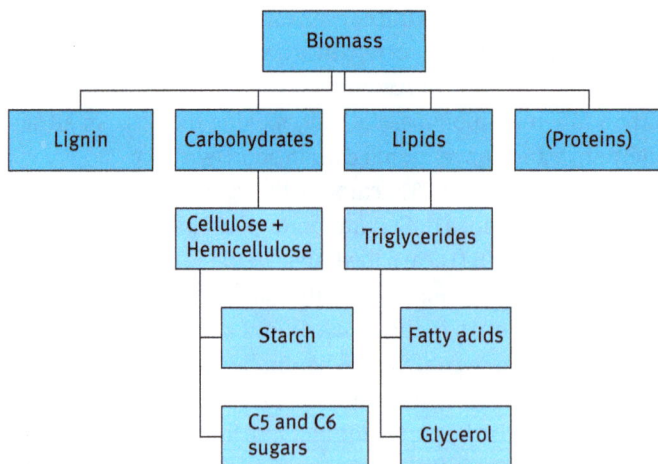

Fig. 4.1: Biomass as feedstock: different basic precursors for the conversion in biorefineries.

The field of bioconversion has reached its present industrially proven level through several waves of technological research and innovations [5]. Examples for established bioconversion processes include the production of ethanol, lactic acid, 1,3-propanediol, and more recently succinic acid [6, 7]. Still, the high feedstock cost poses a major obstacle to large-scale implementation of such bio-based products. Recent research interest has shifted to replacing traditional food-related feedstocks with non-food-related lignocellulosic biomass. Many factors like lignin content, crystallinity of cellulose, and particle size limit the digestibility of hemicellulose and cellulose present in the biomass [8]. In fact, the carbohydrate composition of hydro-lyzate (mainly C6 and C5 sugars) of cellulosic biomass impedes a complete conver-sion by pure cultures in typical bioconversion processes. Intensive system-biological studies to create novel organisms for the complete assimilation of the hydrolyzate car-bohydrate (reviewed in [9, 10]) may remove such obstacles soon and will increase the profitability of the second-generation biorefinery, e.g., for bioethanol from cellulosic materials. In the future, the so-called third-generation biofuels might be produced from algal biomass [11]. Production of biofuels from algae typically relies on the lipid content of the microorganisms. Usually, species such as *Chlorella* are targeted because of their high lipid content (around 60–70% [12]) and their high productivity (7.4 g/L/day for *Chlorella prototothecoides* [13]). However, there are many challenges associated with algal biomass, some are geographical and some are technical.

In this chapter, we first give a brief introduction to the principles of bioconversion at cellular and process levels. The bioproduction of 1,3-propanediol and *n*-butanol are then used as examples to illustrate the principles and practical aspects of bio-conversion in the context of biorefinery. Emphasis is also put on the corresponding downstream processing of bioconversion process as it largely determines the overall bioproduction costs. *n*-Butanol is an excellent example since it can be used directly as fuels or fuel additives or even as a platform chemical in the chemical industry. 1,3-Propanediol is an attractive monomer for new polyesters such polytrimethylene-terephthalate with superior properties [14, 15]. For more examples of possible product groups from bioconversion, the readers are referred to reviews in literature [5, 16–19].

4.2 Principles of bioconversion process

One of the major principles in microbial conversion processes is the coupling of the catabolism and anabolism (Fig. 4.2). Catabolism is the process of converting complex substrates such as sugars $(CH_2O)_n$ into precursors and intermediates (also called monomers such as amino acids), thereby bioenergetic molecules in forms of nicotinamide adenine dinucleotides (NADH) or adenosine triphosphate (ATP) and reducing power ([H]). The monomers and the energy are then used in the anabolic process to form complex molecules such as proteins, deoxyribonucleic acid, ribonucleic acid, and lipids for the synthesis of cellular biomass. The monomers and intermediates can be also converted into different (fermentation) products for example alcohols, organic acids, and amino acids under certain conditions. They are normally the target products of bioconversion processes. The control of the physiological conditions is critical for an efficient production of the target product since microorganisms have been evolved during the evolution primarily for proliferation and survival (maintenance).

The intermediates and energies of a bioconversion process should be well balanced. Of particular importance for bioconversion is the balances of ATP and the

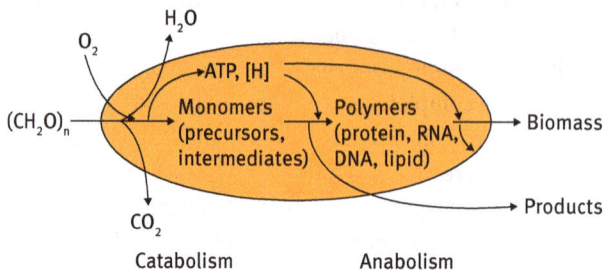

Fig. 4.2: Principles of bioconversion at cellular level

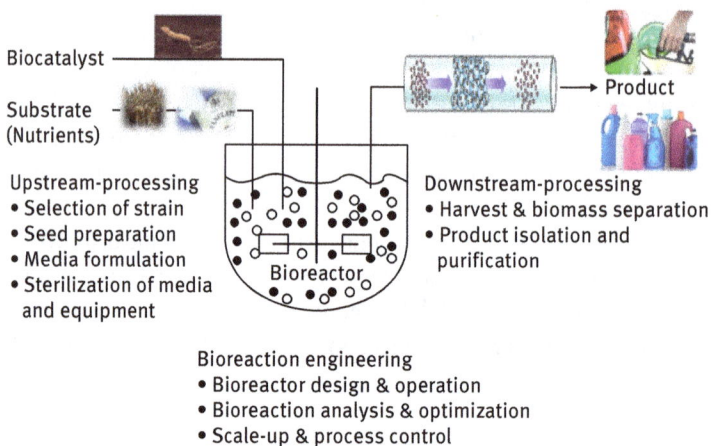

Fig. 4.3: Principles and major steps of bioconversion at process level.

reducing powers in forms of reduced nicotinamide adenine dinucleotides (mainly NADH or its phosphorated form NADPH) of oxidized ones (NAD+ and NADP+) in the catabolism and anabolism. Due to the fact that intracellular NAD+/NADH are not easy to measure, the concept of balance of "reductance degree (κ)" of substrate, biomass, and products is introduced. It will be illustrated below how this can be used to analyze and optimize bioconversion processes.

At process level, a bioconversion process consists of three major steps: upstream processing, bioreaction engineering, and downstream processing (Figure 4.3). The tasks of upstream processing include the selection and optimization of biocatalysts (microbes and enzymes), preparation and sterilization of substrate and medium (nutrients), and finally, seed preparation. Bioreaction engineering is the heart of a bioconversion process that includes bioreactor design and operation, bioreaction (bioconversion) analysis and optimization, scale-up, and process control. Downstream processing deals with harvest and biomass separation, product isolation, and purification. The principles and practical aspects of bioconversion will be illustrated in the following with two important bioprocesses in more details.

4.3 Examples of bioconversion processes

4.3.1 Microbial production of 1,3-propanediol from glycerol

Glycerol, especially crude glycerol as a by-product directly from plant-oil processing and biodiesel production, is an interesting substrate for bioconversion. Crude glycerol normally contains impurities such as methanol, fatty acids, salts, and heavy metals and may need to be purified for certain fermentation processes with pure culture [20]. Glycerol can only be metabolized by microorganisms that need no external electron acceptors, but provide internal electron sinks. The degree of reduction (κ) of typical microorganisms (with an average cellular formula $CH_{1.9}O_{0.5}N_{0.2}$; $\kappa = 4.3$) is lower than that of glycerol ($\kappa = 4.67$), and therefore, reducing equivalents needs to be released for the balance of reducing power, leading to the production of more reduced products such as 1,3-propandiol (PDO, $\kappa = 5.33$) or ethanol ($\kappa = 6$) [21].

More specifically, for the production of 1,3-propanediol, glycerol is fermented by a dismutation process that comprises a simultaneous oxidation and reduction of the substrate to maintain a redox balance: the reductive pathway (A) yields in 1,3-PDO and the oxidative way (B) channels glycerol into glycolysis pathways leading to different by-products (Figure 4.4).

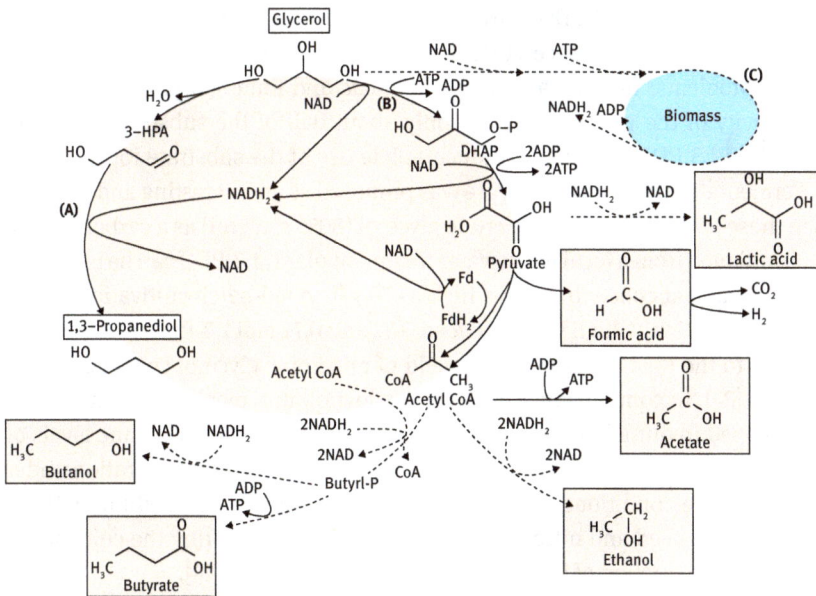

Fig. 4.4: Metabolic pathway of glycerol fermented into 1,3-propanediol.

In the reductive pathway (A), glycerol is metabolized into 3-hydroxypropional-dehyde (3-HPA) by the co-enzyme B12-dependent enzyme glycerol dehydratase. Subsequently, 3-HPA is reduced to PDO via 1,3-PDO oxidoreductase with consumption of $NADH_2$. $NADH_2$ is generated in the oxidative pathway B. In the oxidative pathway (B), glycerol is converted into dihydroxyacetone (DHA) by NAD-dependent glycerol dehydrogenase. Afterward, DHA is phosphorylated into DHA phosphate (DHAP), catalyzed by the enzyme dihydroxyacetone kinase. DHAP is then converted with triosphosphate isomerase into glyceraldehyde-3-phosphate (G3P), which is subsequently metabolized by glycolytic reactions into pyruvate (as described above). The 2-mol ATP generated is utilized for biomass production (C). For maintenance of intracellular redox balance, pyruvate is further converted into different side products, like organic acids or solvents, depending on the type of microorganism. The involvement (selectivity) of individual steps and energy demand varies between the microorganisms and also with the different cultivation conditions.

Conventionally, microbial production of 1,3-PDO is carried out using a single microorganism, either natural strains with glycerol as substrate or genetically engineered strains with glucose as a substrate [22, 23]. The bioconversion of crude glycerol into 1,3-PDO is of particular interest as a part of a biorefinery concept either together with biodiesel or bioethanol production (Figure 4.5). In both cases, glycerol can be obtained as a by-product with impurities as mentioned above. The microbial conversion of glycerol into 1,3-PDO is always associated with the production of organic acids as by-product, due to the necessity of balancing the reducing power. These results in two major problems: first, organic acids are toxic and limit cell growth and thus limit productivity of the process. Second, only about half of the substrate (glycerol) is converted into 1,3-PDO, leading to an incomplete use of the substrate for the target product. Fermentation with mixed culture was proposed as an interesting and effective solution to these problems [23]. Using crude glycerol (80% glycerol) as a carbon source and inocula adapted from a local wastewater treatment plant, 1,3-PDO can be produced as the main product at concentration as high as 70 g/L in fed-batch cultivation with a productivity of 2.6 g/L h. A high yield between 0.57 and 0.72 mol 1,3-PDO/mol glycerol, which is close to the theoretical maximal yield of anaerobic glycerol conversion, has been achieved [24]. In comparison to 1,3-PDO production in typical pure cultures, the process developed in our laboratory with a mixed culture achieved the same levels of product titer, yield, and productivity, but has the decisive advantage of operation under complete non-sterile conditions. Moreover, a defined fermentation medium without yeast extract can be used and nitrogen gassing can be omitted during the cultivation, leading to a strong reduction of investment and production costs [24].

4.3.2 Bioproduction of *n*-butanol

The solvent *n*-butanol is known as an important precursor for bulk chemicals like butyl acrylate as well as diluents for pharmaceutical industry [25]. A very promising application is its utilization as fuel and kerosene additive. In comparison to ethanol, butanol has

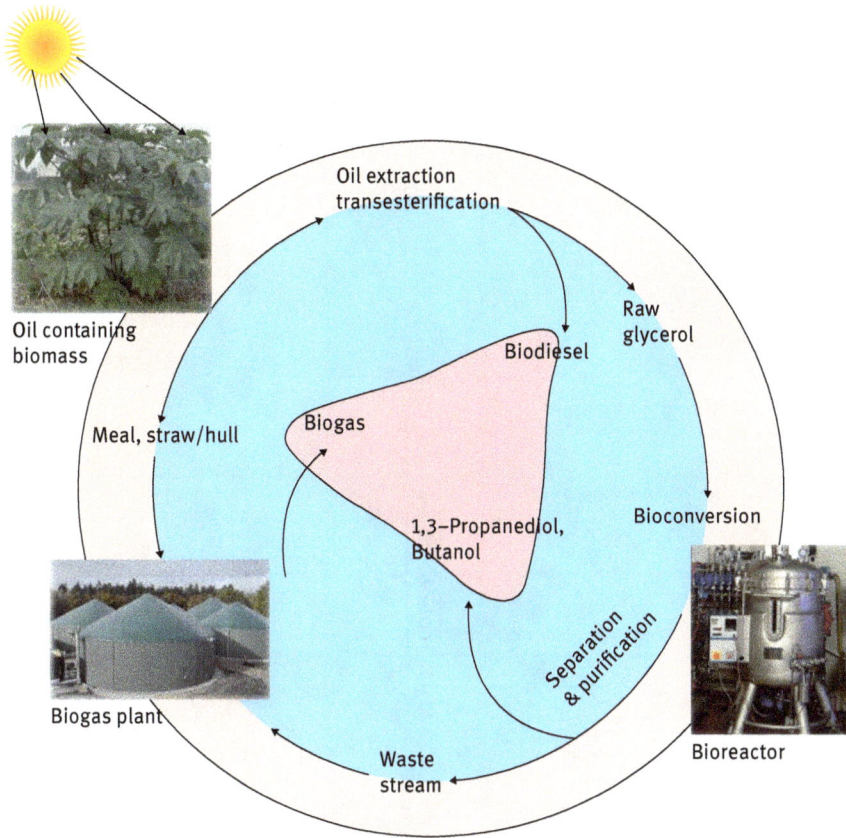

Fig. 4.5: Oil plant biorefinery with production biodiesel, as well as 1,3-propanediol and butanol from raw glycerol, together with energy integration due to biogas production from waste streams.

higher energy content, is less volatile, less corrosive, has a higher flash point, and is less hygroscopic; thus, it can be mixed in higher concentrations with gasoline [26, 27]. Therefore, many countries/companies have made large efforts to make biobutanol a commercial reality. China plans to implement 0.21 million tons per year of ABE solvent capacity, which is expected to increase to 1 million tons per year [28]. British petroleum (BP) and DuPont are working in collaboration on a project to improve butanol fermentation [27]. BP has also launched a subsidiary named BP Biofuels to commercialize butanol fermentation in a higher range. Other major players in the development of butanol fermentation are Butalco (Switzerland), Syntec (Canada), Green Biologics (UK), Gevo (USA), Cobalt Technologies (USA), Tetravitae Bioscience (USA), and ButylFuel LLC (USA). In 2008, the global market for n-butanol was 2.8 million tons, estimated to be worth approximately $5 billion. The average growth is expected to be 3.2% pa with demand concentrated in North America (28%), Western Europe (23%), and North East Asia (35%) [28]. More interestingly, biobutanol has the potential to substitute for both ethanol and biodiesel in the biofuel market estimated to be worth $247 billion by 2020 [29].

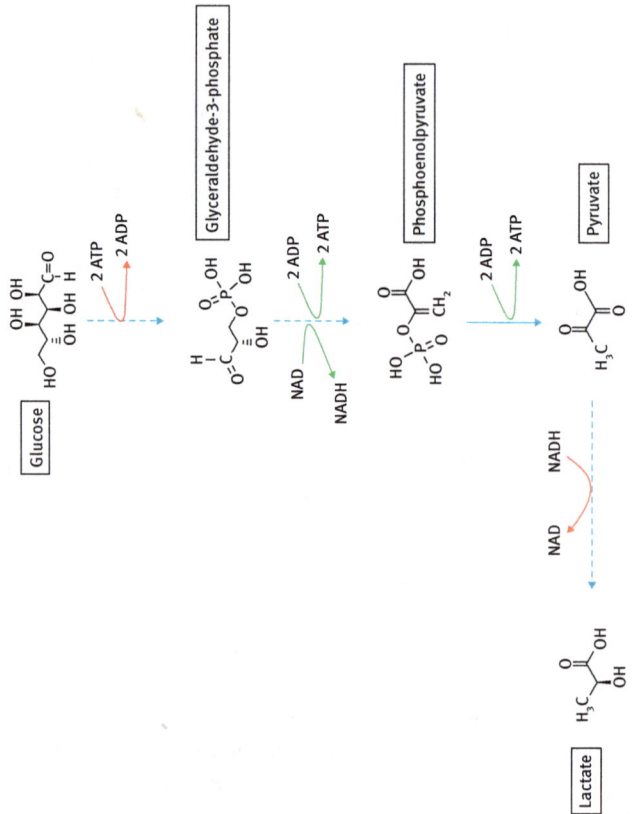

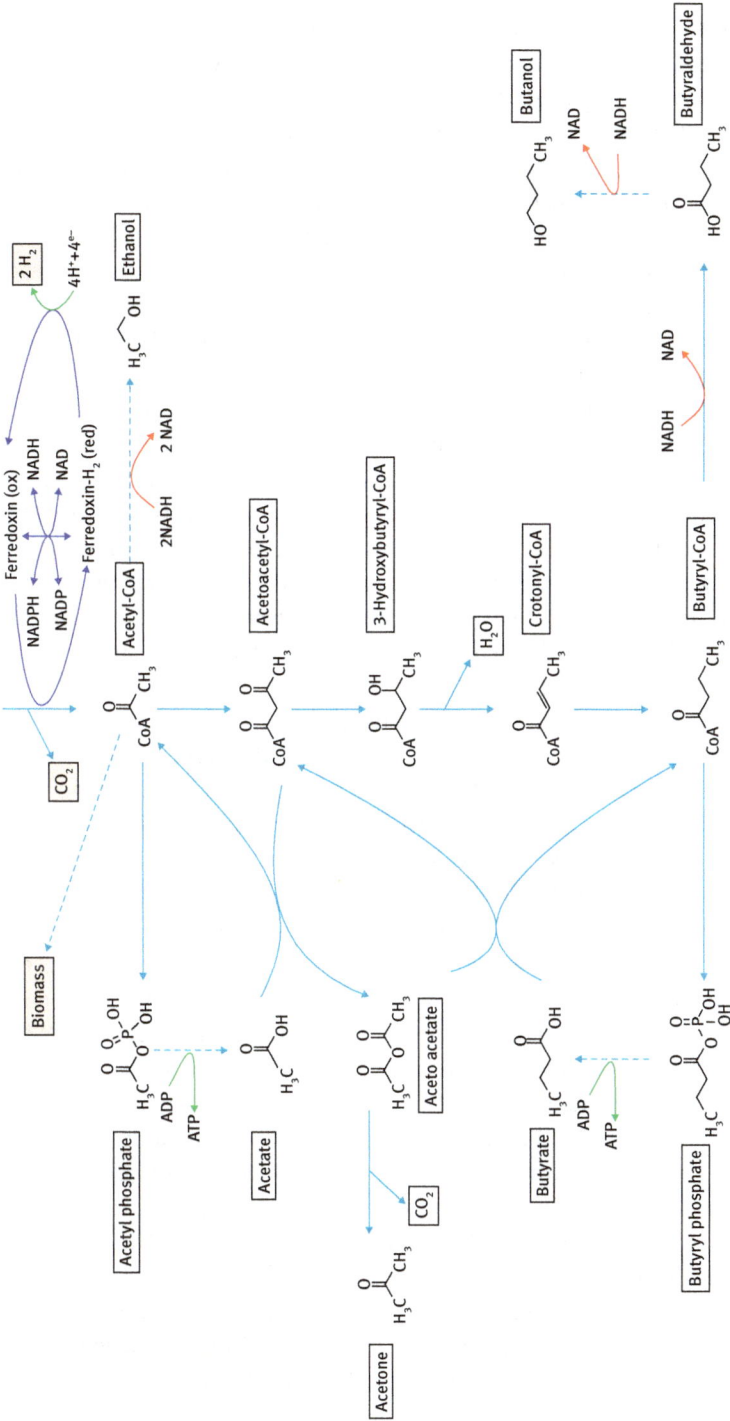

Fig. 4.6: Metabolic pathway of glucose conversion into acetone, butanol, and ethanol.

Renewable *n*-butanol is produced from the fermentation of carbohydrates in a process often referred to as the ABE fermentation, named after its major chemical products: acetone, butanol, and ethanol. The ABE fermentation is a proven industrial process that uses solventogenic *Clostridia* to convert sugars, starches, or hydrolyzate into solvents [28, 30, 31]. To date, *Clostridium acetobutylicum* ATCC 824 remains the best studied and manipulated strain [28]. Figure 4.6 shows the general metabolic pathway from glucose to ABE in *Clostridia*. Hexose sugars namely glucose, fructose, and galactose are catabolized to pyruvate via Embden-Meyerhof-Parnas pathway. Pyruvate is subsequently converted to acetyl CoA by oxidative decarboxylation using pyruvate-ferredoxin oxidoreductase. During the formation of acetyl CoA, a reduced ferredoxin molecule is formed that further acts as an electron donor to reduce NAD+/NADP+. The formation of NADH/NADPH is catalyzed by the enzymes NADH/NADPH-oxidoreductase, respectively. These are key enzymes that also produce energy-rich molecules for biomass growth. Acetaldehyde is then converted to ethanol by acetaldehyde dehydrogenase and ethanol dehydrogenase or is converted to acetoacetyl CoA by acetyl CoA acetyltransferase, which is subsequently converted to 3-hydroxybutyryl CoA using dehydrogenase enzyme. The product is dehydrated by enzyme crotonase to form crotonyl CoA, which is later converted to butyryl CoA by a corresponding dehydrogenase enzyme. Pyruvate, acetyl CoA, and butyryl CoA are the important precursors for the production of acids and solvents of industrial interest. The acids acetate and butyrate are formed from acetyl CoA and butyryl CoA, respectively, via analogous pathways, with corresponding acyl-phosphate as intermediate. ATP is generated during the process. Butanol is formed from butyryl CoA with butyraldehyde as intermediate, whereas ethanol is formed from acetyl CoA with acetaldehyde as intermediate. Reducing equivalents NADH/NADPH are utilized during this process. Acetone is formed via decarboxylation of acetoacetate, which was formed from acetoacetyl CoA.

Even though *Clostridia* are mostly known for *n*-butanol production from glucose, they are also able to produce butanol form glycerol as single substrate or in mixture with glucose [32, 33]. With glucose as the main carbon source, the fermentation profile of most solventogenic *Clostridia* is divided into two distinct phases: acidogenic phase, in which acids and cell biomass are first produced, followed by solventogenic phase, in which most of the acids are converted to solvents. With glycerol as sole substrate for *n*-butanol production, no such a generic phase separation is observed [33]. Interestingly, *Clostridium pasteurianum* was reported to utilize glycerol and convert it into 1,3-PDO and butanol as the main fermentation products [33, 34]. Moreover, we demonstrated recently that the blend of both glucose and glycerol stream favored the highest butanol productivity by *C. pasteurianum*, and limitation of either substrate will inhibit butanol production [33]. The fermentation stopped and the cell growth was inhibited after a butanol concentration of 21 g/L was reached. Using *in situ* removal of butanol, cell growth was retrieved and a process with simultaneous production of butanol and 1,3-PDO was developed in laboratory scale and is being scaled up in a pilot plant (see below).

In order to penetrate the larger biofuel market, biobutanol needs to compete in costs (priced on energy basis) with ethanol despite its superior fuel properties. Specific application as a component of aviation gasoline and jet fuel may open a new market for biobutanol. Reduction in feedstock cost offers the best opportunity especially since *Clostridia* are well suited for sugars derived from cellulosic material. Indeed, *Clostridia* have broad substrate ranges (including pentose sugars) and display superior tolerance to typical feedstock inhibitors. The main commercial challenges for the conventional butanol fermentation have been extensively reviewed [28, 35, 36]. In general, there is a need for cheaper feedstocks, improved fermentation performance, and more sustainable process operations for solvent recovery and water recycling. Feedstocks contribute most to the production costs followed by the recovery cost. *n*-Butanol titer rarely exceeds 20 g/L due to product inhibition. Therefore, product removal within fermentation, so-called *in situ* removal, seems mandatory. Engineering approaches for improved product separation are given in the downstream processing part below. An alternative approach for the biofuel market resides with iso-butanol using a synthetic microbe [37]. However, it is still unclear how robust and economic these processes are at commercial scale and whether they can accommodate cellulosic feedstocks.

4.4 Downstream processing

The strong link between bioconversion and downstream processing should always kept in mind when a biorefinery process is to be established. Conversion of biomass substrate by microbes is a complex process and many organisms produce a variety of by-products in addition to the main product that need to be separated. Prior to using the bioproduct in chemical or pharmaceutical industry, it has to be separated from water, by-products, and other fermentation residues. A clean product is only obtained by suitable separation and subsequent purification steps. Downstream processing is one of the most important factors in determining the overall production costs of the final product. Even if low-priced substrates are converted, it can make up 50–70% of the total production costs, mainly due to low product concentration and high energy demands for recovery [38]. Thus, the determination of a reasonable downstream process is of major economic interest. In the following, a general procedure for the treatment of fermentation broth is given, and different downstream methods are explained with the examples of bioproduction of *n*-butanol and 1,3-PDO.

4.4.1 General scheme of downstream processing of fermentation broth

After cultivation, the final fermentation broth is a multicomponent mixture requiring a stepwise recovery and purification of the desired product. In Figure 4.7, the general procedure to separate and purify a typical product like 1,3-PDO from fermentation broth is given. The main constituents of the broth are liquids like water, the desired product, residual substrate, dissolved salts, and side-products (e.g., other alcohols, organic acids). Additionally, there are solids, such as cells and cell debris, and undissolved proteins. The solid parts require an initial separation step; otherwise, cells and proteins can cause foaming within distillation [39]. Furthermore, the sugar and protein complexes may react in a Maillard reaction and cause plaque on heating devices. The solid-liquid separation is normally performed by centrifugation or filtration. The chosen method depends on the size of organisms or molecules that should be excluded [40]. Figure 4.8 shows different filtration types, depending on the compounds that should be separated. A stepwise filtration might be necessary, to avoid the plugging of fine membrane pores by large molecules or agglomerates [38].

Another way of separation, which is mainly used in wastewater treatment, is flocculation. Flocculation agents, e.g., chitosan or polyacrylamid, are added to form large agglomerates with the bacteria, cell debris, proteins, nucleic acids, and other broth components. After a certain flocculation time, the heavy agglomerates sank to the bottom of the vessel and can be excluded via decantation or by centrifugation [41]. The next step is the removal of salts, resulting from media components and pH regulation (addition of base or acids). If they remain in the broth, the ions can cause deactivation of catalysts or encrusting on heating devices in the distillation steps. However, if a subsequent salting-out process (see below) for product removal follows, this step might be not necessary. The most common methods applied here are electrodialysis or ion exchange chromatography. The main component of fermentation broth is water, since the desired product is typically produced in a concentration range of 15–150 g/L. The water content can be reduced by evaporation, to minimize the liquid stream amount, and accordingly pumping capacity in subsequent purification steps. Moreover, the desired product can be separated by extraction or adsorption on active carbon or resins. But this requires additional steps for recovery of extractants and desorption processes, which further increases the process costs. Hence, it should only be utilized if the amount of other pollutants is low and the product is very valuable. To achieve a high purification grade of the product (e.g., >99% purity), more purification steps are necessary. Mostly, this is performed by distillation/rectification or chromatography processes. Figure 4.9 shows a possible process route for the separation and purification of 1,3-PDO suggested by Kaeding *et al.* [39], containing all the explained general steps.

The main drawback of many methods is the fouling of filtration, pervaporation, and electrodialysis units, which require regular maintenance work and exchange of the membranes. The fouling is caused by plugging of the membranes from residual proteins, salts, or cell debris. In addition to this, electrodialysis is very susceptible to product loss in the saline effluent. Meanwhile, evaporation and distillation processes require high energy input, which also increases operation expenses. Therefore, a

Broth

Filtration
- Filtration
- Centrifugation
- Flocculation

Cells,
cell debris ←—— Solid-liquid
separation

- Electrodialysis, ion exchange
  chromatography
- Evaporation, extraction

Salt,
water ←—— Primary
recovery

- Distillation
- Preparative chromatography

Impurities, ←—— Final
water purification

Final product

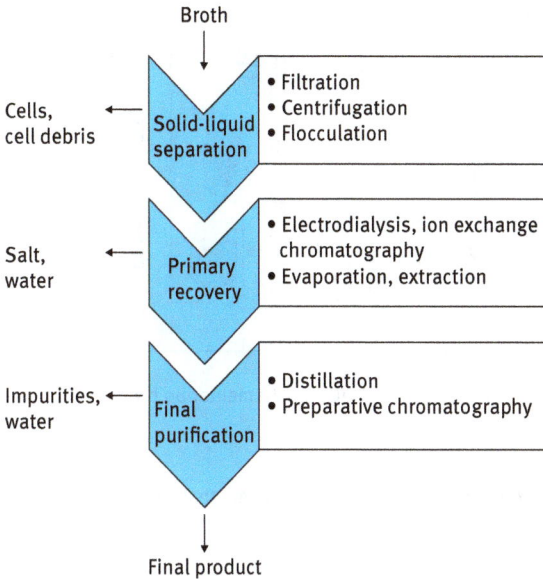

Fig. 4.7: General scheme for the recovery of a typical product like 1,3-PDO from fermentation broth (according to Xiu and Zeng [38]).

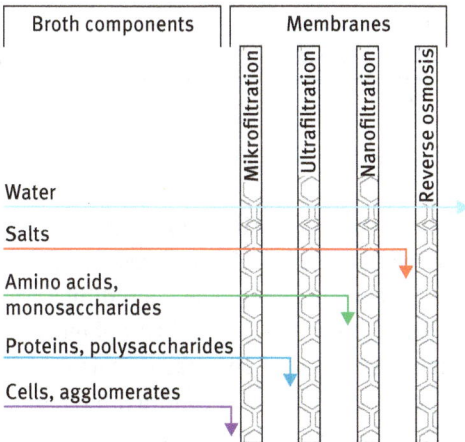

Broth components		Membranes			
	Mikrofiltration	Ultrafiltration	Nanofiltration	Reverse osmosis
Water				
Salts				
Amino acids, monosaccharides				
Proteins, polysaccharides				
Cells, agglomerates				

Fig. 4.8: Filtration methods for the exclusion of components with different sizes.

precise decision of downstream methods is mandatory for an economical large-scale process, even though different types are possible to achieve high product purity. The choice of a suitable method is determined by the highest difference in chemical or physical properties of the broth components (see table 4.1). Depending on the composition of the broth, a combination of more than one process might be necessary.

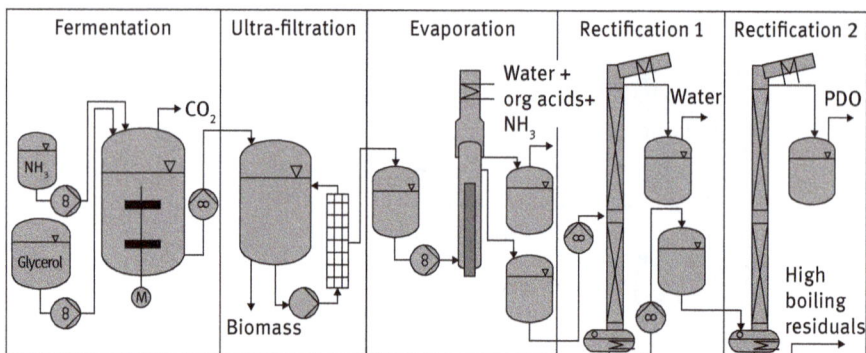

Fig. 4.9: Flow scheme for separation and purification of 1,3-PDO from fermentation broth (from Kaeding et al. [39] with permission).

Table 4.1: Downstream methods and different properties of the components that should be separated [42].

Different properties	Methods
Vapor pressure	Evaporation, distillation, pervaporation, and gas stripping
Solubility	Extraction, absorption, and crystallization
Sorption behavior	Adsorption and chromatography
Membrane permeation	Filtration and pervaporation
Behavior on field forces	Electrophoresis and centrifugation

4.4.2 Downstream processing methods for 1,3-PDO and n-butanol

Many efforts have been done in recent years to realize significant reductions in energy intensity and process costs for purification technologies in bioprocesses. Nevertheless, downstream processing for 1,3-PDO is fairly challenging. The main difficulties are due to the high boiling point (213°C), less volatility, and high hydrophilic character. 1,3-PDO is further processed into fine chemicals and pharmaceuticals, which requires a high purification grade of ~99.9% or even higher. In contrary, butanol is rather volatile and has a lower boiling point. However, so far, only low product concentrations around 20 g/L could be achieved by fermentation, causing high costs for evaporation and other downstream processes. Next to this, butanol above 5 g/L causes a serious growth inhibition. The integration of product recovery in the fermentation process (*in situ* separation) is one option to (i) avoid product inhibition and thus increase product yield and productivity and (ii) save costs for downstream processing by transferring the product into a more easily. This shows that not only knowledge about the bioconversion is mandatory, but also engineering concepts for product separation to develop successful and economic biorefinery processes.

4.4.2.1 Evaporation and Distillation

Evaporation and distillation processes are the common methods in downstream processing, either at the beginning or at the end to achieve high purification grades of the product. Evaporation comprises the simple technique of heating up a fluid to critical temperature and evaporating volatile components, normally water. This technique is performed to reduce the water amount in fermentation broth and increase the product concentration, respectively. Meanwhile, distillation (lat. *destillare*=dripping down) works with the exchange between liquid and gas phases. Continuous distillation uses columns with a temperature range inside. The separation occurs through selective evaporation, condensation, and mass transfer between these phases on several separation trays. At the bottom the temperature is higher and thus the less volatile compound remains, whereas the higher volatile fraction raises to the top and can be excluded (Figure 4.10).

However, evaporation and distillation require a large energy input to achieve the high boiling temperatures. This could be overcome by including a catalytic reaction step within the column, which is called reactive distillation. Nevertheless, distillation allows so far the best product recovery, and high purities, even from low concentrated feeds. It is also advantageous that distillation methods are well studied, easy to scale up, and offering high potentials for optimizations, with available simulation software [43].

Aqueous solutions with butanol above 7 wt% form a binary heterogeneous azeotropic mixture, where the vapor phase is in equilibrium with two liquid phases: a butanol-rich organic phase (up to 70 wt%) and a butanol-depleted watery phase. Both components can be separated by a system of two distillation columns and a decanter. The energy demand of a plain distillation process for butanol separation exceeds the energy content of butanol itself (36 MJ/kg), which makes such a process unfeasible. For example, the plain distillation of a fermentation broth with 5 g/L requires 79 MJ/kg [44]. Matsumura *et al.* [45] used an *in situ* pervaporation process

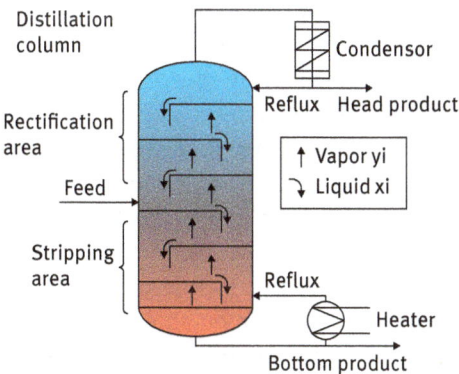

Fig. 4.10: Scheme of separation in a distillation column.

(see below) with membranes containing oleyl alcohol. Afterwards, the condensate is purified *via* distillation, achieving 99 wt% pure butanol, with a total energy demand of 7.4 MJ/kg. Kraemer *et al.* [46] suggested a combined extraction-distillation method for the separation of butanol from an acetone-butanol-ethanol fermentation by *C. acetobutylicum*. At first, butanol is extracted with mesitylene (1,3,5-trimethylbenzene), which is subsequently recovered in a distillation process. In silico modeling results in a reduced energy demand of 4.8 MJ/kg butanol. If butanol is an intermediate for further chemical reactions, the efficiency of a distillation processes can be increased by using reactive distillation. In this case, a reaction zone is located between the stripping and the rectification area of the column. The catalysts are immobilized on grids or packing material and placed on several trays around the column feed. For example, water, butanol, and acetic acid form butyl acetate via an ester-forming reaction, catalyzed by ionic exchange resins. The butanol and water are excluded and recycled at the column head, whereas butyl acetate is removed as bottom product. This could further increase the process economic of a biorefinery; however, it requires pretreated feeding solution to avoid the pollution of catalyst.

4.4.2.2 Gas stripping

Volatile compounds like ethanol or butanol could be successfully removed from fermentation broth by a gas stripping process. Thereby, the broth is sparged with an inert gas that carries the product due to vapor pressure equilibria. The gas (with product and water) is led over a cooling or flash unit, where the product condenses in high concentrations (Figure 4.11). In case of a butanol and water, a biphasic system is formed, which can be further concentrated via distillation. One main advantage of gas stripping is that it does not directly effects the microorganisms, especially when their own produced gases are used. However, the sparging might directly influence the redox potential and pH value of the fermentation broth, due to removal of hydrogen and CO_2. The latter one could further be needed for growth. In general, since only volatile compounds are removed, no loss of nutrients and important intermediates occur, provided that no volatile intermediates are produced within fermentation and no volatile supplements, e.g., like ammonia used for pH regulation, are added. Gas stripping achieves high product recoveries, even at ambient temperatures, as used in cultivation. Furthermore, it can be operated continuously with a simple equipment and could be easily scaled up for industrial processes [47]. However, the energy requirements for cooling and gas recycling as well as the addition of antifoam have a significant effect on operation costs of this method, and even though butanol is removed, other side-products that also inhibit growth at high concentrations accumulate in the broth.

Many research has been performed about gas stripping in butanol and ABE fermentation processes [47–50]. Recently, Xue *et al.* [51] used an intermittent gas stripping to remove butanol in an ABE fermentation of *C. acetobutylicum*. This resulted in a significantly improved productivity and yield for *n*-butanol. The gas stripping was realized outside the bioreactor in a fixed bed reactor with immobilized cells and with

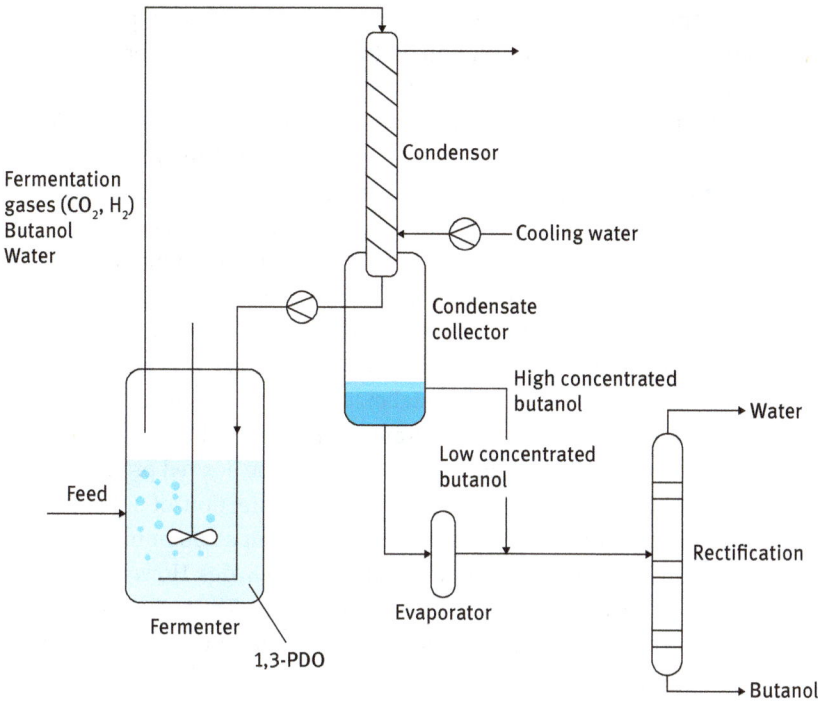

Fig. 4.11: Scheme of gas stripping for in situ recovery of butanol in fermentation processes.

nitrogen sparging. However, immobilized cells are difficult to establish in large-scale processes, and furthermore, nitrogen is quite expensive as stripping gas. Differently from the conventional ABE fermentation of *C. acetobutylicum*, *C. pasteurianum* produces no significant amounts of acetone and ethanol, but higher amounts butanol and 1,3-PDO when glycerol is used as substrate. In a new process, the inhibiting butanol is continuously removed from the fermentation broth by circular sparging of gases (CO_2, H_2) produced by the culture itself. The butanol stripped is collected as a highly concentrated biphasic condensate. At the same time, 1,3-PDO accumulates in the fermenter and reaches a relatively high concentration (above 50 g/L) as well. This facilitates downstream processing and saves operational costs since the main products are already partially separated. Overall, both products were simultaneously produced in higher concentrations, higher yields, and increased productivity compared to conventional processes without gas stripping.

4.4.2.3 Extraction methods

Extraction (lat. *extrahere*=pull sth. out) is defined as a separation method, where the desired compound is dissolved into an immiscible extractant (gas or liquid) out of a mixture of different compounds (liquid or solid). The type of liquid-liquid extraction is the most common one because of its high selectivity toward the desired product. The contact between extractant and fermentation broth can either be direct or in separate mixing devices or columns. A successful separation is based on a careful screening of suitable extractant. The main performance criteria are high distribution coefficient D (mass fraction in extractant over mass fraction in water/broth) and high selectivity (D of product over D of water) as well as simple regeneration of the extractant, e.g., *via* heat induction. Further important is the immiscibility of the extractant in fermentation broth (mainly water) and accordingly the ability to form different phases. This is crucial for the removal of the extractant. Also preferred are easy availability at low costs, low viscosity, simple recovery, and recyclability of the extractant (Friedl and Qureshi, 2001). If extraction is used as an *in situ* process, the extractant has to be non-toxic and non-influencing on the microbial metabolism. Therefore, a simulation software like ESP (Extractant Screening Programme) is a helpful tool in finding an adequate extractant with high distribution coefficients for the desired product.

Malinowsky screened different extractants for the extraction of 1,3-PDO, like series of pentanol to nonanol and hexanal to decanal as well as other organics [52]. The most suitable ones seem to be aliphatic alcohols and aldehydes. However, 1,3-PDO shows a generally low distribution in all these extractants. Thus, in a large-scale process high amounts of extractants would be necessary, additionally increasing the cost of downstream processing [52]. A new approach in the field of extractants is the utilization of ionic liquids (ILs), which are salts in liquid state below 100°C. Müller and Górak [53] used different types of ILs on the basis of 1-butyl-3-methylimidazolium trifluoromethansulfonate with varying anions for 1,3-PDO extraction. However, until now, ILs are very expensive in production, limiting their success as broad applied extractant. Furthermore, this method is not suitable *in situ* because ILs are highly toxic for the bacteria.

Butanol shows an excellent solubility in organic solvents and is therefore a good candidate for extractions. Many screenings have been performed to find suitable extractants among alcohols, alkanes, esters, different oils, as well as surfactants, ILs, and supercritical carbon dioxide [54–60]. The most common extractants for butanol that fulfills many screening criteria are dodecanol and oleyl alcohol. Evans and Wang [61] used a mixture of both to separate butanol *in situ* from a fermentation broth of *C. acetobutylicum* and thereby increased the butanol titer by 72%. A problem in liquid-liquid extraction is the formation of emulsions or the fouling of the extractant, due to direct contact with the fermentation broth. This can be overcome by membrane extraction (perstraction), where a membrane excludes solid broth components. However, for this technique, the membrane must ensure high (selective) permeation rates that allow sufficient mass transfer of the desired product. Another drawback is the backflow of the extractant into the fermentation broth, causing loss of extractant and, in case of toxicity, inhibition of bacterial growth. Furthermore, large membrane areas and high viscous extractants are difficult to operate in large-scale process [54].

In reactive extraction processes, the desired compound reacts with an additional component (reactant) into the desired product or an intermediate. This new product is then extracted with a solvent, which could be the reactant at the same time. The chemical reaction aims to increase the solubility or distribution efficiency of the product in the extractants. Malinowski [62] transformed 1,3-PDO with formaldehyde or acetaldehyde into 2-methyl-1,3-dioxane and extract it subsequently with o-xylene, toluene, or ethyl benzene. One drawback of this method is the occurrence of side-reactions with other broth components and forming additional undesired pollutants.

Salting-out processes base on a slightly different principle such as extraction, but it is often combined with that. By the addition of salts in high concentrations (>25%), water-soluble constituents are displaced from a watery phase, due to the increased ionic strength of the liquid. Therefore, the excluded liquids form a new organic phase. This technique is also used to remove proteins, which precipitate in a solid phase. Suitable salts are, for example, K_2HPO_4, K_3PO_4, K_2CO_3, KF, $K_3C_6H_5O_7$, or $(NH_4)_2SO_4$, which are added to the fermentation broth. However, sometimes, salt addition alone is not efficient enough. Meanwhile, the usage of extractive solvents could also be insufficient because 1,3-PDO shows a low distribution coefficient. Thus, Li *et al.* [63] combined both methods for the separation of 1,3-PDO produced by *Klebsiella pneumoniae*. They found out that the extraction efficiency increases with stronger polarity of the solvent and higher charge of ions. A mixture of methanol and potassium phosphate achieved the best results, and the 1,3-PDO recovered was 98%. Recently, they have been able to establish a continuous removal process in the laboratory scale using ethanol and dipotassium phosphate, recovering 90 wt% of 1,3-PDO from pre-filtered fermentation broth within 11 h [64]. However, if an extractant like ethanol is already present in the cultivation broth, process economics can be enhanced [65]. Not only the product, but other by-products (in this case, acetoin and 2,3-butandiol) and cells can be removed at the same time. Nevertheless, in all cases, a subsequent distillation of the extraction mixture is required.

4.4.2.4 Pervaporation/Membrane separation

In a pervaporation process, the separation occurs through a selective diffusion of the product through a membrane, while holding back water and undesired components (Figure 4.12). The product molecules absorb into the membrane, permeate through it, and evaporate. On the other side of the membrane, a vacuum is applied, which causes the evaporation. Thus, the main driving force is not only the permeability through the membrane, but also the differences in partial pressure of the different broth components. In a second step, the vapor phase is condensed simply by removing the vacuum. A broad variety of membranes is available and tested: either hydrophobic membranes for product recovery from fermentation broth or hydrophilic membranes for the dehydration of high concentrated alcohols [66, 67]. Hydrophilic membranes are made from polyvinylalcohol, polyacrylic acid, or polyacrylonitrile. The most

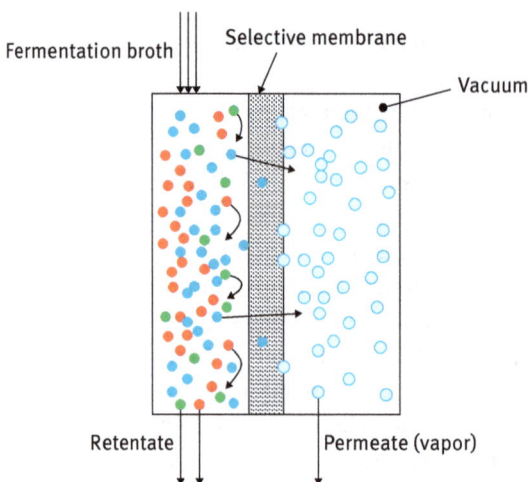

Fig. 4.12: Scheme of pervaporation process as an *in situ* method for production recovery from fermentation broth.

common hydrophobic membrane consists of polydimethylsilocaxe. In terms of sustainability, biopolymers, like cellulose [68, 69], chitosan [70], or sodium alginate [71, 72], are of increasing interest, especially for the dehydration of aqueous alcohol mixtures.

Li *et al.* used a Na-ZSM-5 zeolite membrane for 1,3-PDO removal from an aqueous glycerol-glucose mixture. But both flux and selectivities were far too low for an economical separation procedure [73]. Another novel approach is the incorporation of ILs in the membranes, which successfully increased the selectivity, but further decreased the flux due to higher permeation resistance [74]. Nevertheless, pervaporation is suitable for the case of inhibitory products like butanol, since they can be removed *in situ*, thereby enhancing the process productivity. Heitmann *et al.* [75] used pervaporation membranes, with immobilized ILs to increase the selectivity. The separation was tested on a water-butanol mixture with low butanol concentrations and at cultivation temperature of 37°C. So far, these membranes exhibit good selectivities but insufficient permeability fluxes (560 g/m² h) due to high membrane thickness. This led to a final butanol concentration of 55 wt% in the condensate, which requires an additional distillation step. Thus, compared to conventional polymeric membranes IL membranes need further improvement in permeability. The main bottleneck of this technique is that the membranes are quite susceptible for fouling, and in large-scale processes, immense membrane areas are necessary.

4.4.2.5 Adsorption and Chromatography

In adsorption processes, the desired molecules within a fluid detach on solid surface by means of van der Waals forces (physisorption), ionic exchange, or chemical reaction (chemisorption). It is frequently applied for the cleaning of gases and multicomponent liquids, due to its high selectivity. The adsorbents can be placed in form of pellets or powder directly into the fermentation broth or immobilized in a separated column. Adsorption processes normally achieve a high purity, exhibit simple design and operation, and can be considered as environmental friendly because the adsorbance material is recyclable [76]. Recently, Wang *et al.* [76] reported about the adsorption of 1,3-PDO from crude glycerol on a cheap cation exchange resin. The process showed a high adsorption capacity of 360 mg/g at moderate temperature of 45°C. Other adsorbance materials for 1,3-PDO are, for example, polymeric resins [77] or active char [38]. However, every adsorption process demands a desorption step for product recovery. Physisorption is reversible, whereas chemisorption requires a higher energy input for the recovery of the desired product. So far, this method is not economical for biorefinery processes in terms of low-value high-volume products like fuels or bulk chemicals. Qureshi *et al.* [78] reported about the adsorption of butanol on silicalite, a hydrophobic molecular sieve out of SiO_2, which can be regenerated by heat. The energy requirement to for adsorption and desorption processes is around 8.2 MJ/kg, which is far below the demand for distillation.

To avoid fouling of adsorbents in an *in situ* process by cells or debris, a filtration membrane could be applied between bioreactor and absorption column. Another problem is the adsorption of nutrients on the adsorbents. Less substrate leads to lower productivity and increases fermentation costs [79]. Furthermore, the adsorption of by-products and intermediates should be avoided because it decreases the adsorption capacity of the material. Additionally, they could contaminate the final product or impede the desorption process. The fixation of intermediates like butyric acid is especially problematic because they are needed for the bioconversion into butanol; thus, a capture on the adsorbents further decrease the productivity [78]. Nielsen *et al.* [79] could overcome the drawback of butyric acid adsorption by changing the pH of the fermentation broth, but, on the one hand, this interferes with microbial growth and, on the other hand, it is not applicable for other acids or adsorbents.

Chromatography is not only an analytical method, but it is also a very useful tool for separation. The broth components flow in a mobile phase over a stationary phase. Different constituents attach or adsorb on the stationary phase by different extents, according to their physical or chemical properties. Some molecules leave the device very quickly with the mobile phase; retarded molecules are excluded on later time point, thus allowing the separation. Chromatography is a highly selective method, but the setup of large-scale devices is difficult because large exchange surfaces are necessary and high pressure loss occurs. The stationary phases are mainly thin membranes or fixed bed bodies covered with ion exchange resins, molecular sieves, or (reversible) chemical reaction agents. Like other methods, they are very susceptible to fouling; thus, the devices must be regenerated frequently [38]. Recent methods are successful, but due to low 1,3-PDO and butanol concentrations, they are not suitable for large-scale productions like biorefineries [80, 81].

4.5 Concluding remarks

Despite impressive advances, there remain major challenges to fully harness the advantages of bioconversion. In general, compared to chemical industries, there are only a few industrially important products that are being produced via biotechnology. Besides ethanol, *n*-butanol, lactic acid, succinic acid, and 1,3-propanediol, there are mainly traditional fermentation products such as amino acids (e.g., L-lysine, L-glutamate), organic acids (e.g., acetic acid) as well as vitamins and antibiotics. In fact, bioconversion has been intensively studied for numerous products that are of great interest as chemicals and fuels. Currently, the competitiveness of bioconversion of renewable resources is still limited in comparison to chemical synthesis routes based on fossil resources. Solutions and several engineering strategies were previously proposed and discussed [5, 19, 82, 83]. Enzyme engineering, cost-effective pretreatment of biomass, use of alternative cheaper feedstock, constructing new hyperproductive strains, or applying innovative fermentation and downstreaming processes should strongly improve the competitiveness of bioconversion processes.

Among others, the availability and costs of feedstock play a vital role. For most of the bioprocesses to produce biofuels and bulk chemicals, the substrate costs of biomass today is far too high to compete with conventional carbon sources like petroleum or natural gas [84]. The natural resistance of plant cell walls to microbial and enzymatic deconstruction, collectively known as "biomass recalcitrance", is largely responsible for the high cost of lignocellulose bioconversion. To achieve a sustainable energy production, it will be necessary to overcome the chemical and structural properties that have evolved in biomass to prevent its disassembly. For future biorefineries, research overcoming biomass recalcitrance may primarily target the co-engineering of new cell walls to be degraded by newly engineered enzymes designed for this role. To reach this goal, new findings from plant science and carbohydrate chemistry must be translated and integrated into the conversion processes [83].

Furthermore, future microbial cells should be able to tolerate inhibitors normally found in raw substrates and conduct multiple conversion reactions with extended substrate spectrum for an efficient utilization of carbohydrates. Here, research on metabolic engineering, systems, and synthetic biology will play central roles. It should be mentioned that cost benefits of using genetic engineering exist only if the investment and operating expenses (use of antibiotics, sterile conditions, safety, biomass disposal, etc.) required to run such processes on an industrial scale are compensated by the costs benefit of the product. This is, of course, the case with high-value products (pharmaceuticals, specialties), but can be challenging for biofuels and bulk chemicals, especially in the low-cost sector.

Alternatively, other approaches have been proposed to increase the economic efficiency of bioconversion processes and deserve more attention in the future, of which mixed culture and unsterile fermentations (illustrated above for the production of 1,3-PDO) are worth mentioning [85]. For example, the newly developed glycerol-based and low-cost unsterile mixed-culture fermentation process for the production of 1,3-PDO could be economically and ecologically attractive and even competitive to the glucose-based process using recombinant *E. coli*, which can be only operated as pure culture under sterile conditions [24]. In our opinion, the use of complex substrates in a biorefinery approach necessitates the use of mixed microbial cultures, especially because microbial consortia can perform complicated functions and are more robust to environmental fluctuations than individual populations. Interest has recently emerged in engineering microbial consortia, but studies with the use of synthetic engineered consortia are just in their infancy and still in have not left laboratory studies [86, 87]. Nevertheless, in a short- to middle-term perspective, the use of defined or minimal microbial consortia involving a few species seems to be promising. Moreover, waste stream treatment has to be forced to go beyond its original restricted environmental mandate and seeks ways and means to turn wastes into useful materials rather than merely eliminate their health hazards and nuisance value.

Most techniques in downstream processing are well known from the petrochemical industry, and much progress has been done in the last years for their optimization. However, purification of products from fermentation broths is more challenging, primarily due to their complex and varying composition. Today, the relatively low product concentrations achieved in bioconversion are the main barrier for a more wide application of bioconversion processes over chemical synthesis for fuels and chemicals. Therefore, *in situ* product removal of growth-inhibiting compounds is a key technology to overcome this drawback. Promising approaches include pervaporation techniques or membrane techniques, since they work efficiently under fermentation conditions, like low temperature. For example, Xue *et al.* [88] used membranes with incorporated carbon nanotubes instead of toxic and expensive ILs. Next to this, the heat and energy integration in a biorefinery is of major importance to create an autonomic and sustainable process.

Ultimately, for more advanced and cost-effective biorefinery processes, the general path along the biofuel and biochemical production route will generally rely on consolidation of processing steps (both bioconversion and downstream processing). It is clear that in terms of biorefinery, there will be no general process steps or similar refineries as in fossil-based chemistry. The construction of a biorefinery process, with bioconversion, pretreatment, and downstream processing, strongly depends on the availability and choice of substrates and microorganisms as well as location and transportation possibilities. Therefore, it should be always a specific concept that requires knowledge in interdisciplinary fields to combine nature and technology in a successful and sustainable manner.

Bibliography

[1] Kamm, B., Kamm, M., Biorefinery-systems, Chem Biochem Eng Q **18** (2004) 1–7.

[2] Serrano-Ruiz, J. C., Ramos-Fernández, E. V., Sepúlveda-Escribano, A., From biodiesel and bioethanol to liquid hydrocarbon fuels: new hydrotreating and advanced microbial technologies, Energy Environ Sci **5** (2012) 5638–5652.

[3] Huber, G. W., Corma, A. Synergies between bio- and oil refineries for the production of fuels from biomass, Angew Chem Int Ed **46** (2007) 7184–7201.

[4] Preisig, H. A., Wittgens, B., Thinking towards synergistic green refineries, Energy Procedia **20** (2012) 59–67.

[5] Lee, R. A., Lavoie, J.-M., From first- to third-generation biofuels: challenges of producing a commodity from a biomass of increasing complexity, Animal Front **3** (2013) 6–11.

[6] Wang, Y., Tashiro, Y., Sonomoto, K., Fermentative production of lactic acid from renewable materials: recent achievements, prospects, and limits, J Biosci Bioeng **119** (2015) 10–18.

[7] Max, B., Salgado, J. M., Rodríguez, N., Cortés, S., Converti, A., Dominguez, J. M., Biotechnological production of citric acid, Braz J Microbiol **41** (2010) 862–875.

[8] Abbas, C. A., Bao, W. L., Beery, K. E., Corrington, P., Cruz, C., Loveless, L., Sparks, M., Trei, K., Manual of Industrial Microbiology and Biotechnology: Bioethanol Production from Lignocellulosics: Some Process Considerations and Procedures, 3rd ed., ASM Press, Washington, DC, 2010.

[9] Zaldivar, J., Nielsen, J., Olsson, L., Fuel ethanol production from lignocellulose: a challenge for metabolic engineering and process integration, Appl Microbiol Biotechnol **56** (2001) 17–34.

[10] Klinke, H. B., Thomsen, A. B., Ahring, B. K., Inhibition of ethanol-producing yeast and bacteria by degradation products produced during pre-treatment of biomass, Appl Microbiol Biotechnol **66** (2004) 10–26.

[11] Brennan, L., Owende, P., Biofuels from microalgae – a review of technologies for production, processing, and extractions of biofuels and co-products, Renew Sustain Energ Rev **14** (2010) 557–577.

[12] Liang, Y., Sarkany, N., Cui, Y., Biomass and lipid productivities of Chlorella vulgaris under autotrophic, heterotrophic and mixotrophic growth conditions, Biotechnol Lett **31** (2009) 1043–1049.

[13] Chen, C. Y., Yeh, K. L., Aisyah, R., Lee, D. J., Chang, J.S., Cultivation, photobioreactor design and harvesting of microalgae for biodiesel production: a critical review, Bioresource Technol **102** (2011) 71–81.

[14] Chatzifragkou, A., Dietz, D., Komaitis, M., Zeng, A., Papanikolaou, S., Effect of biodiesel-derived waste glycerol impurities on biomass and 1,3-propanediol production of Clostridium butyricum VPI 1718, Biotechnol Bioeng **107** (2010) 76–84.

[15] Zeng, A.-P., Biebl, H., Bulk chemicals from biotechnology: the case of 1,3-propanediol production and the new trends, in Tools and Applications of Biochemical Engineering Science, Springer Berlin/Heidelberg, 2002, 239–259.

[16] Schmid, A., Dordick, J. S., Hauer, B., Kiener, A., Wubbolts, M., Witholt, B., Industrial biocatalysis today and tomorrow, Nature **409** (2001) 258–268.

[17] Thiel, K. A., Biomanufacturing, from bust to boom . . . to bubble? Nat Biotechnol **22** (2004) 1365–1372.

[18] Hatti-Kaul, R., Törnvall, U., Gustafsson, L., Börjesson, P., Industrial biotechnology for the production of bio-based chemicals – a cradle-to-grave perspective, Trends Biotechnol **25** (2007) 119–124.

[19] Willke, T., Vorlop, K.-D., Industrial bioconversion of renewable resources as an alternative to conventional chemistry, Applied Microbiology and Biotechnology **66** (2004) 131–142.

[20] Chatzifragkou, A., Aggelis, G., Komaitis, M., Zeng, A.-P., Papanikolaou, S., Impact of anaerobiosis strategy and bioreactor geometry on the biochemical response of *Clostridium butyricum* VPI 1718 during 1,3-propanediol fermentation, Bioresource Technol **102** (2011) 10625–10632.

[21] Drożdżyńska, A., Katarzyna, L., Katarzyna, C., Biotechnological production of 1,3-propanediol from crude glycerol, J Biotechnol Comput Biol Bionanotechnol **92** (2011) 92–100.

[22] Bizukojc, M., Dietz, D., Sun, J., Zeng, A.-P., Metabolic modelling of syntrophic-like growth of a 1,3-propanediol producer, Clostridium butyricum, and a methanogenic archeon, Methanosarcina mazei, under anaerobic conditions, Bioprocess Biosyst Eng **33** (2010) 507–523.

[23] Friedmann, H., Zeng, A. P., Process and apparatus for the microbial production of a specific product and methane, 2008, Patent DE 102007001614 A1 and WO 2006/021087 A1 3/2006.

[24] Dietz, D., Zeng, A. P., Efficient production of 1,3-propanediol from fermentation of crude glycerol with mixed cultures in a simple medium, Bioprocess Biosyst Eng **37** (2014) 225–233.

[25] Dürre, P., Biobutanol: an attractive biofuel, accessed 23 July 2014. Biotechnol J **2** (2007) 1525–1534.

[26] García, V., Päkkilä, J., Ojamo, H., Muurinen, E., Keiski, R. L., Challenges in biobutanol production: how to improve the efficiency? Renew Sustain Energ Rev **15** (2011) 964–980.

[27] Lee, S. Y., Park, J. H., Jang, S. H., Nielsen, L. K., Kim, J., Jung, K. S., Fermentative butanol production by clostridia, Biotechnol Bioeng **101** (2008) 209–228.

[28] Green, E. M., Fermentative production of butanol – the industrial perspective: energy biotechnology – environmental biotechnology, Curr Opin Biotechnol **22** (2011) 337–343.

[29] pikeresearch, available from: http://www.pikeresearch.com/research/biofuels-markets-and-technologies, accessed on 25/09/2014.

[30] Grobben, N. G., Eggink, G., Cuperus, F. P., Huizing, H. J., Production of acetone, butanol and ethanol (ABE) from potato wastes: fermentation with integrated membrane extraction, Appl Microbiol Biotechnol **39** (1993) 494–498.

[31] Madihah, M. S., Ariff, A. B., Sahaid, K. M., Suraini, A. A., Karim, M. I. A., Direct fermentation of gelatinized sago starch to acetone-butanol-ethanol by Clostridium acetobutylicum, World J Microb Biot **17** (2001) 567–576.

[32] Biebl, H., Fermentation of glycerol by Clostridium pasteurianum – batch and continuous culture studies, J Ind Microbiol **27** (2001) 18–26.

[33] Sabra, W., Groeger, C., Sharma, P. N., Zeng, A.-P., Improved n-butanol production by a non-acetone producing Clostridium pasteurianum DSMZ 525 in mixed substrate fermentation, Appl Microbiol Biotechnol **98** (2014) 4267–4276.

[34] Jensen, T., Kvist, T., Mikkelsen, M., Westermann, P., Production of 1,3-PDO and butanol by a mutant strain of Clostridium pasteurianum with increased tolerance towards crude glycerol, AMB Express **2** (2012) 44.

[35] Zheng, Y.-N., Li, L.-Z., Xian, M., Ma, Y.-J., Yang, J. M., Xu, X., He, D. Z., Problems with the microbial production of butanol, J Ind Microbiol Biot **36** (2009) 1127–1138.

[36] Ezeji, T. C., Qureshi, N., Blaschek, H. P., Bioproduction of butanol from biomass: from genes to bioreactors, Curr Opin Biotechnol **18** (2007) 220–227.

[37] Trinh, C. T., Li, J., Blanch, H. W., Clark, D. S., Redesigning Escherichia coli metabolism for anaerobic production of isobutanol, Appl Environ Microbiol **77** (2011) 4894–4904.

[38] Xiu, Z.-L., Zeng, A.-P., Present state and perspective of downstream processing of biologically produced 1,3-propanediol and 2,3-butanediol, Appl Microbiol Biotechnol **78** (2008) 917–926.

[39] Kaeding, T., DaLuz, J., Kube, J., Zeng, A.-P., Integrated study of fermentation and downstream processing in a miniplant significantly improved the microbial 1,3-propanediol production from raw glycerol. Bioprocess Biosyst Eng **38** (2015) 575–586.

[40] Willke, T., Vorlop, K., Biotransformation of glycerol into 1,3-propanediol. Eur J Lipid Sci Technol **110** (2008) 831–840.

[41] Hao, J., Xu, F., Liu, H., Liu, D., Downstream processing of 1,3-propanediol fermentation broth, J. Chem Technol Biotechnol **81** (2006) 102–108.

[42] Mersmann, A., Scholl, S., Thermische Verfahrenstechnik, in Grote, K.-H., Feldhusen, J., editors, Dubbel, Springer, Berlin, Heidelberg, 2011, N9.

[43] Vane, L. M., Separation technologies for the recovery and dehydration of alcohols from fermentation broths, Biofuels Bioprod Bioref **2** (2008) 553–588.

[44] Friedl, A., Qureshi, N., Maddox, I. S., Continuous acetone-butanol-ethanol (ABE) fermentation using immobilized cells of Clostridium acetobutylicum in a packed bed reactor and integration with product removal by pervaporation, Biotechnol Bioeng **38** (1991) 518–527.

[45] Matsumura, M., Kataoka, H., Sueki, M., Araki, K., Energy saving effect of pervaporation using oleyl alcohol liquid membrane in butanol purification, Bioprocess Biosyst Eng **3** (1988) 93–100.

[46] Kraemer, K., Harwardt, A., Bronneberg, R., Marquardt, W., Separation of butanol from acetone-butanol-ethanol fermentation by a hybrid extraction-distillation process, Comput Chem Eng **35** (2011) 949–963.

[47] Qureshi, N., Blaschek, H., Recovery of butanol from fermentation broth by gas stripping, Renew Energ **22** (2001) 557–564.

[48] Ennis, B. M., Continuous product recovery by in-situ gas stripping/condensation during solvent production from whey permeate using Clostridium acetobutylicum, Biotechnol Lett **8** (1986) 725–730.

[49] Lu, C., Dong, J., Yang, S.-T., Butanol production from wood pulping hydrolysate in an integrated fermentation-gas stripping process, Bioresource Technol **143** (2013) 467–475.

[50] Mariano, A. P., Keshtkar, M. J., Atala, D. I. P., Maugeri Filho, F., Wolf Marciel, M. R., Maciel Filho, R., Stuart, P., Energy requirements for butanol recovery using the flash fermentation technology: energy and fuels, Energ Fuel **25** (2011) 2347–2355.

[51] Xue, C., Zhao, J., Lu, C., Yang, S.-T., Fengwu, B., Tang, I.-C., High-titer n-butanol production by clostridium acetobutylicum JB200 in fed-batch fermentation with intermittent gas stripping, Biotechnol Bioeng **109** (2012) 2746–2756.

[52] Malinowski, J., Evaluation of liquid extraction potentials for downstream separation of 1,3-propanediol, Biotechnol Tech **13** (1999) 127–130.

[53] Müller, A., Górak, A., Extraction of 1,3-propanediol from aqueous solutions using different ionic liquid-based aqueous two-phase systems, Separ Purif Technol **97** (2012) 130–136.

[54] Groot, W. J., Soedjak, H. S., Donck, P. B., Lans, R. G. J. M., Luyben, K. Ch. A. M., Timmer, J. M. K., Butanol recovery from fermentations by liquid-liquid extraction and membrane solvent extraction, Bioprocess Biosyst Eng **5** (1990) 203–216.

[55] Dhamole, P. B., Wang, Z., Liu, Y., Wang, B., Feng, H., Extractive fermentation with non-ionic surfactants to enhance butanol production, Biomass Bioenerg **40** (2012) 112–119.

[56] Ishizaki, A., Michiwaki, S., Crabbe, E., Kobayashi, G., Sonomoto, K., Yoshino, S., Extractive acetone-butanol-ethanol fermentation using methylated crude palm oil as extractant in batch culture of Clostridium saccharoperbutylacetonicum N1-4 (ATCC 13564), J Biosci Bioeng **87** (1999) 352–356.

[57] Roffler, S. R., Blanch, H. W., Wilke, C. R., In-situ recovery of butanol during fermentation 1 + 2, Bioprocess Biosyst Eng **2** (1987) 1–12.

[58] Santangelo, F., Stoffers, M., Górak, A., Extraction of butanol from aqueous solutions and fermentation broth using ionic liquids, in Valenzuela, L. F., Moyer, B. A., editors, Proceedings of the 19th International Solvent Extraction Conference, Gecamin, Santiago, 2011.

[59] Laitinen, A., Kaunisto, J., Supercritical fluid extraction of 1-butanol from aqueous solutions, J Supercrit Fluids **15** (1999) 245–252.

[60] Dooley, K. M., Cain, A. W., Carl, Knopf, F., Supercritical fluid extraction of acetic acid, alcohols and other amphiphiles from acid-water mixtures, J Supercrit Fluid **11** (1997) 81–89.

[61] Evans, P. J., Wang, H. Y., Enhancement of butanol formation by Clostridium acetobutylicum in the presence of decanol-oleyl alcohol mixed extractants, Appl Environ Microb **54** (1988) 1662–1667.

[62] Malinowski, J. J., Reactive extraction for downstream separation of 1,3-propanediol, Biotechnol Progress **16** (2000) 76–79.

[63] Li, Z., Teng, H., Xiu, Z., Extraction of 1,3-propanediol from glycerol-based fermentation broths with methanol/phosphate aqueous two-phase system, Process Biochem **46** (2011) 586–591.

[64] Fu, H., Sun, Y., Xiu, Z., Continuous countercurrent salting-out extraction of 1,3-propanediol from fermentation broth in a packed column, Process Biochem **48** (2013) 1381–1386.

[65] Li, Z., Jiang, B., Zhang, D., Xiu, Z., Aqueous two-phase extraction of 1,3-propanediol from glycerol-based fermentation broths, Separ Purif Technol **66** (2009) 472–478.

[66] Huang, H. J., Ramaswamy, S., Tschirner, U., Ramarao, B., A review of separation technologies in current and future biorefineries, Separ Purif Technol **62** (2008) 1–21.

[67] Vane, L. M., A review of pervaporation for product recovery from biomass fermentation processes, J Chem Technol Biot **80** (2005) 603–629.

[68] Mochizuki, A., Sato, Y., Ogawara, H., Yamashita, S., Pervaporation separation of water/ethanol mixtures through polysaccharide membranes, I. The effects of salts on the permselectivity of cellulose membrane in pervaporation. J Appl Polym Sci **37** (1989) 3357–3374.

[69] Dubey, V., Saxena, C., Singh, L., Ramana, K. V., Chauhan, R. S., Pervaporation of binary water-ethanol mixtures through bacterial cellulose membrane, Separ Purif Technol **27** (2002) 163–171.

[70] Uragami, T., Takigawa, K., Permeation and separation characteristics of ethanol-water mixtures through chitosan derivative membranes by pervaporation and evapomeation, Polymer **31** (1990) 668–672.

[71] Mochizuki, A., Amiya, S., Sato, Y., Ogawara, H., Yamashita, S., Pervaporation separation of water/ethanol mixtures through polysaccharide membranes, IV. The relationships between the permselectivity of alginic acid membrane and its solid state structure, J Appl Polym Sci **40** (1990) 385–400.

[72] Kanti, P., Srigowri, K., Madhuri, J., Smitha, B., Sridhar, S., Dehydration of ethanol through blend membranes of chitosan and sodium alginate by pervaporation, Separ Purif Technol **40** (2004) 259–266.

[73] Li, S., Tuan, V. A., Falconer, J. L., Noble, R. D., X-type zeolite membranes: preparation, characterization, and pervaporation performance. *Microporous and Mesoporous Materials* 2002, **53**, 59–70.

[74] Izak, P., Köckerling, M., Kragl, U., Stability and selectivity of a multiphase membrane, consisting of dimethylpolysiloxane on an ionic liquid, used in the separation of solutes from aqueous mixtures by pervaporation, Green Chem **8** (2006) 947.

[75] Heitmann, S., Krings, J., Kreis, P., Lennert, A., Pitner, W. R., Górak, A., Schulte, M. M., Recovery of n-butanol using ionic liquid-based pervaporation membranes, Separ Purif Technol **97** (2012) 108–114.

[76] Wang, S., Dai, H., Yan, Z., Zhu, C., Huang, L., Fang, B., 1,3-Propanediol adsorption on a cation exchange resin: adsorption isotherm, thermodynamics and mechanistic studies, Eng Life Sci **14** (2014) 485–492.

[77] Luerruk, W., Shotipruk, A., Tantayakom, V., Prasitchoke, P., Muangnapoh, C., Adsorption of 1,3-propanediol from synthetic mixture using polymeric resin as adsorbents, Front Chem Eng China **3** (2009) 52–57.

[78] Qureshi, N., Hughes, S., Maddox, I., Cotta, M., Energy-efficient recovery of butanol from model solutions and fermentation broth by adsorption, Bioprocess Biosyst Eng **27** (2005) 215–222.

[79] Nielsen, L., Larsson, M., Holst, O., Mattiasson, B., Adsorbents for extractive bioconversion applied to the acetone-butanol fermentation, Appl Microbiol Biotechnol **28** (1988) 335–339.

[80] Roturier, J.-M., Fouache, C., Berghmans, E., Process for the purification of 1,3-propanediol from a fermentation medium, Google Patents, 2002.

[81] Hilaly, A., Binder, T., Method of recovering 1,3-propanediol from fermentation broth, Google Patents, 2002, http://www.google.com/patents/US6479716.

[82] Himmel, M. E., Ding, S. Y., Johnson, D. K., Adney, W. S., Nimlos, M. R., Brady, J. W., Foust, T. D., Biomass recalcitrance: engineering plants and enzymes for biofuels production, Science **315** (2007) 804–807.

[83] Bornscheuer, U. T., Huisman, G. W., Kazlauskas, R. J., Lutz, S., Moore, J. C., Robins, K., Engineering the third wave of biocatalysis, Nature **485** (2012) 185–194.

[84] Bozell, J. J., Feedstocks for the future – biorefinery production of chemicals from renewable carbon, Clean Soil Air Water **36** (2008) 641–647.

[85] Sabra, W., Zeng, A.-P., Industrial Biocatalysis: Mixed Microbial Cultures for Industrial Biotechnology: Success, Chance, and Challenges, Pan Stanford Publishing, Boca Raton, FL, USA 2014.

[86] Brenner, K., You, L., Arnold, F. H., Engineering microbial consortia: a new frontier in synthetic biology, Trends Biotechnol **26** (2008) 483–489.

[87] Sabra, W., Dietz, D., Tjahjasari, D., Zeng, A., Biosystems analysis and engineering of microbial consortia for industrial biotechnology, Eng Life Sci **10** (2010) 407–421.

[88] Xue, C., Du, G.-Q., Chen, L.-J., Ren, J.-G., Sun, J. X., Bai, F. W., Yang, S. T., A carbon nanotube filled polydimethylsiloxane hybrid membrane for enhanced butanol recovery, Sci Rep 2014, 4.

Gennaro Agrimi, Isabella Pisano, Maria Antonietta Ricci,
and Luigi Palmieri

5 Microbial strain selection and development for the production of second-generation bioethanol

Abstract: Ethanol is the most important renewable fuel in terms of volume and market value. Today, the so called first-generation bioethanol is primarily produced from the fermentation of sugars deriving form food crops. This has raised several ethical, environmental and political concerns. Second-generation biofuels can be produced using waste lignocellulosic biomass such as agricultural and forestry residues, or dedicated crops grown on marginal lands. Lignocelluloses are composed of cellulose, hemicellulose, lignin, and several inorganic materials. The process for producing ethanol and other biofuels from lignocellulosic biomass includes: biomass pretreatment, enzymatic hydrolysis and detoxyfication, fermentation and distillation-rectification-dehydration. In this chapter we describe advancements in ethanol production from lignocellulosic biomass with a specific focus on the selection and development of the biocatalysts used in the hydrolysis and fermentation steps.

5.1 Introduction

Transportation fuel supply is currently dominated by fossil fuels. The increase in the price of fossil fuels and concerns about the CO_2 emission-driven climate change have prompted a dramatic increase in demand for a sustainable alternative transportation fuels based on renewable resources.

Ethanol is currently the most important renewable fuel in terms of volume and market value. In 2010, worldwide bioethanol production reached 113 billion liters, with the USA and Brazil as the world's top producers, accounting together for 80% of global production. By 2022, world ethanol production is projected to increase by almost 70% compared to the average of 2010–2012 and reaching some 168 bnl [1]. Currently, ethanol is industrially produced from sweet juice (e.g., sugarcane, sugar beet juice, molasses) and starch (e.g., corn, wheat, barley, cassava); biodiesel is generated from vegetal and animal oils. These are the so-called first-generation biofuels.

The major drawbacks of first-generation biofuels are:

- Competition with crops for land that could be used for food production.
- Rising cost of food due to increased demand of grain crops for first-generation biofuel production. These crops represent in fact the staple grains in the diets of many people, especially in less developed countries.

- Necessity to irrigate these crops, which in some regions adds more stress on groundwater sources.
- High input of fertilizers and pesticides necessary for putting poor quality land into production.

Second-generation ethanol can be produced from lignocellulosic biomass such as agricultural and forestry residues, weeds and waste paper, some municipal and industrial wastes, etc. Lignocellulose is the most abundant renewable material on earth. Currently, the most promising and abundant lignocellulosic feedstocks derived from plant residues are from corn stover, sugarcane bagasse, rice, and wheat straws. In addition, dedicated energy crops are being developed. Second-generation biofuel crops have many advantages over first-generation biofuels:
- Lack of direct competition between feeding and fueling.
- Dedicated energy crops can grow on marginal lands that are unsuited for food crops and require less water, fertilizer and pesticide inputs.

Lignocellulosic biomass is composed of three polymers, cellulose, hemicellulose, lignin, and several inorganic materials [2]. Their relative proportion varies according to the specific biomass. Cellulose is the most abundant organic polymer on earth; it is a linear polysaccharide polymer of glucose. The cellulose chains are packed by hydrogen bonds in microfibrils [3] covered by lignin. Hemicellulose is a heteropolymer consisting of xylose-linking compounds like arabinose, glucose, mannose, and other sugars through an acetyl chain. In contrast to cellulose, which is crystalline, hemicelluloses have a random and amorphous structure with little resistance to hydrolysis [2]. Lignin is a very complex molecule mainly constructed of phenylpropane units linked in a three-dimensional structure with hemicellulose and cellulose. Lignin is particularly resistant to chemical and enzymatic degradation. Generally, softwoods contain more lignin than hardwoods and most of the agriculture residues.

The process for producing ethanol and other biofuels from lignocellulosic biomass includes four main steps:
- Pretreatment – breaking bonds between lignocellulose constituents.
- Enzymatic hydrolysis and detoxification – transforming cellulose to glucose and removing fermentation inhibitors originated during pretreatment processing.
- Fermentation – obtaining the desired biofuel through microbial catabolism of sugars.
- Distillation-rectification-dehydration – separating and purifying the fermentation product.

In this chapter, we describe the identification and development of biocatalysts used in the enzymatic hydrolysis and fermentation of lignocellulosic biomass for ethanol production. Other steps of the production process such as biomass pretreatment are treated in chapter 11 of this book.

5.2 Enzymatic hydrolysis

For the production of lignocellulosic ethanol, sugars must be released from cellulose and hemicellulose. The close association of cellulose fibrils with hemicellulose and lignin makes lignocellulosic biomass highly stable and recalcitrant to enzymatic or chemical hydrolysis. Pretreatment is required to alter the biomass structure and chemical composition in order to facilitate the hydrolysis of the carbohydrate fraction; it allows to decrystallize and partially depolymerize cellulose [4, 5]. Pretreatment can be carried out on the basis of mechanical, physical, chemical, physico-chemical, and biological actions or a combination of them. In some pretreatment process, lignin can be separated from hemicellulose and cellulose. Most pretreatments described in other chapters of this book, solubilize hemicellulose, either totally or in a very significant quantity, in an oligomeric form. After pretreatment, two main processes can be used to hydrolyze cellulose and the remaining part of hemicellulose into monomeric sugar constituents required for fermentation into ethanol: acid (dilute or concentrated) and enzymatic treatments.

Enzymatic hydrolysis of cellulose and hemicellulose is carried out by cellulase and hemicellulase enzymes. The hydrolysis takes place under mild conditions (e.g., pH 4.5–5.0 and temperature 40–50°C). The main advantages of this process, compared to acid hydrolysis, are lower corrosion problems, low utility consumption, and absence of fermentation inhibitors release [6, 7].

Enzymatic hydrolysis of cellulose can be divided in three phases: (1) cellulase adsorption onto the surface of the cellulose, (2) hydrolysis of cellulose to fermentable sugars and (3) desorption of the cellulase. Cellulases and hemicellulases can be classified into at least 15 protein families, which are in turn divided into subfamilies. Cellulase cocktails consist of three major classes of enzymes: endoglucanases (EG), exoglucanases (also called cellobiohydrolases, or CBH), and β-glucosidases (βG). These enzymes are commonly called together cellulase or cellulolytic enzymes [7]. The endoglucanases attack the low-crystallinity regions of the cellulose fibers and create free chain-ends. The exoglucanases further degrade the sugar chain by removing cellobiose units (dimers of glucose) from the free chain-ends. The produced cellobiose (a disaccharide formed by two glucose molecules linked by a β-1-4 glycosidic bond) is then cleaved to glucose by βG. This enzyme cannot be considered a cellulase, but its action is very important for the complete depolymerization of cellulose to glucose. If βG activity is too low, cellobiose accumulates and inhibits cellulases. Recent studies support the importance of non-hydrolyzing enzymes in cellulose depolymerization. For example, the expansin protein family contributes to loosening cellulose microfibrils, which determine the swelling of the polysaccharidic polymer. As a result, cellulose crystallinity decreases and its accessibility to hydrolyzing enzymes increases. Also, enzymes that catalyze the oxidative cleavage of cellulose such as polysaccharide monooxygenases contribute to the overall cellulose depolymerization process. Since hemicellulose contains different sugar units, the hemicellulytic enzymes are more complex and include endo-1,4-β-D-xylanases, exo-1,4-β-D-xylosidases, endo-1,4-β-D-mannanases, β-mannosidases, acetyl xylan esterases, α-glucuronidases, α-L-arabinofuranosidases and α-galactosidases [6].

Bacteria and fungi are good sources of cellulases and hemicellulases. The enzymatic cocktails that are usually employed comprise mixtures of several hydrolytic enzymes as well as non-hydrolyzing proteins. Different enzyme cocktails are today available. The optimal composition of the cocktails depends on the specific biomass and the fermentation process used.

5.2.1 Identification and development of biocatalysts for cellulose and hemicellulose hydrolysis

Despite recent advancements, cellulose and hemicellulose enzyme cocktail costs are still considered to be one of the major impediments for an economically sustainable production of biofuels form lignocellulose. The low cellulase activity, which is approximately 10- to 100-folds lower than those of amylases (the enzymes used in the production of first-generation bioethanol) depending on the cellulose pretreatment and hydrolysis process conditions, implies that substantial amounts of lignocellulolytic enzymes must be added for an efficient hydrolysis. Much research efforts have been focused on lowering the cost of enzymes, trying to improve the enzyme-producing strains, screening new organisms, producing more active enzymes, engineering enzymes and improving the enzymatic hydrolytic process parameters [9].

Cellulases and hemicellulases are produced by several lignocellulolytic microorganisms, mainly fungi and bacteria. Nevertheless, mostly *Trichoderma reesei* and its mutants are employed for the commercial production of hemicellulases and cellulases, with a few also produced by *Aspergillus niger*. The fungus *T. reesei* was one of the first cellulolytic organisms isolated in the 1950s. Since then, extensive strain improvement and screening programs have been carried out, and cellulase industrial production processes have been developed in several countries [9].

An important area of biotechnology research for the lignocellulosic biomass treatments has been directed to discovering and identifying organisms that produce novel lignocellulolytic enzymes. The novelty is referred not only to the identification of new enzymes that are able to hydrolyze pretreated cellulose and hemicellulose more rapidly but also to enzymes that can withstand extreme pH, different temperatures and inhibitory agents which can be found in the pretreated biomass.

Several strategies can be employed to identify new enzymes (Figure 5.1).

1. Classical screening of microorganisms. The decomposition of plant biomass is promoted by various microorganisms. Only a few of them are currently known. Some natural habitat (for example forest soil, compost) can be considered a huge source of new (ligno)cellulolytic microbes. These microbes provide a variety of novel cellulases, hemicellulases and accessory proteins. Relevant lignocellulosic

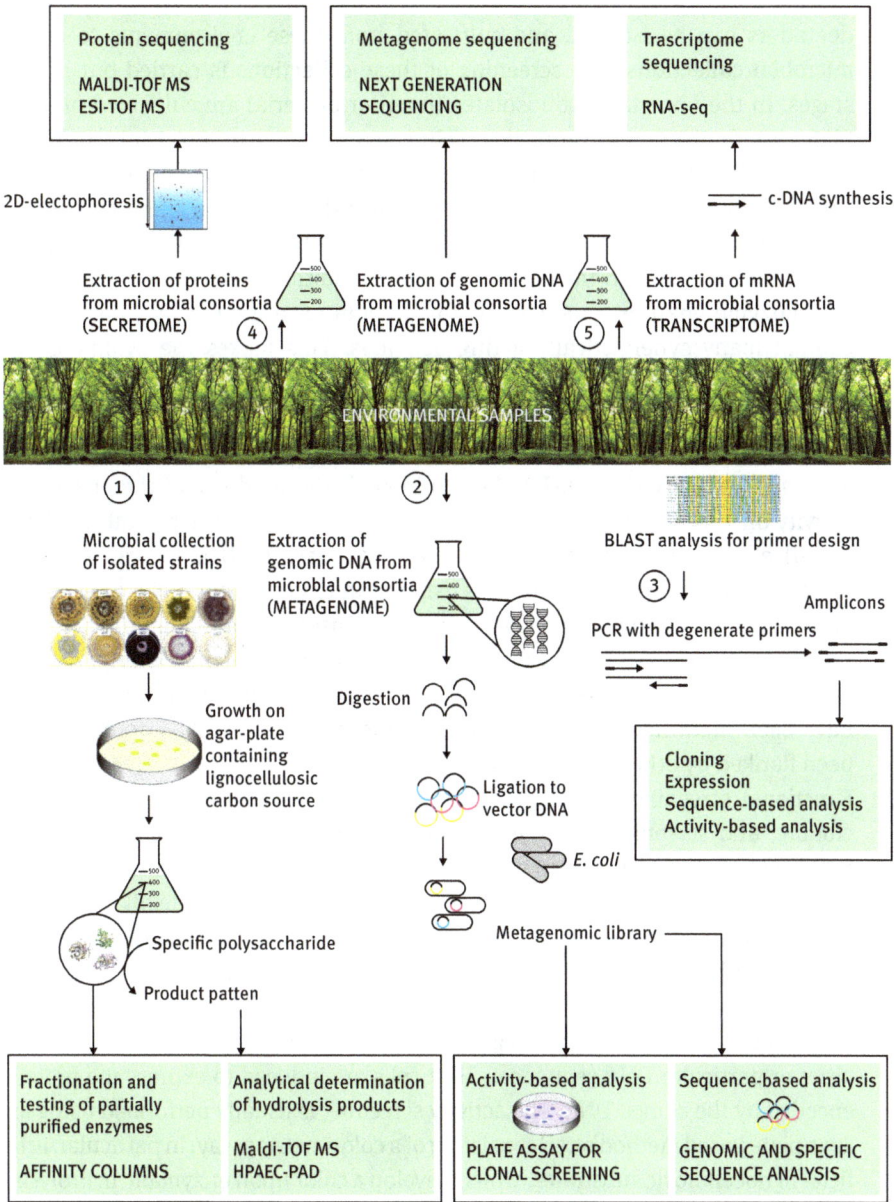

Fig. 5.1: Schematic representation of the strategies used for the identification of novel biocatalysts for cellulose and hemicellulose hydrolysis. (1) Screening of microorganisms in microbial collections. (2) Screening of metagenomic libraries. (3) Screening of enzyme database. (4) Functional screening of proteomes/metaproteomes secreted from lignocellulosic microbial consortia (secretome). (5) Identification and characterization of lignocelullosic enzyme by sequencing microbial consortia genomic DNA (metagenome) or their mRNA (transcriptome).

degraders can be isolated and cultivated from these environments to create microbial collections. The screening of these collections is carried out in two stages. In the first stage, soil isolates (fungi or bacteria) are cultivated on agar plates containing cellulose or lignocellulose as a sole carbon source to select isolates allegedly rich of (hemi)cellulases. In the second stage, microorganisms showing the best growth are cultivated in semi-liquid media again containing cellulose or lignocellulose as the carbon source. At this stage, growth conditions can be varied to match the desired reaction conditions (e.g. high temperature) [10]. Enzyme mixtures produced by fungal cultures are typically complex and contain many synergistically acting activities. This makes the evaluation of different enzyme components in the mixture a challenging task. The enzyme mixture produced by the selected microorganism can be analyzed for various hemicellulase and cellulase activities using specific analytical methods such as mass spectrometry (MALDI-TOF MS) to identify the products of the enzymatic activity on specific lignocellulosic substrates and obtain a fingerprinting of the (hemi)cellulolytic activities present in the crude enzyme [11]. Moreover, a fractionation of partially purified enzymes using affinity chromatography and testing of the enzymatic activities can be carried out. Based on the screening, the most promising strains are further analyzed and subjected to gene cloning/genome sequencing. Because less than 1% of the potential microbes in the biosphere have been identified using traditional methods of culture, this approach has been flanked by others [12].

2. Functional screening of metagenomic libraries. It consists in the construction, cloning, and screening of a metagenomic library (a collection of DNA derived from environmental samples containing microbial consortia with an interesting phenotype) by activity- or sequenced-based methods. In detail, total DNA from all microorganisms living in a defined habitat is isolated, digested, and cloned in a host organism, such as *Escherichia coli*. The result is a collection of microbial cells that contain pieces of foreign DNA, a metagenomic library. These libraries can be used in screening experiments to find the desired enzyme activities or particular DNA sequences. Host cells are induced to express the protein encoded by the cloned DNA. The activity screening is usually performed using an agar-plate based method often consisting of a colorimetric assay; in particular, artificial lignocellulosic substrates, which develop a color upon enzymatic hydrolysis, are used. Alternatively, enzymatic activities can be assayed in cell lysates. Through this metagenomic approach, several lignocellulolytic enzymes, some of which display high activity or novel features, have been identified from different environments (anaerobic digesters, cow rumen, forest soil, animal guts, etc.) [13]. Some limitations of this approach are represented by the complexity of these systems, the difficulty to correlate the identified enzymatic activity to the source organism, the availability of synthetic chemical substrates for the screening assays, the poor quality of the isolated DNA, and the limitations of the expression system (often *E. coli*) that greatly restricts enzyme discovery from eukaryotic species [14].

3. Screening of enzyme database. This approach is based on the availability of annotated sequences of enzymes with specific functions previously identified in different microorganisms. The enzymatic activity of interest is selected in enzyme databases. Biochemical properties and sequences of lignocellulolytic enzymes are included in the BRENDA database (http://www.brenda-enzymes.org); fungal enzyme sequences are collected in the mycoCLAP database (https://mycoclap.fungalgenomics.ca). The CAZy database (https://www.cazy.org) gathers families of structurally related catalytic and carbohydrate-binding functional domains of enzymes that degrade, modify, or create glycosidic bonds. The sequences of the selected enzymes are aligned using specific homology alignment algorithms (Basic Local Alignment Search Tool, or BLAST). Conserved regions are identified and used to design degenerate primers (primers in which one or more positions can be occupied by one of the several possible nucleotides in order to account for small mutations in the conserved regions). These primers can be used for the amplification (PCR) and cloning of the target enzyme from the (meta)genome. The cloned sequence can then be overexpressed in a heterologous host and its activity can be tested [10].

4. Functional screening of proteomes/metaproteomes. This approach consists in the preparation and analysis of protein samples, with either the secretome of microbes grown in specific growth conditions or in proteins isolated from relevant environments (metaproteome) [14]. Two-dimensional (2D) gel electrophoresis is used to separate single proteins and is the primary tool for obtaining a global picture of the expression profile under various cultural conditions. In this method, proteins are first separated based on their net charge by isoelectric focusing and then in the orthogonal direction by molecular mass using a denaturing electrophoresis (SDS-PAGE). The sample preparation and the 2D gel electrophoresis are quite complex, and the reproducibility and resolution still remains a challenge. Once the proteins are separated, it is possible to determine their sequences by mass spectrometric techniques (MALDI-TOF or ESI-TOF MS). Data acquired such as peptide mass fingerprints (PMF) are utilized for protein identification, which is done by analyzing the sizes of tryptic fragments via search engines using protein databases (e.g., NCBI protein database). This approach requires the availability of adequate information about the sequences (either of the coding genes or amino acidic) of the studied enzymes [10].

5. Sequencing-based enzyme discovery. The modern next-generation sequencing (NGS) techniques (Illumina, Roche 454, etc.) allow massive sequencing of nucleic acids in a cost- and time-saving manner, providing information on the coding sequences of lignocellulolytic enzymes found in one organism or in a microbial consortium. Analysis of the genomic data allows the identification of sequences similar to already discovered enzymes. These technologies can allow the identification of a huge number of candidate lignocellulolytic genes in one experiment [15]. Also, next-generation cDNA sequencing (RNA-seq) makes it possible to sequence complete transcriptomes (mRNAs) in a population. It results usable and

informative overcoming some problems related to eukaryotic gene identification [16] such as the occurrence of gaps in the genetic coding sequences (introns). Despite advances in gene discovery through next-generation sequencing, there remains a considerable gap between the growing number of new gene sequences and the number of characterized lignocellulose-active enzymes. The discrepancy between gene discovery and novel enzyme applications is associated with the limitations in functional expression of heterologous enzymes and in the availability of model polysaccharides and substrates in screening experiments. Consequently, once potential lignocellulosic enzymes are identified, their expression and biochemical characterization remains a fundamental requirement [10].

Besides the research for new enzymatic activities, much effort has been put in place in order to lower production costs and improve the existing enzyme preparations. In particular, researchers have sought to improve the following properties: activity, thermal stability, pH stability, salt tolerance.

In addition, the composition of the enzymatic cocktails has been thoroughly investigated. Until a few years ago, the lignocellulosic hydrolytic process suffered the accumulation of short gluco- and xylo-oligosaccharydes, which acted as end-product inhibitors on cellulases and hemicellulases. Consequently, new enzyme preparations have been formulated. They are characterized by a higher content of βG activity. This enzyme acts on cellobiose (4-O-β-D-glucopyranosyl-D-glucose) and on short glucose oligosaccharides, yielding glucose monomers, thus lowering cellulase inhibition [17].

The most common source of lignocellulolytic enzymes, *T. reesei*, naturally secretes low levels of βG. Newer enzyme cocktails supplement the enzyme mixtures produced by *T. reesei* with additional βG from other species. With these improved cocktails, separate hydrolysis and fermentation (SHF) (see the following paragraphs), which typically suffered from end-product inhibition of the enzymatic activities, has witnessed a renewed interest. A further improvement has been the introduction of xylanolytic activities (xylanase and β-xyloxidase) in the enzymatic cocktails since xylo-oligosaccharydes, originating from partial hemicellulose hydrolysis, also have an inhibitory effect on cellulases. In the latest cocktails, oxidative enzymes have also been introduced. They act synergistically with cellulases and hemicellulases to improve the saccharification efficiency [18].

The improvement of the enzymes, enzymatic cocktails, and production process has been carried out by several means. Traditional methods such as random mutagenesis using UV irradiation or chemical mutagens followed by screening procedures have allowed the selection of mutants of *T. reesei* displaying a higher cellulase activity. Genome shuffling approaches (recombination of whole genomes) have been successfully carried out to isolate high producers. Also, target genetic engineering have been employed. For example, Zhang et al. [19] replaced the promoter of the weakly expressed *T. reesei* βG gene with four copies of the much more active promoter of the *cbh1* (cellobiohydrolase) gene, obtaining a strain secreting an enzymatic cocktail with an increased βG activity. Protein engineering has also been employed to obtain cellulases with improved biochemical properties (e.g., higher turnover rate, increased thermal stability) [18].

In order to optimize cellulase production, process engineering approaches have also been tackled: the effect of different carbon sources and different fermentation conditions have been tested, obtaining significant increase in the yield of the process. Moreover, improved cocktails have allowed the decrease of the enzymatic loading and hence the overall costs [17].

5.3 Fermentation inhibitors in lignocellulosic hydrolyzate

The harsh pretreatment methods (physical, chemical, or biological) needed to increase the susceptibility of the raw material to subsequent hydrolysis can generate significant amounts of by-products that inhibit hydrolytic enzymes and microbial metabolism, reducing sugar and ethanol yields and productivities. If highly concentrated hydrolyzates are used in order to achieve higher ethanol concentrations, inhibitory compounds levels increase accordingly. The inhibitory compound pattern is variable depending on the conditions used during pretreatment and hydrolysis (i.e., temperature, pH, and residence time). Consequently, to obtain robust microorganisms tolerating high concentration of inhibitors and an optimal design of the fermentation process, the inhibitory compounds and the molecular mechanisms of inhibition have to be known. Lignocellulose-derived inhibitors have traditionally been divided into three general classes: furan aldehydes, weak acids, and phenolics [20]. Lignocellulosic hydrolyzates vary in their degree of inhibition, and different microorganisms have different inhibitor tolerances. Most studies have been performed on *Saccharomyces cerevisiae* as reported below.

5.3.1 Weak acids

Weaks acids in the context of pretreated lignocellulosic materials are mostly represented by acetic acid, formic acid, octanoic acid, and levulinic acid [21]. Acetic acid is released during pretreatment as a result of the deacetylation of hemicellulose chains, whereas formic and levulinic acid are degradation/breakdown products of furfural and hydroxymethylfurfural (HMF), which in turn derive from the degradation of pentoses and hexoses, respectively [22]. The inhibitory effects of weak acids are strongly correlated with pH and have been attributed to two factors: intracellular acidification and accumulation of anions [23]. Intracellular acidification is believed to result from the passive influx of weak acids in their undissociated form from the extracellular medium (usually pH 5) over the plasma membrane into the cytosol (pH ~7). This phenomenon is particularly increased when the pH of the growth medium is near or lower than pKa of the weak acid in question. The pKa values of the acids relevant in the context of the fermentation of lignocellulosic feedstocks (formic, levulinic, acetic acids) are in the range 3.75–4.75, which means that at pH 5, nearly half of the acetic acid is present in the undissociated form. The decrease in intracellular pH is compensated by the plasma membrane H+-ATPase, which pumps protons out of the cell at the expense of ATP hydrolysis [24]. Consequently, less ATP is available for biomass formation. At higher acid concentrations, ATP demand can be so high that cells cannot avoid acidification of the cytosol. When the proton efflux capability of the cells is exhausted, intracellular acidification occurs, leading to alteration of cellular processes. An alternative theory explaining the weak acid toxicity is based on the fact that carboxylic anions accumulate at high concentration in the cytoplasm impacting cell turgor and amino acids pools [24].

5.3.2 Furan derivatives: furfural and HMF

The furan compounds 5-hydroxymethyl-2-furaldehyde (HMF) and 2-furaldehyde (furfural) are formed by dehydration at high temperatures and low pH from hexoses and pentoses, respectively. The level of furans varies according to the type of raw material and the pretreatment procedure. These compounds have been demonstrated to have numerous adverse effects on microbial metabolism such as reducing the fermentation rate, increasing the lag phase of growth, reducing viability, inhibiting several enzymes in the glycolytic and ethanol production pathways, breaking down DNA, and damaging cellular membranes [25–30]. Inhibitory effects have been demonstrated both in eukaryotes and prokaryotes. Many microorganisms reduce or oxidize furanic aldehydes to their respective alcohol or carboxylic acid forms to detoxify them [31]. The reactivity of the furan derivatives has been attributed to the aldehyde group, which can bind to and inhibit several proteins. The detoxification process itself (reduction of aldehydes to less toxic alcohols) can also cause a depletion of NAD(P)H, which is needed by a high number of intracellular reactions with negative consequences on the overall cellular metabolism. In general, the effects of furans can be explained by a re-direction of cellular bioenergetics in order to fix the damage caused by furans leading to reduced intracellular levels of ATP and NAD(P)H [31].

5.3.3 Phenolic compounds

Phenolic compounds are generated mainly as a consequence of lignin breakdown and carbohydrate degradation during pretreatment processes for hydrolyzate production [7, 8]. The amount and type of phenolics is dependent on the pretreatment technology and the feedstock used, since lignin has different degrees of methoxylation, internal bonds, and association with hemicellulose and cellulose in different lignocellulosic biomasses. Despite the variability of phenolics compounds, the most frequently occurring among them are 4-hydroxybenzaldehyde, 4-hydroxybenzoic acid, vanillin, dihydroconiferyl alcohol, coniferyl aldehyde, syringaldehyde, and syringic acid. Most of the studied phenolic compounds increase the lag phase of growth and reduce the specific growth rate and volumetric ethanol productivity in *S. cerevisiae* [32]. Removing most of the phenolics by enzymatic treatment significantly increases the fermentability of hydrolyzates, demonstrating the significance of these compounds as inhibitors [33]. The chemical nature of the phenolics determines their toxicity, as phenolic aldehydes and ketones are generally regarded as the most inhibitory while phenolic acids and alcohols tend to be seen as the least toxic [31]. In *S. cerevisiae*, distinct physiological responses are observed, suggesting that phenolic compounds have different cellular targets. The inhibitory mechanisms of the phenolic compounds remain to be fully elucidated. It has been hypothesized that they cause loss of membrane integrity, destroying the membrane capability to act as selective barrier [34]. Since phenolic aldehydes, such as the furan aldehydes, are converted to their corresponding alcohol forms in metabolism, it can be also hypothesized that phenolic and furan aldehydes cause the perturbation of redox metabolism. Weakly acidic phenolic compounds may also act as uncouplers of the electrochemical gradient by transporting protons back across the mitochondrial membranes, thus impairing cellular energetic metabolism [35].

5.3.4 Other inhibitors

Lignocellulosic materials may contain a high concentration of inorganic salts. Moreover, inorganic ions are also added with chemicals used in pretreatment, during pH adjustment of the pretreated raw material and are plausibly also leached off from process equipment [21, 26]. High concentration of salts reduces the volumetric ethanol production rate and the degree of the inhibitory effect is dependent on the nature of the ion, with Ca^{2+} being the most inhibitory [28, 29].

The combined effect of several inhibitors is higher than the summed effects exerted by individual inhibitors (synergistic effect). Furfural and acetic acid interact, reducing synergistically the specific growth rate, ethanol yield, and biomass yield [28, 29]. Syringaldehyde and acetovanillone interact synergistically in *S. cerevisiae*, reducing the volumetric ethanol productivity when added to a wheat straw hydrolyzate pretreated by wet oxidation [37]. Furfural acts synergistically with HMF and aromatic aldehydes (4-hydroxybenzaldehyde, vanillin, and syringaldehyde) in *E. coli*, reducing the specific growth rate and ethanol yield. Furfural interacts synergistically with its conversion product furfuryl alcohol, decreasing the *E. coli*-specific growth rate and ethanol yield [38].

5.3.5 Detoxification

Detoxification is particularly important when strongly inhibiting hydrolyzates are fermented, when high concentrations of inhibitors accumulate in the fermentation unit due to recirculation of streams, or when a fermenting organism with low inhibitor tolerance is used. Detoxification has, however, some disadvantages: additional costs, more complicated process of biomass-to-ethanol conversion, and additional waste generation [39]. Three main types of treatments are employed to reduce the inhibitory compounds to non-inhibitory levels: physical, chemical, and biological.

5.3.5.1 Physical and chemical detoxification methods

As demonstrated by Palmqvist et al. [40], roto-evaporation of lignocellulose hydrolyzate can decrease the levels of volatile inhibitory compounds such as acetic acid, furfural, and vanillin in a lignocellulosic hydrolyzate. Using this method, however, non-volatile fermentation inhibitors are retained in the hydrolyzates. Another physical method is the separation of inhibitory compounds from hydrolyzates through membrane extraction. Microporous polypropylene hollow fiber membranes were used for the extraction of sulfuric acid, acetic acid, 5-hydroxymethyl furfural, and furfural from corn stover hydrolyzed with dilute sulfuric acid. In this case, ethanol yields from detoxified hydrolyzates were about 10% higher than those from hydrolyzates detoxified using ammonium hydroxide treatment [41].

Hydrolyzates can be detoxified using several chemical approaches:
- Solvent extraction – solvents such as diethyl ether or ethyl acetate are employed to remove acetic acid furfural, vanillin, and 4-hydroxybenzoic acid.
- Activated carbon/charcoal adsorption – this treatment removes lignin-derived inhibitors. It does not require pH adjustments. Differently from the inhibitors, sugars have a relatively low affinity toward charcoal. This application is expensive since powdered activated charcoal cannot be regenerated, and a 10% loss generally occurs during each thermal reactivation cycle.
- Use of ion exchange resins to remove acetic acid.
- Overliming/alkali treatment – increasing the pH to 9–10 with $Ca(OH)_2$ (overliming) and readjustment to 5.5 with H_2SO_4 leads to the precipitation of toxic components and to the degradation of some inhibitors due to their instability at high pH. This treatment, however, leads to extensive loss of fermentable sugars [20].

5.3.5.2 Biological detoxification methods

Biological methods of detoxification involve the use of microorganisms and/or enzymes that act on specific toxic compounds present in the hydrolyzates, converting them into less toxic substances. The advantages of biological detoxification over physical or chemical methods include little waste production, mild reaction conditions, no use of toxic and corrosive chemicals, less toxic products generated by side-reaction, and a lower energy demand. Biological treatments can be performed directly in the fermentation vessel prior to fermentation or simultaneously with the fermentation [42]. Disadvantages of using biological methods are long incubation periods required for detoxification, high enzyme production costs, and the loss of sugars [43]. These treatments can be carried out using either whole cells or isolated enzymes. For example, treatment of hydrolyzates with the filamentous soft rot fungus *T. reesei* led to the removal of acetic acid, furfural, and benzoic acid derivatives [40]. Treatment with the enzymes peroxidase and laccase, isolated from the ligninolytic fungus *Trametes versicolor*, led to selective and complete removal of phenolic monomers and phenolic acids [44]. The detoxifying mechanism was suggested to be oxidative polymerization of low-molecular-weight phenolic compounds. Immobilization would facilitate enzyme recovery and reduce cost. Laccases have been proven to be particularly successful in the detoxification processes. Their industrial use would require, however, the production of large quantities of enzyme at low cost. Genetic engineering of *S. cerevisiae* for production of active laccase from the white rot fungus *Trametes versicolor* allowed the in situ detoxification of phenolic inhibitors and dilute acid spruce hydrolyzate fermentation at a faster rate than the parental strain [45].

5.3.6 Development of biocatalysts with enhanced resistance to fermentation inhibitors

Two strategies have been employed to make a biocatalyst more resistant to fermentation inhibitors: metabolic engineering and evolutionary engineering. Metabolic engineering is the purposeful modification of metabolism in order to improve the traits of a cell and hence the production of a fermentation product. An important part of metabolic engineering is the genetic modification of the biocatalyst; in this case, genes encoding enzymes, transporters, or transcriptional regulators are deleted or overexpressed based on biochemical information previously acquired, in order to improve the fitness of the biocatalysts in the presence of the fermentation inhibitors. Insights about the genetic traits involved in inhibitor tolerance or detoxification can be obtained by systems biology approaches (genomics, transcriptomics, metabolomics, fluxomics) where cells are treated with a single inhibitor or with a mixture of different inhibitors and the cellular response is measured in terms, for example, of changes in gene expression. Metabolic engineering can thus be defined a rational approach [46]. For example, to counteract redox unbalance that can take place in yeast in the presence of high concentration of furan derivatives (furfural and HMF), *GSH1*, a gene involved in the glutathione (GSH) synthetic pathway, was overexpressed, obtaining a better performance in terms of ethanol yield and conversion of furfural and HMF in pretreated spruce saccharification and fermentation (SSF; see paragraph 5.4.2) process [47].

Meanwhile, in evolutionary engineering (also called directed evolution), cells are exposed to selective pressure (in this case fermentation inhibitors) in prolonged cultures and the best growing mutants, mimicking in a much shorter time and accelerating what happens in the natural evolution process, are selected. Several cultivation methods can be applied; the most used ones are serial batch cultures with increasing inhibitor concentrations and continuous cultivations (chemostats) using increasing dilution rates (growth rates) in the presence of inhibitors. To increase the mutation rate, and consequently the evolution rate, random mutagenesis protocols can be employed using a chemical mutagen such as EMS (ethyl methansulfonate) or UV irradiation [48]. For example, to improve the resistance of the *S. cerevisiae* strain TMB3400 to fermentation inhibitors, Koppram et al. [49] employed two evolutionary engineering protocols comprising either long-term adaptation in repetitive batch cultures in shake flasks using a cocktail of 12 different inhibitors or a long-term chemostat adaptation using spruce hydrolyzate. The isolated evolved strains displayed significantly improved growth performance over TMB3400 when cultivated in spruce hydrolyzate under anaerobic conditions with a significant increase of the ethanol productivity. Eventually, the obtained evolved strains can give very important insights in inverse metabolic engineering experiments; in this case, a systems biology approach is used to find out the molecular basis of the improved traits (in this case, improved tolerance). The information obtained from these experiments can represent the basis for a following metabolic engineering of the biocatalysts. In many cases, the outlined approaches (metabolic and evolutionary engineering) have been used in concert for strain development [46].

5.4 Fermentation

The production of lignocellulosic ethanol can be divided into four main steps: pretreatment, hydrolysis, fermentation, and distillation. Based on the integration of the above-mentioned steps, four process formats, characterized by the increase of process integration, can be used:
1. separate hydrolysis and fermentation (SHF)
2. simultaneous saccharification and fermentation (SSF)
3. simultaneous saccharification and co-fermentation (SSCF)
4. consolidated bioprocessing (CBP)

Currently, the trend is to go toward an integrated process. However, since some of the steps have different optimal operational conditions, it must be pointed out that a reduction of the costs can also be obtained, optimizing the individual steps. In addition, hybrid processes are being developed. In these cases, individual steps take place under their optimal conditions; before the conclusion of the first process, the following step is started in the same vessel. For example, cellulose can be partially hydrolyzed under its optimal conditions; afterward, operational conditions are changed, cells are added, and fermentation begins [50].

5.4.1 Separate hydrolysis and fermentation (SHF)

In this process, pretreated lignocellulose is hydrolyzed to monosaccharides and sub-sequently fermented to ethanol in separate vessels (Figure 5.2).

This mode of operation allows to perform hydrolysis and fermentation at their own optimal conditions. The optimal temperature for cellulase activity is usually between 45°C and 50°C, whereas that of fermenting microorganisms is between 30°C and 37°C [51].

This process has two main drawbacks. First, inhibition of cellulase activity by the released sugars, mainly cellobiose; meanwhile, βG, the enzyme hydrolyzing cellobi-ose is inhibited by released glucose. Second, SHF is prone to microbial contamina-tions during the hydrolysis process, with cellulase preparations being not sterile. The hemicellulose hydrolyzate produced after the pretreatment and containing a large fraction of pentose sugars can be fermented in a separate tank [7, 8] Alternatively, hexoses and pentoses can be fermented simultaneously. This process is called sepa-rate hydrolysis and co-fermentation (SHCF).

5.4.2 Simultaneous saccharification and fermentation (SSF)

In SSF, glucose produced by the hydrolyzing enzymes is consumed immediately by the fermenting microorganism present in the same vessel (Figure 5.3).

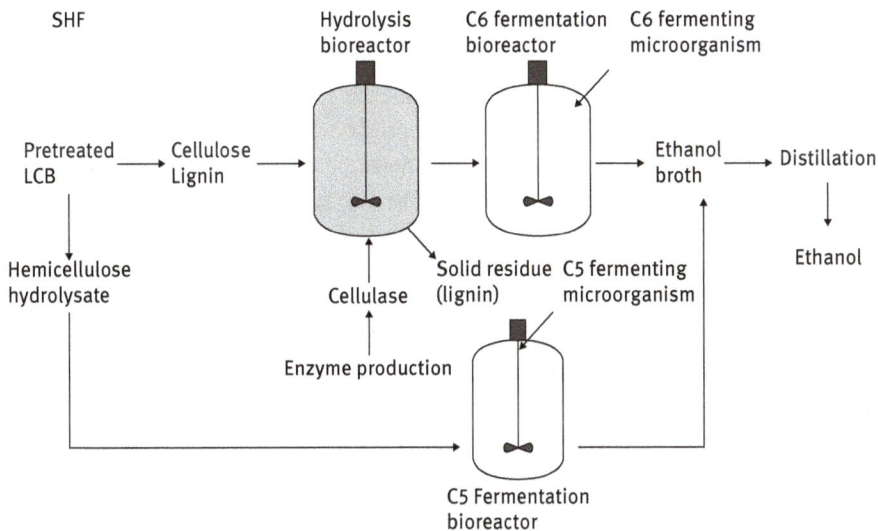

Fig. 5.2: Schematic flow sheet for bioethanol production from lignocellulosic biomass using SHF. LCB, lignocellulosic biomass.

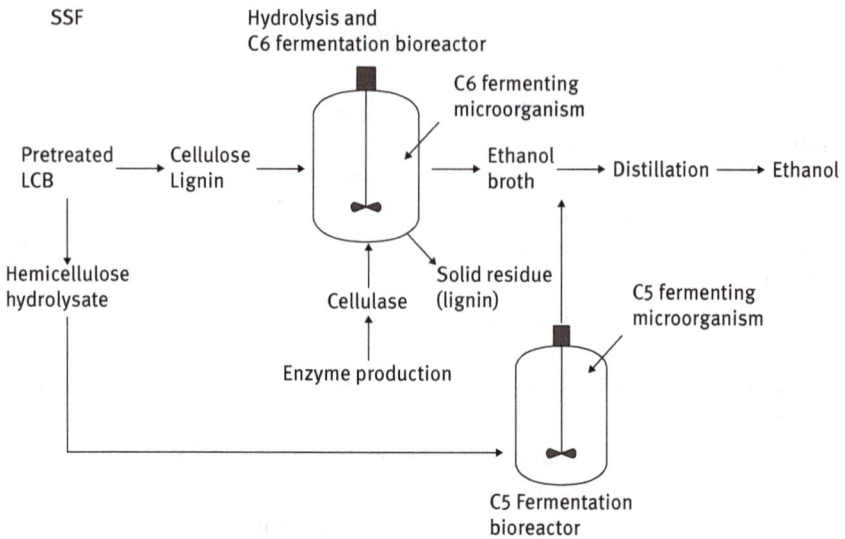

Fig. 5.3: Schematic flow sheet for bioethanol production from lignocellulosic biomass using simultaneous enzymatic saccharification (hydrolysis), and fermentation (SSF). LCB, lignocellulosic biomass.

There are several advantages of SSF compared to SHF. SSF avoids end-product inhibition since the inhibition effects of cellobiose and glucose on the enzymes are minimized by keeping a low concentration of these sugars in the media. Consequently, higher ethanol yields from cellulose have been reported for SSF compared to SHF. Moreover, SSF requires lower amounts of enzyme [4] and shows a lower risk of contamination due to the presence of the produced ethanol. Finally, SSF requires a lower number of separate vessels in comparison to SHF, resulting in lower capital costs.

Despite all these advantages, several problems remain to be solved. The most important problem is the difference between optimal temperatures of the hydrolyzing enzymes and fermenting microorganisms. The optimal temperature for hydrolytic enzymes, whose activity represents the rate-limiting step of SSF, is usually between 45°C and 50°C, whereas fermenting microorganisms like *S. cerevisiae* have an optimal temperature between 30°C and 35°C and are virtually inactive at more than 40°C. As a compromise, SSF is usually carried out at 34–37°C. Several thermotolerant bacteria and yeasts, such as *Kluyveromyces marxianus* have been proposed for use in SSF to raise the temperature close to the optimal temperature of hydrolysis. Another problem is the difficulty to achieve high substrate loads as a consequence of mechanical mixing problems and insufficient mass transfer. Moreover, cells produced during SSF cannot be easily separated from the solid residue (mainly lignin) and thus cannot be recycled, as in the case of SHF. As for SHF, as well as for SSF, the pentose-rich hemicellulosic stream can be fermented by a pentose-fermenting microorganism in a separate tank [7, 8].

5.4.3 Simultaneous saccharification and co-fermentation (SSCF)

SSCF is a variation of SSF in which the fermentation of both five- and six-carbon sugars to ethanol (co-fermentation) is carried out in the same vessel (Figure 5.4). A substantial portion of hemicellulose is hydrolyzed during most types of pretreatment. Cellulose is not separated from hemicellulose after the pretreatment and is hydrolyzed and converted to ethanol together with hemicellulose sugars [7, 8].

Saccharomyces cerevisiae recombinant strains capable of fermenting both hexose and pentose sugars are used in this process. Using a recombinant xylose-fermenting *S. cerevisiae* industrial strain, ethanol concentrations reaching 40 g/L, and yields up to 80% of the theoretical based on xylose and glucose have been achieved [52]. Also, the recombinant bacterium *Zymomonas mobilis* and the natural pentose-fermenting yeast *Pichia stipitis* have been used in SSCF. However, yields obtained were lower and a thorough detoxification of the hydrolyzate was necessary [53].

5.4.4 Consolidated bioprocessing (CBP)

CBP is a process in which cellulase production, substrate hydrolysis, and fermentation to ethanol takes place in a single process step by a cellulolytic microorganism (Figure 5.5). It is based on utilization of mono- or co-cultures of microorganisms that ferment cellulose to ethanol without the addition of cellulases. Although still immature, CBP could become the choice technology for producing ethanol commercially and is the logical end-point in the evolution of ethanol production from lignocellulosic materials. This mode of operation has potentially lower costs and higher efficiency than processes featuring dedicated cellulase production. This results from

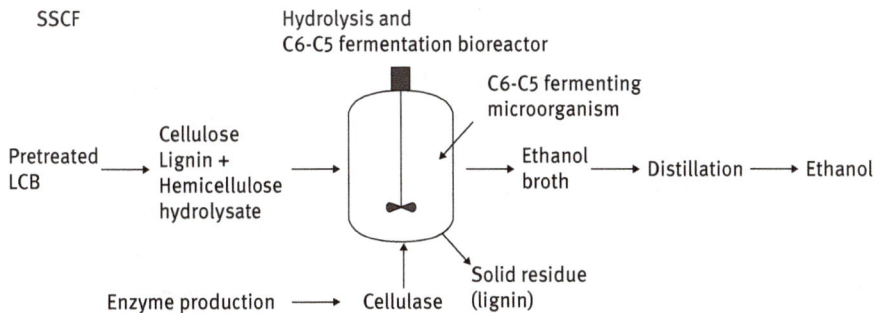

Fig. 5.4: Schematic flow sheet for bioethanol production from lignocellulosic biomass using simultaneous saccharification (hydrolysis) and co-fermentation (SSCF). LCB, lignocellulosic biomass.

CBP   Cellulase production, hydrolysis and
                                C6-C5 fermentation bioreactor

Cellulose
Pretreated → Lignin +   Ethanol → Distillation → Ethanol
LCB   Hemicellulose   broth
         hydrolysate

Cellulase producing   Solid residue
C6-C5 fermenting   (lignin)
microorganism

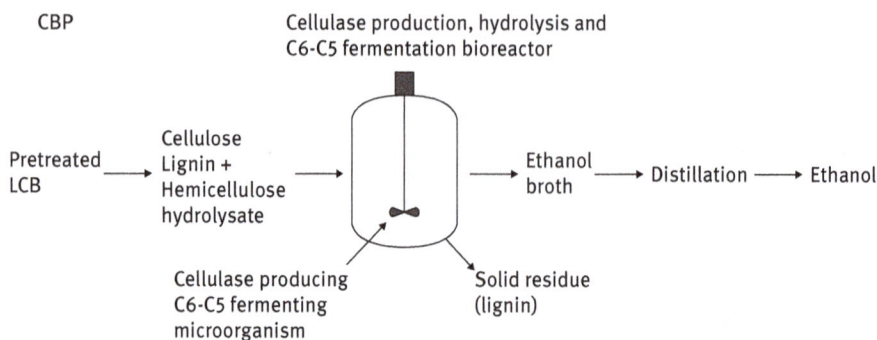

Fig. 5.5: Schematic flow sheet for bioethanol production from lignocellulosic biomass using CBP. LCB, lignocellulosic biomass.

avoided costs for capital, substrate, media, and utilities associated with cellulase production. Moreover, higher hydrolysis rates are potentially achievable using thermophilic organisms expressing complexed cellulase systems or cellulose-adherent cellulolytic microorganisms. This can in turn reduce reactor volume and capital investment [54].

5.4.5 High-gravity fermentation

In sugar/starch-based ethanol production, the fermentation step is carried out in the presence of very high sugar levels (up to 350 g/L). This process yields 10–15% ethanol, resulting in savings of capital cost and energy input (e.g., in the distillation costs) and is called high-gravity fermentation (HGF). A current trend is to apply this kind of process to lignocellulosic biomass in order to improve the overall economy of the process. To obtain at least 4% ethanol (the benchmark in HGF), at least 8% sugar levels are necessary (the theoretical yield is about 0.5 g of ethanol/g of sugar); this implies the utilization of a lignocellulosic feedstock with an initial dry matter content (comprising both soluble and insoluble solids) of >20%. Since the lignocellulosic materials are highly hygroscopic, this results in very viscous slurries with problems in the hydrolysis and in the fermentation steps [17].

HGF of lignocellulosic material implies the increase of the concentration of the fermentation inhibitors released; also, other factors that hamper cellular growth such as high osmolarity and ethanol concentration are increased in HGF. As a consequence, the need for a robust industrial ethanol-producing microorganism is critical in HGF. To partly overcome this problem, HGF is often carried out in a fed-batch mode where the lignocellulosic biomass is gradually added to the fermentation tank. These conditions allow the biocatalyst to gradually detoxify some inhibitors.

In HGF, the high solids content and viscosity hinders efficient mixing; this hampers the enzyme-substrate interaction. To overcome this problem, new pretreatment conditions, reactor configurations, and mixing strategies are being developed [17].

5.5 Microbial biocatalysts

An optimal microbial catalyst involved in biofuel fermentation should (1) ferment sugars to bioethanol or other biofuels with high yields and productivity, (2) be able to ferment both hexoses and pentoses, which are highly concentrated in hemicellulose making second-generation biofuels profitable, (3) have a high tolerance to fermentation inhibitors and ethanol, (4) be resistant to contamination (resistance to low pH and to high temperatures are helpful in this regard), and (5) use inexpensive media for growth. Other important properties of the "perfect biocatalyst" are capability of producing (hemi)cellulolytic enzymes and thermostability. Some of these features can be natural in most microorganism; no microorganism, however, displays all of them. Hence, metabolic engineering and directed evolution have been employed to improve the biocatalysts (Figure 5.6).

5.5.1 *Escherichia coli*

Escherichia coli is a widely used microorganism in industrial processes; it can ferment hexoses and pentoses; it does not have complex nutritional requirements, and it has a long history as an industrial cell factory [55]. However, it is not able to produce ethanol. The introduction in the chromosome of *E. coli* B of *Zymomomnas mobilis* pyruvate decarboxyalse (*pdc*) and alcohol dehydrogenase II genes generated the strain KO11, which was able to produce ethanol from all the sugars of the hemicellulose hydrolyzate with more than 95% theoretical yield [55].

A limitation of these recombinant *E. coli* strains is that they are not capable of fermenting simultaneously hexoses and pentoses, preferring glucose over other sugars. The major problem that hampers the use of these ethanologenic *E. coli* strains is their lower resistance to fermentation inhibitors, ethanol, and process stress (change in pH, salts, temperature) compared to *S. cerevisiae*. Random and rational approaches have been carried out in order to improve this trait [56], but the overall robustness is still not comparable to that of *S. cerevisiae*. Another disadvantage of *E. coli* is its narrow and neutral optimal pH range (pH 6.0–8.0). Furthermore, some *E. coli* strains are considered human pathogens, and hence, the use of residual cell as mass for animal feed is not accepted. All these factors led us to consider *E. coli* unsuitable for industrial ethanol production [55].

Introduction of *E. coli* pentose utilization pathways
Improvement of resistance to fermentation inhibitors

Z. mobilis

Introduction of *Z. mobilis* pyruvate decarboxylase and alcohol dehydrogenase
Improvement of resistance to fermentation inhibitors

E. coli

Introduction of bacterial or fungal pentose utilization pathways
Improvement of cofactor unbalance
Overexpression of xylulokinase, transaldolase and transketolase
Introduction of a pentose transporter

S. cerevisiae

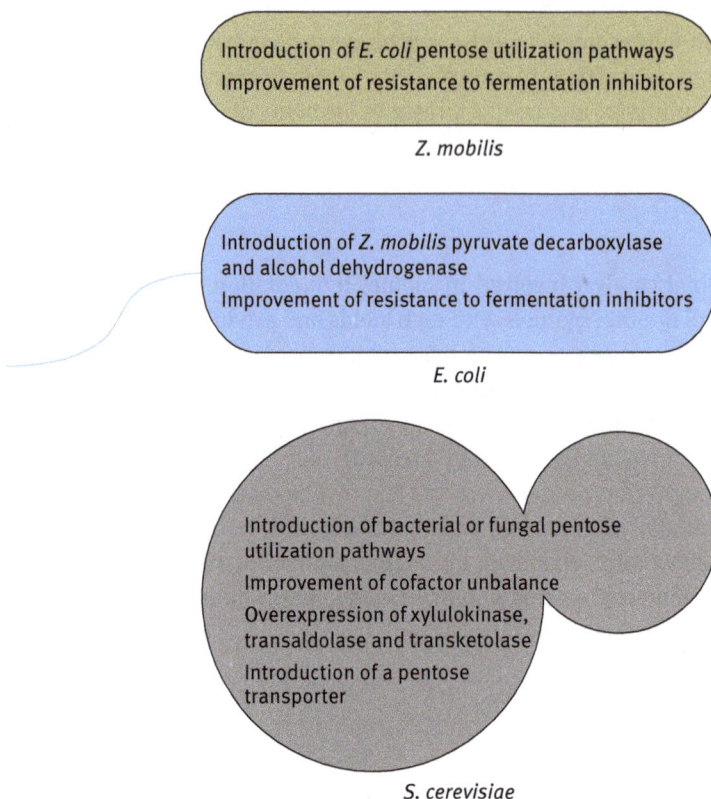

Fig. 5.6: Main metabolic engineering and directed evolution approaches used to enable *Z. mobilis*, *E. coli*, and *S. cerevisiae* to ferment C6 and C5 sugars to ethanol.

5.5.2 *Zymomonas mobilis*

This Gram-negative bacterium is a natural ethanol producer. It is a GRAS (generally regarded as safe) microorganism. Moreover, it shows a high ethanol tolerance (up to 120 g/L), a higher ethanol yield (5–10%), and a higher ethanol productivity (2.5 times) compared to that of *S. cerevisiae* [57].

The high ethanol yield and productivity depend on its unique capability of fermenting glucose using the Entner-Doudoroff (ED) pathway that yields less ATP per glucose than the classical Embden-Meyerhoff-Parnas pathway; consequently, less biomass and more ethanol are formed from the carbon source used.

Zymomonas mobilis cannot, however, ferment pentoses. The introduction of the *E. coli* genes xylose isomerase (*xylA*), xylulose kinase (*xylB*), transketolase (*tktA*), and transaldolase (*talB*) enabled *Z. mobilis* to convert xylose into xylulose-5-phosphate, an intermediate of the pentose-phosphate pathway [58]. The recombinant strain obtained, CP5(pZB5), fermented xylose with an 86% ethanol yield. The subsequent introduction of the *E. coli* arabinose operon (L-arabinose isomerase (*araA*), L-ribulose kinase (*araB*), L-ribulose-5-phosphate-4-epimerase (*araD*)), trans-ketolase (*tktA*), and transaldolase (*talB*) allowed the utilization of arabinose [59]. Although the theoretical yield was close to 100%, the fermentation rate was much lower than that observed with the xylose-fermenting strain. Another recombinant strain, AX101, was generated carrying the genes required to ferment both xylose and arabinose integrated into its genome [60]. Although co-fermentation was obtained by this strain (yield 84%), it showed a sugar preference with glucose being consumed first, followed by xylose and arabinose. *Zymomonas mobilis* has proved to be an efficient biocatalyst in SSCF. Poplar wood chips pretreated with steam explosion followed by over liming to reduce fermentation inhibitor concentration has been used as SSF substrate. The process was carried out at 34°C (a compromise between optimal fermentation and enzymatic hydrolysis temperatures), pH 5.5, for 7 days, obtaining a final ethanol concentration of 30 g/L and achieving 54% conversion of all potentially available biomass sugars [61]. The main disadvantage of *Z. mobilis* recombinant strains is their sensitivity to fermentation inhibitors especially to acetic acid, which is further exacerbated in the presence of ethanol. This problem is particularly difficult to solve for mixed-sugar fermentations since it has been reported that *Z. mobilis* cells display a lower energetic state when grown on xylose [60].

5.5.3 Other bacteria

Some members of the Gram-positive bacteria genus *Clostridium* such as *C. thermocellum* display interesting features for the development of CBP and are described in Section 5.6.

Among other Gram-negative bacteria, the most interesting appears to be *Klebsiella oxytoca*. This bacterium is able to utilize C5 and C6 sugars and cellobiose. Moreover, it tolerates low pH better than *E. coli*. The main disadvantage of *K. oxytoca* is that its fermentation of sugars yields a mixture of products. Metabolic engineering has sought to improve this trait, integrating into the genome of *K. oxytoca* the *Z. mobilis* pyruvate decarboxyalse (*pdc*) and alcohol dehydrogenase II genes. The highest yield obtained has been, however, 0.35 g ethanol/g of sugars [62].

5.5.4 *Saccharomyces cerevisiae*

Saccharomyces cerevisiae has been used for a long time for the production of ethanol in beverages and first-generation bioethanol. It displays a number of desirable features in an industrial setting: it is a GRAS microorganism, highly resistant to fermentation inhibitors, ethanol, and low pH.

However, *S. cerevisiae* can neither ferment pentoses aerobically nor anaerobically. Many other yeasts such as *P. stipitis* are efficient pentose fermenters. However, most of them cannot produce ethanol anaerobically but only in conditions of oxygen restriction (Custers effect). It is very difficult to maintain microaerobic conditions in an industrial fermentation process; if too much oxygen is provided, ethanol yields will be lower, as it will be respired. Hence, it is important that xylose is fermented anaerobically [63].

To make *S. cerevisiae* a good xylose fermenter, metabolic engineering has faced two tasks: (i) expression of heterologous genes to enable *S. cerevisiae* to convert xylose to xylulose and (ii) redirection of the xylulose metabolic flux in order to increase its consumption since *S. cerevisiae* grows on xylulose with a maximum specific growth rate 10 times lower than that on glucose [64].

Two approaches have been followed to extend the *S. cerevisiae* substrate range to xylose:
- introduction of xylose reductase (XR) and xylytol dehydrogenase (XDH) from *P. stipitis* or other natural pentose-fermenting yeasts
- introduction of a xylose isomerase (XI) of fungal or bacterial origin

5.5.4.1 Expression of the fungal xylose utilization pathway

Saccharomyces cerevisiae cannot metabolize xylose but is able to ferment xylulose although with much lower efficiency. Many yeasts convert xylose into xylulose in a two-step process (Figure 5.7b). Xylose is first reduced to xylitol by an NADPH-dependent XR. Xylitol is then oxidized to xylulose by a NAD^+-dependent XDH. *S. cerevisiae* presents both enzymatic activities. They are, however, too low to allow the reaction to proceed. Many xylose-fermenting yeasts have very active XR and XDH. The introduction of the *P. stipitis* XR and XDH into *S. cerevisiae* enabled this yeast to grow on xylose; the main product was, however, xylitol, not ethanol [65]. XR consumes mainly NAD(P)H, and XDH converts NAD^+ to NADH. This leads to a cofactor unbalance resulting in the increase of the NADH concentration (or of the NADH/NAD^+ ratio) that inhibits the action of XDH and decreases the fermentation of xylose to ethanol.

Many studies have demonstrated that in order to have an effective xylose fermentation to ethanol and a glucose-xylose co-fermentation, it is necessary to overexpress xylulokinase (XK). Overexpression of *P. stipitis* XDH and XR and of the endogenous ScXK in an industrial *Saccharomyces* strain generated a biocatalyst able to ferment glucose and xylose to ethanol under aerobic and anaerobic conditions. This yeast was able to produce 22 g/L ethanol from a corn fiber hydrolyzate with a yield of 0.46 g of ethanol/g of sugars (mainly glucose and xylose) [66].

The integration of the *P. stipitis* xylose utilization pathway in an industrial *S. cerevisiae* strain followed by random mutagenesis and selection on xylose led to the strain TMB3400, which was able to ferment a spruce hydrolyzate, obtaining up to 45.2 g/L of ethanol with a yield of 0.43 g/g [67]. The molecular analysis of the genes mutated in TMB3400, as compared to the parent strain, led to the identification of important cellular targets whose overexpression resulted to higher and efficient

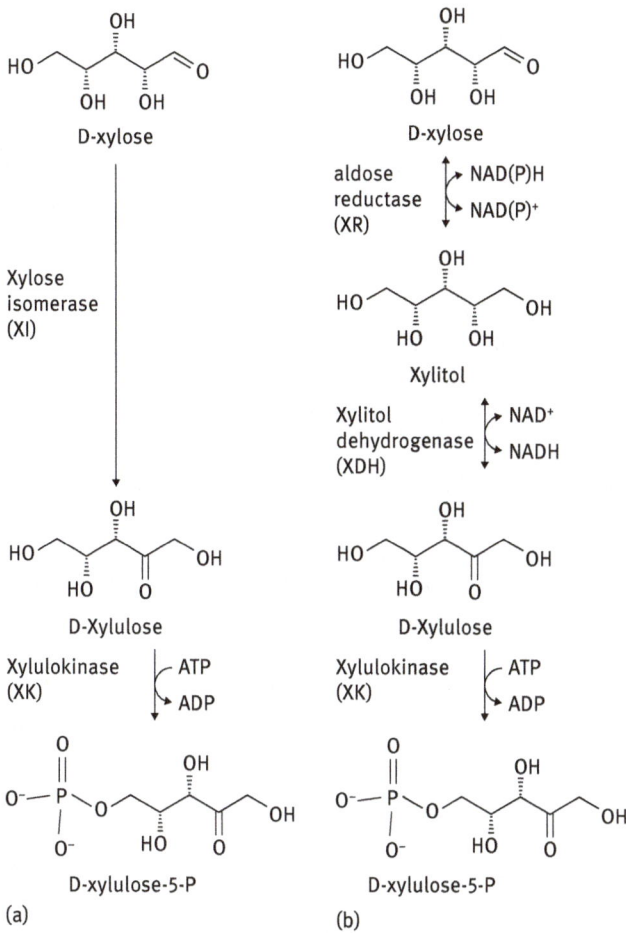

Fig. 5.7: D-Xylose catabolism in (a) bacteria and (b) fungi. D-Xylulose-5-P is further catabolized through the pentose phosphate pathway to fructose-6-P and glyceraldehyde-3-P, which are metabolized in the glycolytic pathway and converted to ethanol.

xylose fermentation: (a) XKS1-encoding xylulokinase; (b) *SOL3*, *GND1*, *TAL1*, and *TKL1*, encoding enzymes in the pentose phosphate pathway; (c) *HXT5*, encoding a proved to be critical for hexose transporter [67].

The first xylose-fermenting strains displayed a very low growth rate. This was partially ascribed to the low activity of the PP pathway in *S. cerevisiae* and particularly transaldolase (TAL) and transketolase (TKL), which are necessary to metabolize xylulose generated from XR+XDH (or from XI, see Section 5.5.4.2).

In *S. cerevisiae*, xylose utilization rate is lower compared to that of natural xylose-utilizing yeasts and is much lower than that of glucose. An explanation of its preference for glucose is that *S. cerevisiae* has many transporters for hexoses (Hxt1, Hxt2, Hxt4,

Hxt5, Hxt7, and Gal2) but none specific for pentoses. Transport of xylose is completely inhibited by glucose; consequently, the pentose is consumed only after the depletion of glucose. Xylose transport limits xylose metabolism especially in fast xylose-metabolizing strain [68]. Several evolutionary engineering approaches aimed at improving xylose-fermenting capability resulted in evolved strains showing improved xylose transport kinetics [69]. Many transporters from plants, bacteria, and xylose-fermenting yeasts have been expressed in *S. cerevisiae* in order to improve cellular xylose uptake.

5.5.4.2 Expression of the bacterial xylose utilization pathway

Bacteria are able to convert xylose into xylulose via a one-step reaction catalyzed by xylose isomerase (XI) (Figure 5.7a). XI expression in *S. cerevisiae* has been sought for a long time. It allows to circumvent the cofactor unbalance problem observed upon the expression of XR and XDH.

Many attempts to express a bacterial XI in *S. cerevisiae* were unsuccessful. The discovery of a eukaryotic XI in the fungus *Pyromyces* sp. E2 and its high-level expression in *S. cerevisiae* (strain RWB202) allowed a very slow growth on 2% xylose under anaerobic conditions [70]. In order to improve xylose utilization, RWB202 was subjected to an evolutionary engineering approach, through a prolonged cultivation on xylose. The obtained mutant strain (RWB-202AFX) grew aerobically and anaerobically on xylose. The fermentative performance of RWB-202AFX was further improved by a metabolic engineering approach: xylulokinase (EC 2.7.1.17), ribulose 5-phosphate isomerase (EC 5.3.1.6), ribulose 5-phosphate epimerase (EC 5.3.1.1), transketolase (EC 2.2.1.1), and transaldolase (EC 2.2.1.2) were overexpressed, and *GRE3*, encoding an aldose reductase, was deleted to further minimize xylitol production [71]. The process was then further improved by an evolutionary engineering protocol in automated sequencing-batch reactors on glucose-xylose mixtures followed by prolonged anaerobic cultivation. A single-strain isolate (RWB 218) was obtained, which rapidly consumed glucose-xylose mixtures anaerobically, in synthetic medium. RWB218 fermented a wheat straw hydrolyzate, obtaining 38 g/L ethanol with the remarkable yield of 0.51 g/g of total sugars, which is close to the theoretical yield [71].

5.5.4.3 Expression of the arabinose utilization pathway

Arabinose is present in some types of lignocellulose hydrolyzates such as corn fiber and many herbaceous crops. Only a few yeasts are naturally able to ferment arabinose to ethanol with low efficiency [72]. The fungal arabinose pathway comprises aldose (xylose) reductase, L-arabinitol 4-dehydrogenase, L-xylulose reductase, D-xylulose reductase, and D-xylulokinase (Figure 5.8b). This pathway consists of two NAD$^+$-linked oxidations and two NADPH-linked reductions, resulting in a redox cofactor imbalance under anaerobic conditions, which reduces the fermentation rate [72]. Overexpression of all the genes of the fungal pathway resulted in the first *S. cerevisiae* strain capable of fermenting L-arabinose to ethanol but at a very low rate (0.35 mg ethanol/g/h under anaerobic conditions) [73].

Also, bacteria are able to ferment arabinose. The bacterial arabinose pathway does not include any redox reaction and comprises L-arabinose isomerase (*araA*), L-ribulokinase (*araB*), and L-ribulose-5-phosphate 4-epimerase (*araC*) (Figure 5.8a). Becker and Boles [74] introduced *E. coli araB* and *araD* and *Bacillus subtilis araA* into *S. cerevisiae*. Moreover, they overexpressed GAL2, encoding a galactose permease that has been shown to catalyze the arabinose uptake. Through an evolutionary engineering protocol, a strain capable of producing ethanol from arabinose under oxygen-limited conditions with a theoretical yield of 60% was selected [74]. More recently, an improvement of the fermentation performances was obtained, replacing the *B. subtilis araA* with the enzyme from *Bacillus licheniformis*, leading to a considerably decreased lag phase. Subsequently, the codon usage of all the genes involved in the L-arabinose pathway was adapted to that of the highly expressed genes encoding glycolytic enzymes in *S. cerevisiae*. The ethanol production rate from L-arabinose could be increased more than 2.5-fold and the ethanol yield could be increased from 0.24 g ethanol/g consumed L-arabinose to 0.39 g ethanol/g consumed L-arabinose [75].

Arabinose and xylose pathways have been co-expressed in *S. cerevisiae*. Wisselink et al. [76] reported the construction of the first yeast strain capable of fermenting mixtures of glucose, xylose, and arabinose with a high ethanol yield (0.43 g/g of total sugar) without formation of the side products xylitol and arabitol. The kinetics of anaerobic fermentation of glucose-xylose-arabinose mixtures were greatly improved using an evolutionary engineering strategy. Yeast cells were successively cultivated in mixtures of glucose, xylose, and arabinose, allowing them to evolve longer periods on the less preferred carbon sources. The evolved strain (IMS0010) showed a significant reduction in the time required to completely ferment a mixture containing 30 g/L glucose, 15 g/L xylose, and 15 g/L arabinose [77].

5.5.5 Other yeasts

Pichia stipitis is a natural pentose-fermenting yeast. It is able to ferment other sugars (e.g., glucose, cellobiose) with good yields (0.31–0.48 g ethanol/g of fermented sugar). However, compared to *S. cerevisiae*, this yeast displays several drawbacks: (1) it produces ethanol only in microaerophilic conditions; (2) sugar consumption rates are slower than in *S. cerevisiae*; (3) it is sensitive to metabolic inhibitors. The sequencing of *P. stipitis* genome opens the door to possible genetic engineering approaches to solve these problems [77].

Two other yeasts have been considered for biofuel production mainly for their ability to ferment many sugars at high temperature (up to 48–50°C), which makes them interesting biocatalysts in an SSF and possibly in CBP process: *Kluyveromyces marxianus* and *Hansenula polymorpha*. Both of them are, however, unable to produce ethanol in the presence of excess sugar and oxygen or in anaerobiosis [77].

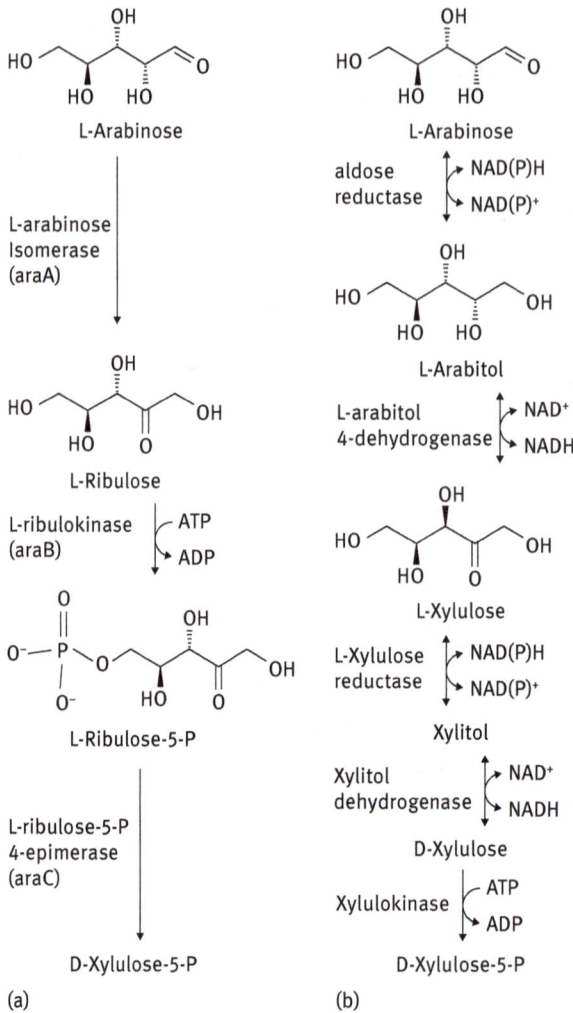

Fig. 5.8: D-Arabinose catabolism in (a) bacteria and (b) fungi. D-Xylulose-5-P is further catabolized through the pentose phosphate pathway to fructose-6-P and glyceraldehyde-3-P, which are metabolized in the glycolytic pathway and converted to ethanol.

5.6 Strain development for CBP

As outlined in Section 5.4.4, CBP is considered by many researchers the best possible option for a profitable second-generation ethanol production. For the development of an efficient CBP, microbial catalyst should show the following features:

1. Excellent ethanol-producing capabilities.
2. Efficient (hemi)cellulase production.
3. Tolerance to ethanol, fermentation inhibitors, and other stresses commonly found in the ethanol production.

4. Since the most efficient cellulases found to date have an optimal operative temperature of 50° C, the ideal microorganism has also to be thermostable. At industrial level, high temperatures are beneficial, to allow the reduction of cooling costs and decrease the contamination risk by mesophils.

Currently, no organism combines all of these properties, although many match some of them [54]. Consequently, researchers are developing metabolic and genetic engineering approaches to generate a recombinant biocatalyst that can be employed in CBP.

Two engineering strategies are being pursued:

1. Converting a cellulolytic microorganism into an efficient ethanol producer. To render a cellulolytic microorganism ethanologenic, researchers have tried to increase ethanol titers by introducing heterologous ethanol synthetic pathways. Most research has been focused on thermophilic Gram-positive bacteria. Many of these bacteria are strict anaerobes and produce mixtures of fermentation products. Generally, they display a low tolerance to ethanol and to fermentation inhibitors. Metabolic engineers have sought to eliminate fermentation by-products and the increase in ethanol tolerance. One of the most studied microorganisms in this regard is *Clostridium thermocellum*. It is a thermophilic anaerobic bacterium that can grow at temperatures as high as 60°C and is capable to naturally producing ethanol, albeit at low concentrations (<3 g/L). As many clostridia, it expresses a high number of different cellulases and hemicellulase on its cell surface. These enzymes are linked to a scaffold protein and constitute a complex known as cellulosome. *Clostridium thermocellum* cellulosome has been proved to be more efficient than free hemicellulases since the multi-enzymatic complex provides a synergistic mechanism of action. Several engineered strains have been developed using directed evolution to improve ethanol or inhibitor tolerance. Despite this, the utilization of cellulolytic microorganisms in an economically feasible CBP is still far from being achieved [48, 78].

2. Developing an ethanologenic microorganism that expresses and secretes heterologous cellulases. The expression of cellulases and hemicellulases in an ethanologenic microorganism is a difficult task since the hydrolysis of the lignocellulosic biomass requires the simultaneous presence of many different enzymatic activities. Two expression strategies have been pursued: (a) secretion of cellulolytic enzymes by recombinant ethanologens and (b) expression and display of these enzymes on the cell surface.

The latter approach appears to be superior since the activity of the enzymes displayed on the cell surface is retained as long as the expressing cells keep their viability. Moreover, in this way, it is possible to produce a large amount of biomass prior to its addition to the pretreated lignocellulosic material that has to be fermented. A number of ethanologenic hosts have been chosen for the expression, but much research has been focused on *S. cerevisiae* for its known properties as ethanol producer (see Section 5.5.4). However, as outlined before, thermostability is an important feature in CBP; hence, other thermotolerant yeasts such as *Kluyveromyces marxianus* and *Hansenula polymorpha* have also been employed. Despite the efforts and the promising results achieved, an industrial (hemi)cellulose-producing ethanologenic microorganism capable of converting pretreated raw lignocellulosic biomass into ethanol in an economically feasible process has still not been obtained [48, 54].

Bibliography

[1] OECD-FAO Agricultural Outlook 2014.
[2] Hon, D. N. S., Shiraishi, N., Wood and Cellulosic Chemistry, Revised and Expanded, CRC Press, New York, 2000.
[3] Ha, M. A., Apperley, D., C., Evans, B. W., Huxham, I. M., Jardine, W. G., Vietor, R., J., Reis, D., Vian, B., Jarvis, M. C., Fine structure in cellulose microfibrils: NMR evidence from onion and quince, Plant J **16** (1998) 183–190.
[4] Sun, Y., Cheng, J., Hydrolysis of lignocellulosic materials for ethanol production: a review, Bioresour Technol **83** (2002) 1–11.
[5] Mosier, N., Wyman, C., Dale, B., Elander, R., Lee, Y. Y., Holtzapple, M., Ladisch, M., Features of promising technologies for pretreatment of lignocellulosic biomass, Bioresour Technol **96** (2005) 673–686.
[6] Chandel, A. K., Chan, E. S., Rudravaram, R., Narasu, M. L., Rao, L. V., Ravindra, P., Economics and environmental impact of bioethanol production technologies: an appraisal, Biotechnol Mol Biol Rev **2** (2007) 14–32.
[7] Taherzadeh, M. J., Karimi, K., Acid-based hydrolysis processes for ethanol from lignocellulosic materials: a review, BioResources **2** (2007) 427–499.
[8] Taherzadeh, M. J., Karimi, K., Enzymatic-based hydrolysis processes for ethanol, BioResources **2** (2007) 707–738.
[9] Howard, R. L., Abotsi, E., Jansen van Rensburg, E. L., Howard, S., Review – Lignocellulose biotechnology: issues of bioconversion and enzyme production, Afr J Biotechnol **2** (2003) 12.
[10] Liu, G., Qin, Y., Li, Z., Qu, Y., Development of highly efficient, low-cost lignocellulolytic enzyme systems in the post-genomic era, Biotechnol Adv **31** (2013) 962–975.
[11] Van Gool, M. P., Vancso, I., Schols, H. A., Toth, K., Szakacs, G., Gruppen, H., Screening for distinct xylan degrading enzymes in complex shake flask fermentation supernatants, Bioresour Technol **102** (2011) 6039–6047.
[12] Debnath, R., Sarma, R., Saikia, R., Yadav, A., Bora, Y., Metagenomics: a hunting expedition in microbial diversity, in Gaur, R., Gautam, H., editors, Molecular Biology of Bacteria, Nova Science Publishers, New York, 2013, 19–30.
[13] Duan, C. J., Feng, J. X., Mining metagenomes for novel cellulase genes, Biotechnol Lett **32** (2010) 1765–1775.
[14] Warnecke, F., Luginbuhl, P., Ivanova, N., Ghassemian, M., Richardson, T. H., Stege, J. T., Cayouette, M., McHardy, A. C., Djordjevic, G., Aboushadi, N., Sorek, R., Tringe, S. G., Podar, M., Martin, H. G., Kunin, V., Dalevi, D., Madejska, J., Kirton, E., Platt, D., Szeto, E., Salamov, A., Barry, K., Mikhailova, N., Kyrpides, N. C., Matson, E. G., Ottesen, E. A., Zhang, X., Hernandez, M., Murillo, C., Acosta, L. G., Rigoutsos, I., Tamayo, G., Green, B. D., Chang, C., Rubin, E. M., Mathur, E. J., Robertson, D. E., Hugenholtz, P., Leadbetter, J. R., Metagenomic and functional analysis of hindgut microbiota of a wood-feeding higher termite, Nature **450** (2007) 560–565.

[15] Hess, M., Sczyrba, A., Egan, R., Kim, T. W., Chokhawala, H., Schroth, G., Luo, S., Clark, D. S., Chen, F., Zhang, T., Mackie, R. I., Pennacchio, L. A., Tringe, S. G., Visel, A., Woyke, T., Wang, Z., Rubin, E. M., Metagenomic discovery of biomass-degrading genes and genomes from cow rumen, Science **331** (2011) 463–467.

[16] Takasaki, K., Miura, T., Kanno, M., Tamaki, H., Hanada, S., Kamagata, Y., Kimura, N., Discovery of glycoside hydrolase enzymes in an avicel-adapted forest soil fungal community by a metatranscriptomic approach, PLoS One **8** (2013) 8.

[17] Koppram, R., Tomas-Pejo, E., Xiros, C., Olsson, L., Lignocellulosic ethanol production at high-gravity: challenges and perspectives, Trends Biotechnol **32** (2014) 46–53.

[18] Wang, M., Li, Z., Fang, X., Wang, L., Qu, Y., Cellulolytic enzyme production and enzymatic hydrolysis for second-generation bioethanol production, Adv Biochem Eng Biotechnol **128** (2012) 1–24.

[19] Zhang, J., Zhong, Y., Zhao, X., Wang, T., Development of the cellulolytic fungus Trichoderma reesei strain with enhanced beta-glucosidase and filter paper activity using strong artificial cellobiohydrolase 1 promoter, Bioresour Technol **101** (2010) 9815–9818.

[20] Palmqvist, E., Hahn-Hägerdal, B., Fermentation of lignocellulosic hydrolysates, I: inhibition and detoxification, Bioresour Technol **74** (2000) 17–24.

[21] Klinke, H. B., Thomsen, A. B., Ahring, B. K., Inhibition of ethanol-producing yeast and bacteria by degradation products produced during pre-treatment of biomass, Appl Microbiol Biotechnol **66** (2004) 10–26.

[22] Taylor, M. P., Mulako, I., Tuffin, M., Cowan, D., Understanding physiological responses to pre-treatment inhibitors in ethanologenic fermentations, Biotechnol J **7** (2012) 1169–1181.

[23] Russell, J. B., Diez-Gonzalez, F., The effects of fermentation acids on bacterial growth, Adv Microb Physiol, **39** (1998) 205–234.

[24] Jarboe, L. R., Royce, L. A., Liu, P., Understanding biocatalyst inhibition by carboxylic acids, Front Microbiol **4** (2013) 272–272.

[25] Modig, T., Liden, G., Taherzadeh, M. J., Inhibition effects of furfural on alcohol dehydrogenase, aldehyde dehydrogenase and pyruvate dehydrogenase, Biochem J **363** (2002) 769–776.

[26] Liu, Z. L., Moon, J., Andersh, B. J., Slininger, P., J., Weber, S., Multiple gene-mediated NAD(P)H-dependent aldehyde reduction is a mechanism of in situ detoxification of furfural and 5-hydroxymethylfurfural by Saccharomyces cerevisiae, Appl Microbiol Biotechnol **81** (2008) 743–753.

[27] Heer, D., Sauer, U., Identification of furfural as a key toxin in lignocellulosic hydrolysates and evolution of a tolerant yeast strain, Microb Biotechnol **1** (2008) 497–506.

[28] Palmqvist, E., Almeida, J. S., Hahn-Hagerdal, B., Influence of furfural on anaerobic glycolytic kinetics of Saccharomyces cerevisiae in batch culture, Biotechnol Bioeng **62** (1999) 447–454.

[29] Palmqvist, E., Grage, H., Meinander, N. Q., Hahn-Hagerdal, B., Main and interaction effects of acetic acid, furfural, and p-hydroxybenzoic acid on growth and ethanol productivity of yeasts, Biotechnol Bioeng **63** (1999) 46–55.

[30] Taherzadeh, M. J., Gustafsson, L., Niklasson, C., Liden, G., Conversion of furfural in aerobic and anaerobic batch fermentation of glucose by Saccharomyces cerevisiae, J Biosci Bioeng **87** (1999) 169–174.

[31] Almeida, J. R., Bertilsson, M., Gorwa-Grauslund, M. F., Gorsich, S., Liden, G., Metabolic effects of furaldehydes and impacts on biotechnological processes, Appl Microbiol Biotechnol **82** (2009) 625–638.

[32] Larsson, S., Quintana-Sainz, A., Reimann, A., Nilvebrant, N. O., Jonsson, L. J., Influence of lignocellulose-derived aromatic compounds on oxygen-limited growth and ethanolic fermentation by Saccharomyces cerevisiae, Appl Biochem Biotechnol **84–86** (2000) 617–632.

[33] Larsson, S., Reimann, A., Nilvebrant, N. O., Jönsson, L., Comparison of different methods for the detoxification of lignocellulose hydrolyzates of spruce, Appl Biochem Biotechnol **77** (1999) 91–103.

[34] Heipieper, H. J., Keweloh, H., Rehm, H. J., Influence of phenols on growth and membrane permeability of free and immobilized Escherichia coli, Appl Environ Microbiol **57** (1991) 1213–1217.

[35] Adeboye, P. T., Bettiga, M., Olsson, L., The chemical nature of phenolic compounds determines their toxicity and induces distinct physiological responses in Saccharomyces cerevisiae in lignocellulose hydrolysates, AMB Express **4** (2014) 46.

[36] Maiorella, B. L., Blanch, H. W., Wilke, C. R., Feed component inhibition in ethanolic fermentation by Saccharomyces cerevisiae, Biotechnol Bioeng **26** (1984) 1155–1166.

[37] Klinke, H. B., Olsson, L., Thomsen, A. B., Ahring, B. K., Potential inhibitors from wet oxidation of wheat straw and their effect on ethanol production of Saccharomyces cerevisiae: wet oxidation and fermentation by yeast, Biotechnol Bioeng **81** (2003) 738–747.

[38] Zaldivar, J., Martinez, A., Ingram, L. O., Effect of selected aldehydes on the growth and fermentation of ethanologenic Escherichia coli, Biotechnol Bioeng **65** (1999) 24–33.

[39] Parawira, W., Tekere, M., Biotechnological strategies to overcome inhibitors in lignocellulose hydrolysates for ethanol production: review, Crit Rev Biotechnol **31** (2011) 20–31.

[40] Palmqvist, E., Hahn-Hägerdal, B., Galbe, M., Zacchi, G., The effect of water-soluble inhibitors from steam-pretreated willow on enzymatic hydrolysis and ethanol fermentation, Enzyme Microb Technol **19** (1996) 470–476.

[41] Grzenia, D. L., Schell, D. J., Wickramasinghe, S. R., Membrane extraction for detoxification of biomass hydrolysates, Bioresour Technol **111** (2012) 248–254.

[42] López, M. J., Nichols, N., Dien, B. S., Moreno, J., Bothast, R. J., Isolation of microorganisms for biological detoxification of lignocellulosic hydrolysates, Appl Microbiol Biotechnol, **64** (2004) 125–131.

[43] Yang, B., Wyman, C. E., Pretreatment: the key to unlocking low-cost cellulosic ethanol, Biofuels Bioprod Biorefining **2** (2014) 26–40.

[44] Jönsson, L. J., Alriksson, B., Nilvebrant, N. O., Bioconversion of lignocellulose: inhibitors and detoxification, Biotechnol Biofuels **6** (2013) 16.

[45] Larsson, S., Cassland, P., Jonsson, L. J., Development of a Saccharomyces cerevisiae strain with enhanced resistance to phenolic fermentation inhibitors in lignocellulose hydrolysates by heterologous expression of laccase, Appl Environ Microbiol **67** (2001) 1163–1170.

[46] Ling, H., Teo, W., Chen, B., Leong, S. S. J., Chang, M. W., Microbial tolerance engineering toward biochemical production: from lignocellulose to products, Curr Opin Biotechnol **29** (2014) 99–106.

[47] Ask, M., Mapelli, V., Hock, H., Olsson, L., Bettiga, M., Engineering glutathione biosynthesis of Saccharomyces cerevisiae increases robustness to inhibitors in pretreated lignocellulosic materials, Microb Cell Fact **12** (2013) 87.

[48] Hasunuma, T., Kondo, A., Development of yeast cell factories for consolidated bioprocessing of lignocellulose to bioethanol through cell surface engineering, Biotechnol Adv **30** (2012) 1207–1218.

[49] Koppram, R., Albers, E., Olsson, L., Evolutionary engineering strategies to enhance tolerance of xylose utilizing recombinant yeast to inhibitors derived from spruce biomass, Biotechnol Biofuels **5** (2012) 32.

[50] Jäger, G., Büchs, J., Biocatalytic conversion of lignocellulose to platform chemicals, J Biotechnol **7** (2012) 1122–1136.

[51] Wingren, A., Galbe, M., Zacchi, G., Techno-economic evaluation of producing ethanol from softwood: comparison of SSF and SHF and identification of bottlenecks, Biotechnol Prog **19** (2003) 1109–1117.

[52] Ohgren, K., Bengtsson, O., Gorwa-Grauslund, M. F., Galbe, M., Hahn-Hagerdal, B., Zacchi, G., Simultaneous saccharification and co-fermentation of glucose and xylose in steam-pretreated corn stover at high fiber content with Saccharomyces cerevisiae TMB3400, J Biotechnol **126** (2006) 488–498.

[53] Teixeira, L. C., Linden, J. C., Schroeder, H. A., Simultaneous saccharification and cofermentation of peracetic acid-pretreated biomass, Appl Biochem Biotechnol **84–86** (2000) 111–127.

[54] Lynd, L. R., van, Z. yl, W. H., McBride, J. E., Laser, M., Consolidated bioprocessing of cellulosic biomass: an update, Curr Opin Biotechnol **16** (2005) 577–583.

[55] Ingram, L. O., Gomez, P. F., Lai, X., Moniruzzaman, M., Wood, B. E., Yomano, L. P., York, S. W., Metabolic engineering of bacteria for ethanol production, Biotechnol Bioeng **58** (1998) 204–214.

[56] Alper, H., Stephanopoulos, G., Global transcription machinery engineering: a new approach for improving cellular phenotype, Metab Eng **9** (2007) 258–567.

[57] Rogers, P. L., Jeon, Y. J., Lee, K. J., Lawford, H. G., Zymomonas mobilis for fuel ethanol and higher value products, Adv Biochem Eng Biotechnol **108** (2007) 263–288.

[58] Zhang, M., Eddy, C., Deanda, K., Finkelstein, M., Picataggio, S., Metabolic engineering of a pentose metabolism pathway in ethanologenic Zymomonas mobilis, Science **267** (1995) 240–243.

[59] Deanda, K., Zhang, M., Eddy, C. ., Picataggio, S., Development of an arabinose-fermenting Zymomonas mobilis strain by metabolic pathway engineering, Appl Environ Microbiol **62** (1996) 4465–4470.

[60] Mohagheghi, A., Evans, K., Chou, Y. C., Zhang, M., Cofermentation of glucose, xylose, and arabinose by genomic DNA-integrated xylose/arabinose fermenting strain of Zymomonas mobilis AX101, Appl Biochem Biotechnol **98–100** (2002) 885–898.

[61] McMillan, J. D., Newman, M. M., Templeton, D. W., Mohagheghi, A., Simultaneous saccharification and cofermentation of dilute-acid pretreated yellow poplar hardwood to ethanol using xylose-fermenting Zymomonas mobilis, Appl Biochem Biotechnol, **77–79** (1999) 649–665.

[62] Moniruzzaman, M., Dien, B. S., Ferrer, B., Hespell, R. B., Dale, B. E., Ingram, L. O., Bothast, R. J., Ethanol production from AFEX pretreated corn fiber by recombinant bacteria, Biotechnol Lett **18** (2014) 985–990.

[63] van Maris, A. J. A., Abbott, D. A., Bellissimi, E., van den Brink, J., Kuyper, M., Luttik, M. A. H., Wisselink, H. W., Scheffers, W. A., van Dijken, J. P., Pronk, J. T., Alcoholic fermentation of carbon sources in biomass hydrolysates by Saccharomyces cerevisiae: current status, Antonie Van Leeuwenhoek **90** (2006) 391–318.

[64] Ostergaard, S., Olsson, L., Nielsen, J., Metabolic engineering of Saccharomyces cerevisiae, Microbiol Mol Biol Rev **64** (2000) 34–50.

[65] Kötter, P., Ciriacy, M., Xylose fermentation by Saccharomyces cerevisiae, Appl Microbiol Biotechnol **38** (1993) 776–783.

[66] Moniruzzaman, M., Dien, B. S., Skory, C. D., Chen, Z. D., Hespell, R. B., Ho, N. W. Y., Dale, B. E., Bothast, R. J., Fermentation of corn fibre sugars by an engineered xylose utilizing Saccharomyces yeast strain, World J Microb Biot **13** (1997) 341–346.

[67] Wahlbom, C. F., van, Z. yl, W. H., Jonsson, L. J., Hahn-Hagerdal, B., Otero, R. R., Generation of the improved recombinant xylose-utilizing Saccharomyces cerevisiae TMB 3400 by random mutagenesis and physiological comparison with Pichia stipitis CBS 6054, FEMS Yeast Res **3** (2003) 319–326.

[68] Gardonyi, M., Jeppsson, M., Liden, G., Gorwa-Grauslund, M. F., Hahn-Hagerdal, B., Control of xylose consumption by xylose transport in recombinant Saccharomyces cerevisiae, Biotechnol Bioeng **82** (2003) 818–824.

[69] Kuyper, M., Harhangi, H. R., Stave, A. K., Winkler, A. A., Jetten, M. S. M., de Laat, W. T. A. M., den Ridder, J. J. J., Op den Camp, H. J. M., van Dijken, J. P., Pronk, J. T., High-level functional expression of a fungal xylose isomerase: the key to efficient ethanolic fermentation of xylose by Saccharomyces cerevisiae? FEMS Yeast Res, **4** (2003) 78.

[70] Kuyper, M., Toirkens, M. J., Diderich, J. A., Winkler, A. A., van Dijken, J. P., Pronk, J. T., Evolutionary engineering of mixed-sugar utilization by a xylose-fermenting Saccharomyces cerevisiae strain, FEMS Yeast Res **5** (2005) 925–934.

[71] van Maris, A. J., Winkler, A. A., Kuyper, M., de Laat, W. T., van Dijken, J. P., Pronk, J. T., Development of efficient xylose fermentation in Saccharomyces cerevisiae: xylose isomerase as a key component, Adv Biochem Eng Biotechnol **108** (2007) 179–204.

[72] Dien, B. S., Kurtzman, C. P., Saha, B. C., Bothast, R. J., Screening for L-arabinose fermenting yeasts, Appl Biochem Biotechnol **57–58** (1996) 233–242.

[73] Richard, P., Verho, R., Putkonen, M., Londesborough, J., Penttila, M., Production of ethanol from L-arabinose by Saccharomyces cerevisiae containing a fungal L-arabinose pathway, FEMS Yeast Res **3** (2003) 185–189.

[74] Becker, J., Boles, E., A modified Saccharomyces cerevisiae strain that consumes L-arabinose and produces ethanol, Appl Environ Microbiol **69** (2003) 4144–4150.

[75] Wiedemann, B., Boles, E., Codon-optimized bacterial genes improve L-arabinose fermentation in recombinant Saccharomyces cerevisiae, Appl Environ Microbiol **74** (2008) 2043–2050.

[76] Wisselink, H. W., Toirkens, M. J., Wu, Q., Pronk, J. T., van Maris, A. J. A., Novel evolutionary engineering approach for accelerated utilization of glucose, xylose, and arabinose mixtures by engineered Saccharomyces cerevisiae strains, Appl Environ Microbiol **75** (2009) 907–814.

[77] Weber, C., Alexander, F., Feline, B., Dawid, B., Heiko, D., Thorsten, S., Eckhard, B., Trends and challenges in the microbial production of lignocellulosic bioalcohol fuels, Appl Microbiol Biotechnol **87** (2010) 1303–1315.

[78] Akinosho, H., Yee, K., Close, D., Ragauskas, A., The emergence of Clostridium thermocellum as a high utility candidate for consolidated bioprocessing applications, Front Chem **2** (2014) 66.

Benjamin Katryniok, François Jérôme, Eric Monflier, Sébastien
Paul, and Franck Dumeignil

6 Biomass-derived molecules conversion to chemicals using heterogeneous and homogeneous catalysis

Abstract: This chapter presents the current catalytic technologies and options for upgrading molecules originating from biomass preprocessing, namely natural polymers (here, mainly cellulose and lignin), C1 to C6 platform molecules, and fatty compounds, to higher value added chemicals.

6.1 Introduction

Heterogeneous and homogeneous catalysis (sometimes grouped under the name of "chemo-catalysis") possess a role of the upmost importance in the establishment of biorefineries. Together with thermochemistry and biotechnological transformations (="biocatalysis"), chemocatalysis enables exploring various routes for upgrading raw materials to a variety of chemicals that are further used to ultimately elaborate end-products and commercial goods.

Chemo-catalytic reactions can be directly carried out using as reactants more or less pre-processed natural polymers (cellulose, hemicellulose, lignin, etc.) or using the lignocellulosic-derived "platform molecules" (corresponding to the "commodities" of the petro-based refineries) as raw materials. In biorefineries, the platform molecules are then important chemical intermediates as a selected number of basic entries of complex, integrated chemical transformation networks. Oleaginous plants further give access to other families of starting molecules, notably including triglycerides and fatty compounds (acids and esters). In addition, biomass can provide many other types of molecules, which can be processed using specifically developed advanced catalytic pathways if needed. For example, while still underexploited, proteins can constitute a very important base for chemical transformations.

In this chapter, we will first describe catalytic transformations of selected natural polymers (mainly cellulose and lignin), before discussing the possible reactions from cellulosic biomass-derived C1 to C6 platform molecules. The second main part of this chapter is centered on oleaginous plants, starting with a reminder about the products of oil extraction, before describing the direct transformations of the recovered triglycerides and of the fatty compounds obtained via their cleavage.

6.2 Lignocellulosic biomass

6.2.1 Natural polymers processing

6.2.1.1 Glucidic polymers

In nature, polysaccharides represent a huge reservoir of renewable carbon from which a very rich chemistry can be subsequently derived. Among the different classes of polysaccharides available on earth, one can mention cellulose, hemicellulose, starch as well as chitin and a wide range of functionalized compounds that can be extracted from algae. Cellulose, a biopolymer of glucose, is certainly the polysaccharide with most interest at the industrial level, not only because of its very large availability (45% of the annual production of biomass), but also because it is a non-edible and cheap resource [1–5]. Catalytic depolymerization of cellulose to glucose is a prerequisite step, after which different chemicals, in the form of platform molecules or fuels, can be produced (Figure 6.1). However, the catalytic depolymerization of cellulose is strongly hampered by its highly crystalline structure hindering catalyst accessibility. Hence, in most cases, harsh conditions of temperature and pressure are required, thus making the control of the reaction selectivity to glucose very difficult. In order to enhance its reactivity, cellulose is generally subjected to a pretreatment prior to catalytic hydrolysis, and two main strategies are employed. The first method consists in the solubilization/precipitation of cellulose in ionic liquids, or liquid acids (H_3PO_4, triflic acid) or alkaline solutions (NaOH/urea, ammonia) or mixtures

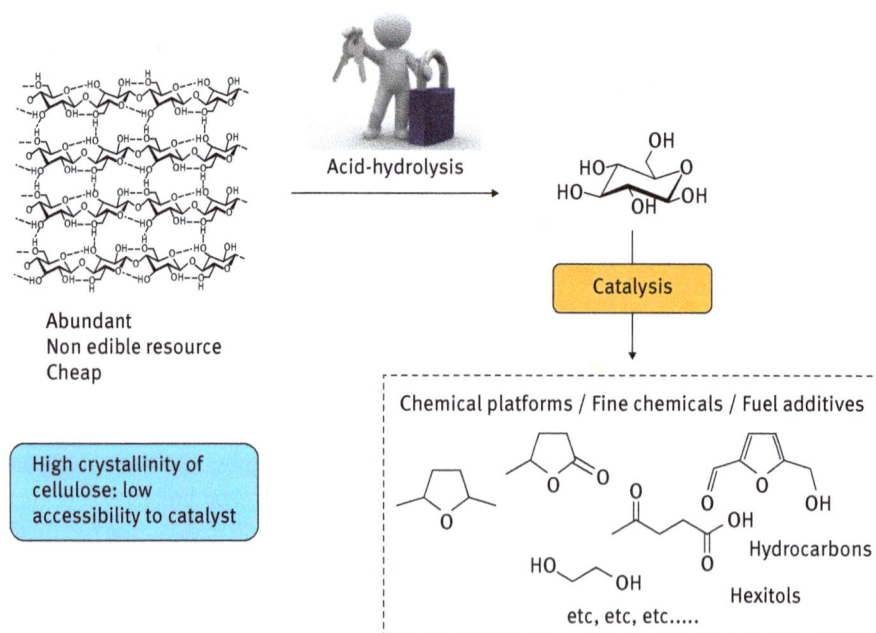

Fig. 6.1: Catalytic conversion of cellulose to value added chemicals (see Section 6.2.2.6 for details).

of dimethylsulfoxide (DMSO)/LiCl or dimethyl acetamide (DMA)/LiCl [6–8]. The second method involves thermochemical, mechanical, or physical treatments of cellulose. All these pre-treatments aim at facilitating a better diffusion of the catalyst within the cellulosic network by modifying the physico-chemical properties of cellulose (i.e., change in its crystalline structure, degree of polymerization, or particle size).

In most cases, homogeneous acid catalysts (H_2SO_4, HCl, etc.) are used to depolymerize cellulose to glucose. These depolymerization routes are industrially achieved either in diluted or concentrated conditions. Diluted conditions generally require high pressure and temperature (>180°C) and unfortunately suffer from a lack in selectivity. In concentrated acidic solutions, catalytic reactions can occur at lower temperature (100°C) and with shorter reaction times (a few minutes), enabling a better control of the reaction selectivity. However, corrosiveness and recyclability of the aqueous effluents represent two important drawbacks. Solid acid catalysts (sulfonated carbon, zeolites, metal oxides, phosphates, etc.) have also attracted a considerable attention in the recent years due to their ease of elimination at the end of the process [9, 10]. However, one should mention that stability of solid acid catalysts in water still remains an important obstacle that has to be overcome. It is noteworthy that a few research groups have recently highlighted the possible catalytic depolymerization of cellulose to glucose under dry conditions, thus limiting the problems inherent to wastewater treatments [11]. In these processes, cellulose is impregnated with a catalytic amount of acid (generally H_2SO_4) or mixed with a solid acid catalyst (e.g., kaolinite) prior to ball milling for 2–3 h. Under these conditions, cello-oligomers with a degree of polymerization lower than 6 can be readily obtained. Contrary to cellulose, these cello-oligomers are soluble in water and can then be quantitatively converted to glucose or directly converted to furanic derivatives. While the energy consumption of these mechanical processes is still too high for a direct industrial commercialization, this route clearly opens a promising process as an entry point into biorefineries. Despite much attention to the catalytic depolymerization of cellulose to glucose, it is clear that the latter is often an intermediate to access higher value-added chemicals (cf. Section 6.2.2.6).

Chitin or polysaccharides extracted from algae are also of great interest at the industrial level because they are already functionalized with amino, phosphate, or sulfate groups, thus widening the application field of polysaccharides. The number of catalytic studies on these polysaccharides is, however, much scarcer than those on cellulose. Actually, these polysaccharides are mostly used for the fabrication of materials (composites, hydrogel, etc.) or as temperature-resistant solid supports for the deposition or grafting of catalytic species.

6.2.1.2 Lignin

This three-dimensional polymer comprises a complex non-periodic arrangement of C9-based aromatic units (phenol units: coumaryl alcohol, coumaric acid, hydroxycinnamic acid; guaiacol units: coniferyl alcohol, ferulic acid; syringol unit: sinapyl alcohol), chemically interweaved in the vegetal cells through biological reaction pathways that are still not yet fully elucidated. Lignin thus constitutes a straightforward access to bio-sourced aromatic molecules, which is a unique feature in biomass. However, even if it is relatively easy to deconstruct cellulosic polymers, selective deconstruction of lignin is still a tremendous challenge. This is also linked with the fact that many different types of lignin with a large variety of structural specificities are present in the nature, depending on the vegetal, so that the possibility of finding an agnostic deep deconstruction process seems to be a rather unrealistic dream. Further, catalytic upgrading of lignin obviously implies the use of lignin fractions received from upstream processing of lignocellulose, and depending on the fractionation process (many physical, chemical, or physicochemical fractionation processes have been developed), the chemical quality, nature (e.g., lignosulfonates), and form (liquid in pulping processes or solid in hydrolysis processes) of the recovered lignin vary in a wide range.

The natural decomposition of wood- and thus of lignin- is performed by fungi and takes months, as everyone can conclude when considering rotten wood. Meanwhile, chemical breakdown of lignin is well established and plays a major role in the paper industry [12]. Since paper is based on cellulose from wood, the key step in the paper industry consists of getting rid of the lignin fraction of biomass. The most common process is the Kraft process, whereby wood is cooked with sulfides (S^{2-} from Na_2S) to cleave the ether bonds. Kraft lignin has a molecular mass in the range of 2000–3000 g/mol, which is still far from the monomers. An alternative process uses organic solvents, notably acetone and ethanol, which is referred as the organosolv process [13]. This technology offers notably the advantage of eliminating the use of strong acids, making it environmentally benign. Meanwhile, its application is limited to non-wooden biomass, such as straw, with a recovered lignin exhibiting a molar mass in the range of 1000–2000 g/mol. Recently, within the PCRD7 EuroBioRef project (eurobioref.org) [14], Borregaard (http://www.borregaard.com) [15] developed a novel biomass-agnostic continuous process for lignocellulosics fractionation (the so-called BALI®' process) specifically intended for implementation in biorefineries. Since the as-obtained lignin fractions are still far from the initial monomers, advanced valorization of lignin is still difficult. At the current state, most of the lignin is thermally converted in the paper mills, which simply means that it is burned to generate steam and electricity for the paper production process.

Potential lignin applications have been classified according to various criteria including the current technology status [16]. Concerning potential transformations involving catalysts [17], the most straightforward pathway passes through thermochemistry to generate syngas, which can be further catalytically processed to various chemicals (see Section 6.2.2.1). Lignin pyrolysis or hydroliquefaction processes yield a shallower deconstruction that gives access to pools of lighter aromatic molecules. These processes can be catalytically assisted to orientate as finely as possible selectivity to the desired molecules, which can then be subsequently upgraded. However, the difficulty here is that the product of lignin deconstruction is a complex mixture, and isolating specific fractions can reveal an unaffordable technological and/or cost hurdle. Figure 6.2 gives some target molecules that can be expected from lignin deconstruction with a direct commercial potential or than can be further upgraded. Note that, in the future, we could have selective deconstruction processes giving back the above-described lignin-constitutive C9 units, even if this seems a very big challenge.

Among the molecules described in Figure 6.2, note that vanillin synthesis from lignin is actually a commercial process based on a copper-catalyzed reaction (Figure 6.3).

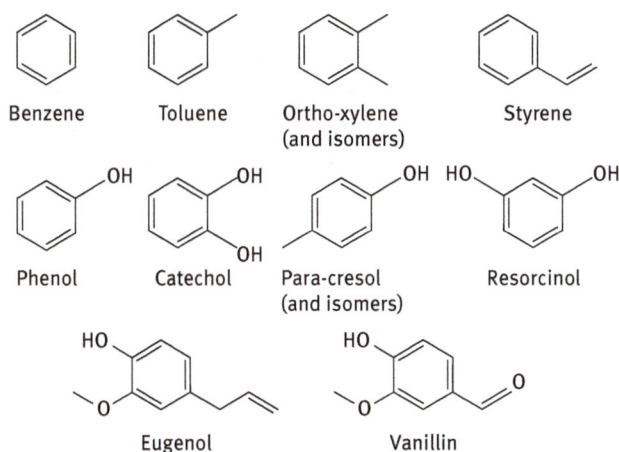

Fig. 6.2: Examples of aromatic molecules of interest that can be derived from lignin.

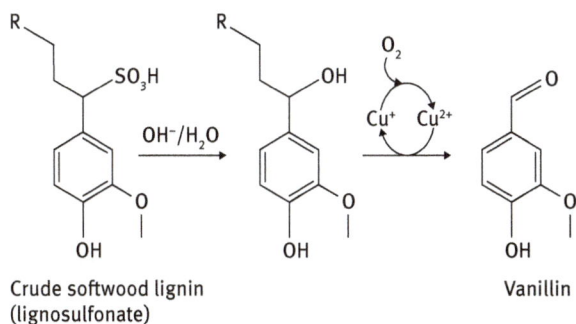

Fig. 6.3: Schematics of the commercial Borregaard (http://www.borregaard.com) process of lignosulfonate conversion to vanillin.

6.2.2 C1–C6 molecules

In this part, we give some details on the transformation of biomass platform molecules derived from cellulosic biomass upstream processing and also glycerol (issued from triglyceride cleavage, see Section 6.3.1). Figure 6.4 gives an overview of the various platform molecules, classified by their number of carbon atoms. Figure 6.4 is based on a DOE list produced in 2004 [18] and its update in 2010 [19] and on some new trends thereafter. The next sections describe typical transformations of some molecules selected in Figure 6.4, but it is possible to find more variations in the extremely abundant literature on the subject.

6.2.2.1 C1 starting materials: biogas and syngas

Using a so-called thermochemical process at high temperature (800°C or more), syngas (CO+H$_2$) can be produced by gasification of any biomass under controlled conditions, which can be optionally catalyst-assisted. Syngas usually contains impurities such as H$_2$S or CO, which can be harmful to downstream chemical processes if the catalysts used therein have not been designed to be, e.g., sulfur-tolerant. Then, after

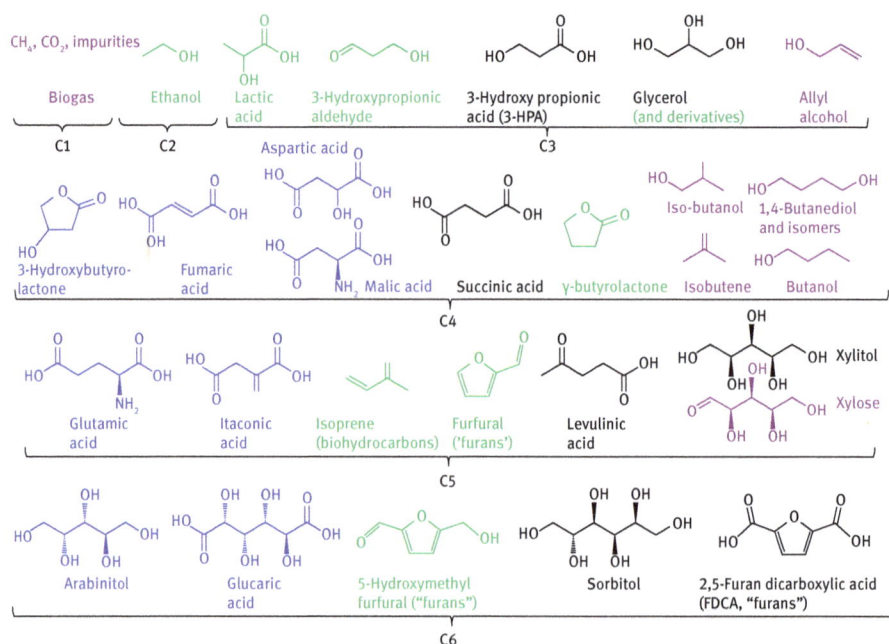

Fig. 6.4: Platform molecules. Blue, molecules only given in the DOE list [18]; green, molecules only given in the 2010 update of the DOE list [19]; black, molecules common to both references [18, 19]; purple, suggestions of other topical platform molecules or of "secondary" important platform molecules derived from mother platform molecules.

purification, and after the H_2-to-CO ratio re-balancing if needed, syngas can be converted to long-chain alkanes (through the well-known Fischer-Tropsch process, which basically yields diesel fractions) or to alcohols (through "syngas conversion"), mainly methanol, as directly obtaining significant quantities of longer alcohols through this technology is still a challenge. Some other variations are also possible, to directly yield, e.g., dimethylether by adding proper acid functionality to the catalysts that enables in situ condensation of the formed methanol molecules. It is even possible to take advantage of the H_2S content in some specific syngas fractions (especially from black liquor issued from paper mills) to synthesize methylmercaptan (CH_3SH) from CO, H_2, and H_2S mixtures, as it was explored in EuroBioRef [14].

Further, anaerobic fermentation of biomass gives biogas, a CH_4 and CO_2 mixture also containing some impurities. While biogas is conventionally used as a fuel, some new, more advanced applications can be envisioned. First, it can be converted to syngas and then subsequently converted in downstream chemicals as mentioned earlier. Then, CH_4 activation, which is still a very big challenge in catalysis, could open perspectives for biogas upgrading to liquid chemicals fractions, facilitating transportation of the as-obtained compounds together with proposing high value-added applications. The presence of CO_2 in the biogas streams could also be profitably used, such as for blocking catalytic basic sites if needed. Papers in this challenging catalytic field still have to be written in the future.

6.2.2.2 C2 molecule – ethanol

Ethanol is by far the most important C2 platform molecule issued from biomass. It can even be considered as a key compound in the frame of the biorefinery concept. Ethanol is one of the rare examples of molecules already massively produced at the commercial scale starting from a renewable feedstock. Indeed, 58.3×10⁶ t of ethanol have been produced worldwide in 2009 [20]. The main part of the production is obtained by a fermentation process using sugars as a feedstock (the so-called first-generation process).

As everybody knows, ethanol is used for the production of alcoholic beverages, but more recently, it is also used massively as a biofuel, either directly blended with gasoline (or used as is in specific flex fuel engines) or after conversion to ethyl *tert*-butyl ether (ETBE) using *iso*-butene as a reactant [21]. Note that, while still not commercial, the conversion of ethanol to hydrocarbon-type fuels has gained a lot of interest in the recent years [22]. Ethanol is also employed as a solvent, a disinfectant/preservative, and, more interestingly, it recently gained a large interest as a platform molecule to synthesize chemicals:

– Ethylene. Bioethylene (also called "green ethylene"), which can be obtained by ethanol dehydration in the presence of acid catalysts, holds a very special position, as it can pave the way to green polymers and other chemicals (as a drop-in product in the current chemical industry). Bioethylene has been first commercially produced by the Brazilian company Braskem in 2010 in order to produce green polyethylene (http://www.braskem.com) [23], but more recently, Axens, Total, and the Institut Français du Pétrole Energies Nouvelles (IFPEN) have also introduced Atol™, a technology for the production of polymer-grade bioethylene via bioethanol dehydration (http://www.axens.net) [24].

– Higher alcohols. Ethanol can also be used to generate higher alcohols, through the so-called Guerbet reaction [25], which needs multifunctional catalysts intimately mixing on their surface acid and basic sites, optionally together with redox sites. A first "dimerization" yields n-butanol from ethanol, but the main issue concerns selectivity, as the formed alcohols can undergo further Guerbet reactions in situ, yielding a blend of higher alcohols.

– Propene. Like ethylene, propene (or 'propylene') is a very important chemical intermediate. It can be used to produce, e.g., polypropylene, as an important polymer in the plastic industry. Surprisingly, direct conversion of ethanol to propene is a quite selective reaction [26] whose mechanism is still quite unclear. The main issue is to control the degree of oligomerization, as higher olefins are also easily formed.

– Butadiene. Ethanol can also be converted to butadiene, an important building block as well. To realize this reaction, ZSM-5, optionally doped with metal(s) or/and phosphorus are the catalytic systems that have driven the greatest attention for this reaction [27].

– Hydrogen. In a biorefinery concept, it is very important to be able to use renewable H_2, e.g., for hydrogenation processes. To this respect, the ethanol reforming reaction is a useful way that is extensively studied to produce sustainable hydrogen from a renewable feedstock [28]. To this respect, $CeNiO_x$ formulations are particularly efficient at low temperatures [29].

– Others. Ethyl esters such as ethyl acetate or ethyl acrylate, ethers, ethylamine/ethylamide, acetaldehyde, and acetic acid are also examples of products of interest obtained from ethanol [30].

Lastly, one should be aware that one of the key points in these reactions is to be able to cope with the bioethanol-specific impurities that could alter the existing downstream processing of the obtained molecules, which have already petro-sourced equivalents. Also, it is crucial to find catalytic systems able to work with more or less large fractions of water. Indeed, this enables shortcutting the complete ethanol distillation chain, thus with substantial cost reductions

6.2.2.3 C3 molecules – glycerol and lactic acid

Glycerol and lactic acid are certainly the most established C3 platform molecules. 3-Hydroxypropionaldehyde (and its acid form, 3-hydroxypropionic acid, or 3-HPA) as well as allyl alcohol might be developed in the future.

– Glycerol. Glycerol is the triple alcohol of propane. It constitutes the backbone of fatty acids in triglycerides, which build up all natural oils and fats. Processing of triglycerides gives a large source of glycerol as a co-product (see Section 6.3.1). Even though several thousands of applications for glycerol are known (to mention, glycerol is added to tobacco and many personal care products as a moistener), the increasing production of biodiesel has generated an oversupply of glycerol. In Europe, the amount of glycerol has increased by 10 times between 2001 and 2010. As a consequence, the valorization of glycerol has become a topical issue in order to increase the economical perspective of the biodiesel units, while, currently, most of the glycerol is still thermally converted. A lot of glycerol upgrading reactions to value-added products have been studied. The most significant one is undoubtedly the commercial synthesis of epichlorohydrine, which has a rather funny history: before the biodiesel boom started at the end of the 1990s, glycerol was a highly demanded feedstock, which was produced from fossil resources (propylene) via epichlorohydrine as an intermediate. Then, with the increasing amounts of glycerol, this process became obsolete and researchers from Solvay discovered that they could reverse their process to then synthesize epichlorohydrine from glycerol in the so-called Epicerol® process (http://www.solvaychemicals.com) [31].

Other pathways for the chemocatalytic valorization of glycerol are still under investigation [32], such as selective reduction to propanediols, dehydration to acrolein [33] (with downstream perspectives, e.g., acrylic acid or acrylonitrile), conversion to glycerol carbonate and subsequent derivatives [34], or oxidation to fine chemicals (notably acids) [35].

Glycerol is also used as a polar head to replace ethylene oxide in the fabrication of non-ionic surfactants. The most well-known application is the catalytic esterification of glycerol with fatty chains yielding the so-called monoglycerides, which have a number of applications in cosmetics and in the food industry. Catalytic etherification of glycerol with fatty alcohol has also received a great deal of attention since it opens an access to surfactants that are much more stable in water. Catalytic esterification or etherification of glycerol with fatty chains generates, however, stringent problems. Indeed, both reactants have opposite polarities, and the reaction media is biphasic, thus inducing mass transfer issues. To achieve an efficient catalytic functionalization of glycerol with fatty chains, catalysts need to exhibit an appropriate hydrophilic/lipophilic balance to favor an optimal contact between both phases. These catalytic systems, the so-called surfactant-combined catalysts, have the ability to emulsify the reaction media while catalyzing the esterification or etherification reaction. For instance, dodecylbenzene sulfonic acid, a homogeneous catalyst, is capable of etherifying glycerol with long alkyl chain alcohol with an unprecedented yield [36, 37].

More recently, glycerol was also proposed as a "green" solvent for performing catalytic reactions in the presence of homogeneous catalysts [38, 39]. In these studies, the homogeneous catalyst is retained in the glycerol phase while reaction products can be continuously extracted with an organic solvent (glycerol, like water, is poorly soluble in common organic solvent including supercritical CO_2). Then, the glycerol phase containing the homogeneous catalyst can be recycled with a maximum of efficiency. Such a concept was particularly efficient for the production 5-hydroxymethylfurfural (5-HMF, see Section 6.2.2.6) from carbohydrates, fungicides of the dithiocarbamate family, fatty esters, etc.

– Lactic acid. Currently, lactic acid is nearly entirely produced at a large scale by fermentation of sugars, even though chemical synthesis pathways have been reported (i.e., from glucose or glycerol). Lactic acid has some direct applications in the food industry, notably as an acidifying agent, as well as in cleaning agents, where it is used due to its antibiotic properties. The largest industrial application of lactic acid is certainly the polymerization to polylactates (polylactic acid, PLA). PLA finds applications as a bio-based "green" packaging material for food, but it still has some technological issues to be solved, including its insufficient thermal stability. Hand in hand with the increasing availability of lactic acid, some possible subsequent downstream conversions have become the focus of research for obtaining value-added products [40, 41]. We can notably mention dehydration to acrylic acid, hydrogenation to 1,2-propanediol, and oxidative dehydrogenation to pyruvic acid.

6.2.2.4 C4 molecules – succinic acid, 1,4-butanediol, γ-butyrolactone, butanol, *iso*-butanol, and olefins

The C4 platform molecules play a very important role in the fossil-based chemical industry. Butadiene is an important intermediate for plasticizers (ABS polymers) and generally finds use in resins for coatings and fiber-based plastics. Meanwhile, bio-sourced C4 platform molecules are much less common than the C3, C5, or C6 ones due to a lower natural occurrence.

– Succinic acid. Succinic acid has driven attention in the chemical industry during the last decades [42]. Its annual production is still low (no more than 30 kt), but several production plants are under construction. BioAmber (http://www.bio-amber.com) [43] and Reverdia (http://www.reverdia.com) [44] are important players in the field, and the most optimistic forecasts predict a five-fold availability in less than 5 years. While succinic acid has nearly no direct application (except for additive in food and drugs), it is a very promising platform molecule since it can be easily converted to 1,4-butanediol, or γ-butyrolactone by catalytic hydrogenation [45]. While not exactly primary products, downstream applications of these two molecules are getting interest, and we decided to include them in the list of platform molecules.

The fermentation process to yield succinic acid needs the neutralization of the acid. When ammonia is used to this purpose, the ammonium salt of succinic acid is obtained and can directly be cyclized to succinimide. In the presence of alcohols, the tertiary amine is formed and can be catalytically reduced to pyrrolidone derivates. Then, succinic acid can also be converted to various valuable pyrrolidone derivates, such as *n*-methyl pyrrolidone, a monomer for polyvinyl pyrrolidones, which finds application in food, drug, and cosmetic industry, and *n*-vinyl pyrrolidone, which finds direct application as a solvent in various processes.

–   1,4-Butanediol. Biobased 1,4-butanediol can be obtained by succinic acid hydrogenation over noble metal catalysts (containing Ru, Pd, and/or Re), and a yield of 90% has been reported. The as-obtained 1,4-butanediol finds direct application as a solvent, but it is generally transformed to tetrahydrofurane (THF), a very widely used solvent in the chemical industry. Other applications of 1,4-butanediol are notably the dehydration to unsaturated alcohols (enols) and butadiene, as a drop-in molecule for the petro-industry. Note that other biosourced butanediol isomers are available and can then also be used as a base for subsequent reactions.

–   γ-Butyrolactone. γ-Butyrolactone is intermediately formed during catalytic hydrogenation of succinic acid to 1,4-butanediol. Selective hydrogenation to γ-butyrolactone is, however, unexpectedly easy over noble metal catalysts containing Ag, Pt, and Pd, with yields up to 95%. In the chemical industry, γ-butyrolactone finds various direct applications, notably as a solvent, but also in pharmaceutics.

–   Butanol and *iso*-butanol. Biotechnologies are increasingly becoming able in accessing butanol (*n*-butanol) and *iso*-butanol from sugar hydrolyzates. These compounds can be, e.g., catalytically dehydrated to their olefinic counterparts, which find direct applications as drop-in biobased molecules in petrorefineries, in which they can be further upgraded. However, specifically designed multifunctional catalysts can enable one-pot upgrading. For example, the reaction of direct conversion of butanol to maleic anhydride has been performed in EuroBioRef [14] over acid/redox bifunctional catalysts, while it is usually performed by 1-butene oxidation in the petrochemical industry.

–   Olefins. In addition to the aforementioned olefins that can be accessed by dehydration of biosourced alcohols (ethylene, 1-butene, *iso*-butene, butadiene) or by chemocatalytic reactions (ethanol to propene [26] or to butadiene [27]), recently, biotechnologies can give direct fermentation access to *iso*-butene (http://www.global-bioenergies.com) 46]. This enables drop-in strategies in petrorefineries, provided the specific products impurities can be correctly handled, which might need specific adaptation of the downstream conventional catalytic formulations.

6.2.2.5 C5 molecules – xylose, furfural, levulinic acid, and isoprene

While C6 monosaccharides can be derived from cellulose, pentoses (C5 sugars) are only accessible by cleaving hemicellulose. Hemicellulose hydrolysis, which is generally based on diluted mineral acids like phosphoric acid or sulfuric acid, notably yields xylose and arabinose.

–   Xylose. Xylose (and arabinose) has only a few direct applications, mainly in the pharmaceutical and the food industries due to its sweetness. However, upgrading of xylose to various chemicals has also been described. Among them, conversion to furfural is well reported in the literature [47]. This reaction is also catalyzed by acids, such as sulfuric acid or solid acid catalysts [48]. Direct one-pot conversion from hemicellulose is also possible, as the intermediately formed xylose is rapidly converted to furfural under the hydrolysis acidic conditions. Furfural finds direct application as a solvent in the chemical industry, but its main potential lies in its role as a key intermediate, thus making itself a platform molecule.

–   Furfural. Various important bulk chemicals can be derived from furfural, such as maleic acid, methyl-tetrahydrofurane (MTHF), THF, or furfuryl alcohol. As mentioned earlier, maleic acid, or maleic anhydride, is an important building block in the synthesis of polyester fibers, whereas THF is a well-established solvent. MTHF has also become a promising solvent in the chemical industry since it exhibits properties close to those of THF or dichloromethane but is considered "greener". The synthesis of maleic acid from furfural is based on selective oxidation. A 50% yield has been reported over heteropolyacid catalysts. This kind of catalysts is also of interest for hydrogenating furfural to furfuryl alcohol, which is of great interest as a building block for resins [49]. Conventionally, furfuryl alcohol is synthesized over copper-chromite catalysts, which causes eco-toxicity issues. As an alternative system, heteropolyacids combined with Raney nickel enables furfuryl alcohol yields over 95%.

–   Levulinic acid. Levulinic acid is a by-product in the dehydration of fructose to 5-HMF (cf. Section 6.2.2.6). It can be obtained in high yields by acidic treatment of glucose, fructose, or cellulose, but also directly from lignocellulose [50]. Nevertheless, being rather considered as an undesired degradation product during 5-HMF synthesis, only a few systematic articles on the selective synthesis of levulinic acid can be found in the literature. This may also be ascribed to the low carbon economy of such a process, since only one carbon atom is lost during the degradation of 5-HMF. Levulinic acid itself has no direct application, but it can be converted to γ-valerolactone (GVL) via hydrogenation over Ru- or Pt-based noble metal catalysts [51]. Similarly to γ-buyrolactone (cf. Section 6.2.2.4), GVL is a promising "green" solvent, but it can also be further converted to MTHF via hydrogenation and successive etherification. Starting from a neat fructose solution, it was even reported that over Ru-based catalysts, the yields in levulinic acid, GVL, and methyl-THF could be tuned by adjusting the reaction conditions.

– Isoprene. Similarly to C4 molecules, biotechnological production of isoprene enables drop-in strategies in petrochemistry (http://www.amyris.com) [52], provided specific impurities are correctly handled, which imposes fine-tuning of the downstream catalytic systems.

6.2.2.6 C6 molecules – 5-hydroxymethylfurfural, 2,5-furandicarboxylic acid, and sorbitol

Glucose and fructose are the most widely known monosaccharides, which are currently derived from sucrose, the table sugar we are consuming every day. The trend is, however, to obtain glucose not only from cellulose, but also from starch, which are natural polymers of D-glucose. Fructose can then be obtained by glucose isomerization.

Processing of sugars by fermentation notably gives ethanol (as well as other compounds, as mentioned earlier; cf. Section 6.2.2.2). However, chemocatalysis gives access to another major product, 5-HMF [47], which is obtained by selective dehydration of fructose or, depending on the catalytic system, directly from glucose or even cellulose.

– 5-Hydroxymethylfurfural. A large variety of processes and catalysts are described in the literature for the synthesis of 5-HMF. The possible catalysts include homogenous acids such as sulfuric and hydrochloric acids and heterogeneous catalysts like aluminosilicates, even involving ionic liquids [53]. From fructose, yields of 80% are widely reported, and from glucose, only slightly lower performance is observed. However, the rapid degradation of 5-HMF to levulinic acid in the presence of water, the natural solvent for monosaccharides, remains the main problem. Thus, efforts have been dedicated to the rapid and efficient extraction of 5-HMF from the aqueous reaction mixture. Many reactor concepts have been described, including counter-courant extraction and continuous stripping of 5-HMF by a nitrogen flow. Nevertheless, the extraction using organic solvents remains the most promising solution. Several extraction solvents such as methyl-*iso*-butylketone, butanol, and DMA have been studied. Alternatively, the reaction was performed in DMSO, which avoids the use of water, whereby 5-HMF degradation is naturally limited. The main drawback when using organic solvents arises from the purification and the separation of the product. Since 5-HMF is highly reactive, distillation is rather difficult to perform. To these purification issues, one must also add the issues related to the general use of organic solvents or ionic liquids, which cannot be always considered "green".

– 2,5-Furandicarboxylic acid. 2,5-Furandicarboxylic acid (FDCA) is the most interesting derivate of 5-HMF, since it is considered as an alternative to terephthalic acid – a main commodity chemical for the polymer industry [54]. Everyone is familiar with polyethylene terephthalate (PET), which is also eponym of the famous PET bottles. Hence, the market potential for FDCA is huge, considering the annual PET world production of around 50 million tons. Large efforts have been then devoted to the development of catalysts for the selective oxidation of 5-HMF to FDCA. Most of the catalytic systems are based on noble metals such as Pt, Pd, and Au, with FDCA yields over 95%. So far, the whole 5-HMF/FDCA value chain has suffered from difficulties in the dehydration reaction, rather than from the oxidation step, and only some attempts to commercial applications are reported, notably by the Dutch company Avantium (http://avantium.com) [55].

– Sorbitol. Sorbitol is another promising intermediate derived from glucose [2]. The sorbitol story is not new at all, as sorbitol is produced at the industrial scale for decades by the hydrogenation of glucose over Raney nickel. Sorbitol is notably used in the food industry for diabetic diet. Contrarily to glucose, sorbitol ingestion does not yield any increase in the blood sugar level, while offering the same sweetness and energy content. More recently, sorbitol has gained attention for its use in the synthesis of isosorbide (http://www.roquette.com) by selective dehydration over acid catalysts [56]. Thanks to the two remaining hydroxyl groups, isosorbide can form, e.g., polyesters, and has also applications in the pharmaceutical industry.

Various catalytic processes have also been developed to directly convert cellulose to commodities without intermediate isolation of glucose. In this context, multifunctional solid catalysts have been proposed. Generally, they exhibit acid and redox catalytic sites located on the same surface. Clearly, this approach is of great interest and allows diversity and complexity to be created from cellulose in a same reactor. Among the recent works, one can note the deposition of transition metals (Ni, Ru, Pd, Pt) over acidic support (e.g., niobium oxide, zeolites). These bifunctional catalysts are capable of depolymerizing cellulose to glucose and reducing glucose to sorbitol, isosorbide, ethylene glycol, diketone derivatives, among others, in a one-pot procedure [1–5]. Note that this strategy is widely employed for the production of fuels such as GVL, hydrocarbons, or alkenes from polysaccharides.

6.3 Oleaginous biomass

6.3.1 Separation and primary products

Prior to refining, seeds are triturated in order to extract vegetable oils. This process enables, obviously, the recovery of triglycerides, but also the recovery of phospholipids, seed meal (containing, e.g., proteins), etc., which can find chemical applications. However, here we will focus on triglyceride-based chemistry. Extracted vegetable oils or triglycerides can be used as collected in catalytic processes (cf. Section 6.3.1.1). They can also be split, and in most cases, the first step of triglyceride refining involves catalytic hydrolysis to fatty acids or alcoholysis (transesterification) to fatty esters aiming at separating the fatty chains from the

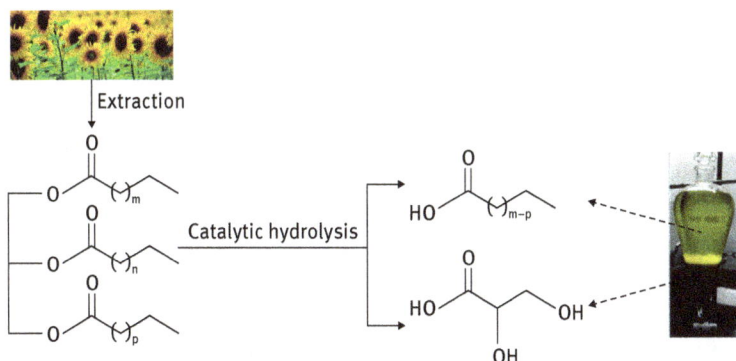

Fig. 6.5: Schematics of catalytic hydrolysis of vegetable oils.

glycerol backbone (which can be used as a platform molecule; cf. Section 6.2.2.3) (Figure 6.5). These reactions are generally performed under basic conditions, but acid catalysts or enzymatic processes are also reported. At the industrial level, homogeneous catalysts such as NaOH or K_2CO_3 are generally employed. Numerous studies have also highlighted the potential of solid basic catalysts such as zeolites, metal oxides, hydrotalcites, etc. However, they generally suffer from a lack in stability when mixed with triglycerides (also containing free fatty acids or, FFAs) and a part of the solid catalyst or of the catalytic phase is dissolved in the reaction mixture, making their long-term recycling difficult and expensive. However, one must mention an industrial plant located in the south of France (IFPEN) involving a heterogeneous catalyst (zinc aluminate) in a continuous way.

As an interesting feature, while it was previously thought that it would be impossible to find biotechnological systems to realize the conversion of sugar hydrolysates to fatty acids, this is now a reality, with high yields, opening the way for diversification of sourcing [57].

6.3.1.1 Triglycerides

While the production of triglycerides represents less than 5% of the worldwide production of biomass, most of the catalytic studies were initially focused on this raw material. This scientific fad for vegetable oils stems from their chemical structure. Historically, catalysis has been developed for the conversion of fossil oils. Vegetable oils comprise long alkyl chains that are quite similar to the hydrocarbons produced and converted in the petrorefineries. Hence, it is clear that conventional solid catalysts specifically designed for the conversion of fossils oils might be potentially used or adapted to the use of vegetable oils without significant technological upheaval. For this reason, numerous well-known catalytic transformations have been carried out on triglycerides. Because catalytic reactions similar to those described hereafter for fatty acid and esters (cf. Section 6.3.1.2) can be generally performed with triglyce-rides (reduction, oxidation, nitration, hydroformylation, etc.), this aspect will not be further described in the present section.

We would like to insist here on the fact that the direct utilization of triglycerides in catalytic processes is, however, more challenging than it looks. The first problem faced by catalysis originates from the very high viscosity of vegetable oils that induces important mass transfer limitations (dispersion of catalysts, diffusion of reactants, solubility of gases, etc.). Triglycerides also contain a large variety of impurities such as water, peroxides, etc. These impurities dramatically impact the stability, activity, and selectivity of catalysts by modifying the nature of catalytic sites or the catalyst support. In addition, the concentration and nature of these impurities as well as the chemical composition of triglycerides (chemical nature of fatty chains) is changing according to its botanical origin. Hence, to this respect, catalytic transformations of refined fatty acids or esters are easier. As a consequence, this also often means that a catalyst has to be specifically designed to a targeted triglyceride.

6.3.1.2 Fatty acids and esters

As mentioned earlier, fatty acids or esters (most conventionally methyl esters) are generated from triglycerides by hydrolysis or transesterification. Various natural fatty acids or esters differ in carbon chain lengths (even numbers) and number of double bonds (never conjugated and largely predominantly in a *cis*- or *Z*-conformation). Typical fatty acids or esters used in industrial processes as feedstocks are shown in Figure 6.6. These compounds are obtained from various types of vegetable oils, but animal fats can also be a source of fatty compounds of interest.

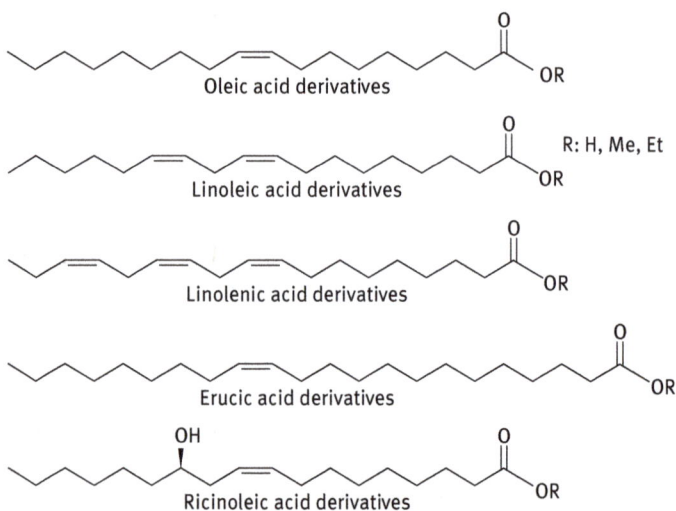

Oleic acid derivatives

Linoleic acid derivatives   R: H, Me, Et

Linolenic acid derivatives

Erucic acid derivatives

Ricinoleic acid derivatives

Fig. 6.6: Some industrially used fatty acid derivatives.

Conventional fatty acids or esters can be mainly functionalized in two ways. The first one consists of a reaction at the carboxylic end of the acid (e.g., conversion into fatty alcohols and fatty amines) and the second one consists of performing reactions on the internal carbon-carbon double bond(s) of the apolar tail [58, 59]. Several reactions involving these double bonds are well established and mostly homogeneously catalyzed. Examples of these reactions are metathesis, hydroformylation, hydroaminomethylation, hydroesterification, silylation, hydrogenation, or oxidation (Figure 6.7). Further, the presence of an OH group on ricinoleic acid and derivatives, which is a quite rare feature, also opens the way for OH functionalization [60]. Note that ricinoleic acid is already used as a base for commercial polymer applications by Arkema (http://www.arkema.com) [61] and Solvay (http://www.rhodia.com) [62].

- Hydroformylation. Catalytic hydroformylation is a possible route to functionalize unsaturated bio-sourced compounds [63]. This reaction takes place under CO/H_2 pressure to yield aldehydes with an atom economy of 100%. The hydroformylation of unsaturated fatty esters has been studied using heterogeneous and homogeneous catalysts in organic solvents or in two-phase aqueous systems. The most active catalytic systems were reported with homogeneous rhodium catalysts

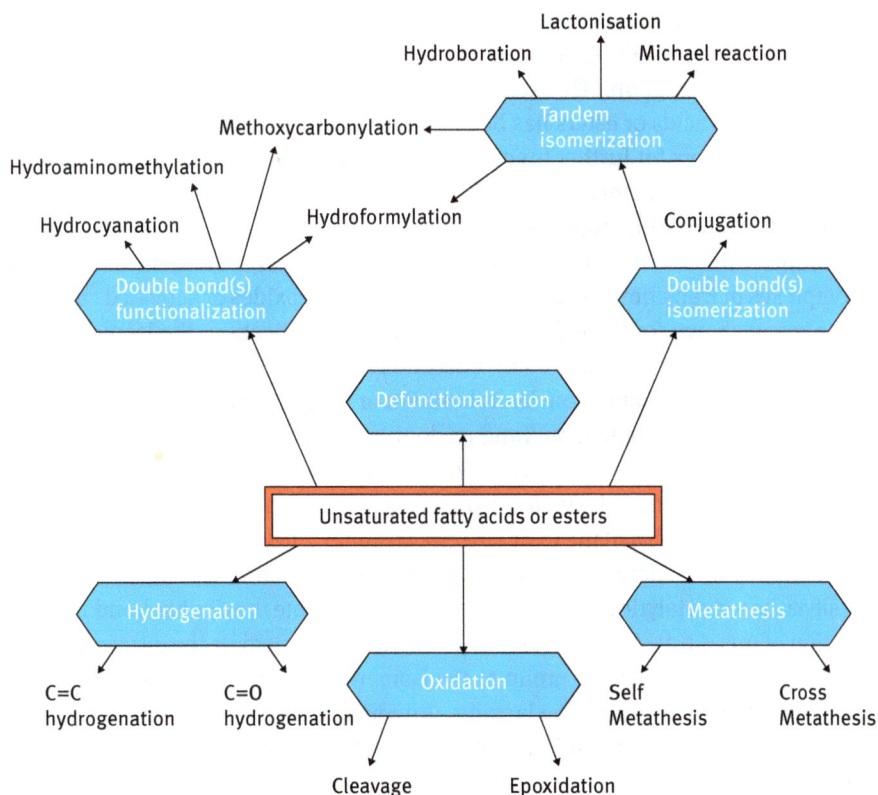

Fig. 6.7: Possible reactions starting from unsaturated fatty acids.

stabilized by phosphines or bulky phosphites. Using the chelating biphosphite ligand biphephos, methyl oleate and ethyl linoleate could even undergo an isomerizing hydroformylation leading to linear aldehydes. However, the yields remained modest due to a very strong hydrogenation side reaction.

- Hydroaminomethylation. Hydroaminomethylation of fatty compounds is a one-pot cascade reaction, starting with the hydroformylation of a double bond, before consecutive condensation of the intermediate aldehyde with the substrate amine, and subsequent hydrogenation of the formed enamine or imine to the desired amine product. This reaction is catalyzed by homogeneous rhodium complexes stabilized by phosphines and allows the introduction of amine functionality on fatty compounds in a one-reaction step.

- Hydroesterification. Hydroesterification (also called alkoxycarbonylation) is an elegant way for generating diesters from carbon monoxide, alcohol, and unsaturated fatty esters. The catalytic system consists of a palladium precursor, a phosphine, and a Brønsted acid. Interestingly, the selective hydroesterification of various unsaturated methyl esters to α,ω-diesters can be realized. The double bond is isomerized along the chain, and the hydroesterification takes place at the terminal carbon atom. This isomerizing hydroesterification requires the presence of a sterically demanding diphosphine ligand such as *bis*(ditertiary-butyl-phosphino-methyl)benzene.

- Selective isomerization. The selective isomerization of carbon-carbon double bonds of fatty acids or esters has been extensively investigated as valuable products or precursors for further functionalization can be obtained through this way. For example, *cis*-9-, *trans*-11-, and *trans*-10, *cis*-12-conjugated linoleic acids that have beneficial health properties can be efficiently synthesized by selective isomerization of linoleic acid. This reaction can be catalyzed by homogenous rhodium complexes or transition metals (Ru, Au) supported on oxides or activated carbons. As already shown for hydroformylation or hydroesterification, the isomerization of double bond can be followed by a second reaction, allowing selective tailored functionalization. Isomerizing hydroboration or trialkylsilylation are other examples of such tandem reactions. Tandem double-bond isomerization and hydroboration of methyl oleate were achieved using a iridium/*bis*(*di*phenylphosphine) ethane catalyst. Methyl oleate can also undergo iridium-catalyzed dehydrogenative silylation with triethylsilane to finally give terminal vinylsilanes. Isomerization of the carbon-carbon double bond can also be directed toward the acid/ester moiety. A silver-based catalytic system could efficiently promote the double-bond isomerization of unsaturated fatty acids and at the same time mediate the intramolecular addition of the carboxylate group to the isomerized double bond, leading to the selective formation of five-ring lactones. Rhodium-phosphite catalysts could also mediate double-bond migrations and catalyze the Michael addition of carbon and nitrogen nucleophiles once the double bond is in conjugation with the carboxylate group. Owing to this isomerizing Michael reaction, β-amino esters or β-arylated products can be easily synthesized depending on the employed nucleophile.

– Metathesis. Self- or cross-metathesis of unsaturated fatty esters is a convenient route to generate diverse types of α,ω-bifunctional molecules. Several highly selective homogeneous or heterogeneous catalytic systems have been reported to perform these reactions. For example, homogeneous ruthenium-based catalysts, such as the Grubbs catalysts, are nowadays commonly used because of their efficiency and especially their simple handling in comparison to less oxygen-stable Schrock catalysts. In addition, the development of new metathesis catalysts like the Hoveyda-Grubbs catalyst allowed cross-metathesis reactions with functionalized olefins such as methyl acrylate.

– Oxidation. The oxidation of the carbon-carbon double bond(s) of unsaturated fatty acids and their derivatives is an important reaction in oleochemistry. For instance, the oxidative cleavage and the epoxidation of unsaturated fatty acid derivatives are performed at the industrial level using ozone and peroxy acids, respectively. The use of transition metal catalysts to perform these reactions is a research topic of high interest, as these catalysts should allow the application of milder reaction conditions and more benign oxidants such as oxygen or hydrogen peroxide. Simple metal salts, metal oxides or peroxides, metal coordination complexes, and heterogeneous catalyst have been proposed to achieve these oxidations. The catalytic systems are generally based on second- and third-row transition metals such as Ru, Re, Mo, Os, or W. Recently, the use of non-toxic and cheap metals such as iron has also been investigated.

– Hydrogenation. Basically, polyunsaturated fatty acids derivatives can be hydrogenated into monounsaturated or saturated esters through carbon-carbon double-bond hydrogenation or into fatty alcohols by the carbon-oxygen double-bond hydrogenation. Furthermore, complete hydrogenation of unsaturated fatty acids derivatives or fatty alcohols can also lead to the removal of oxygen and the formation of saturated hydrocarbons. Homogeneous or heterogeneous catalysts based on VIII group metals such as Ni, Co, Pd, Pt, and Rh are generally used to perform the partial hydrogenation of the carbon-carbon double bond under moderate operating conditions. Among these different metals, Pt appears to be the less active metal toward isomerization, minimizing the selectivity to undesired *trans* monounsaturated fatty acid derivatives. Catalytic hydrogenation of carbon-oxygen double bond requires more drastic conditions and can be achieved using homogeneous or heterogeneous catalysts. For heterogeneous catalysts, it is well recognized that the best catalysts have two different adsorption sites, a metallic center for H_2 adsorption and an electron-deficient center like SnO_x for carbon-oxygen activation. Finally, primary unsaturated fatty amines are produced by the hydrogenation of unsaturated fatty nitriles obtained by reaction of fatty acids or fatty acid esters with ammonia. Most of the current catalytic processes are based on copper-chromite, cobalt, and nickel catalysts.

6.4 Conclusion

Many different catalytic reactions and possibilities, with at least a scientific interest, but increasingly orientated along industrial perspectives have been or are being developed from biomass-derived substrates. However, economic criteria (micro, but also macro), raw materials availabilities, social requirements, Life Cycle Assessment or, LCA, etc. must be the drivers interwoven with these developments to ensure final industrial feasibility and then reality. The as-derived integrated transformation networks are thus determined by numerous parameters and will most probably be constantly refined in the next few decades, due to scientific advances, and also to global context evolution (e.g., shale gas exploitation, development of new crops). Further, chemocatalysis will increasingly work hand in hand with biotechnologies to propose ultra-integrated processes, some of them undoubtedly relying on the novel hybrid catalysis concept, mixing chemocatalysis and biotechnologies in various types of revolutionary cooperative one-pot approaches [64].

Bibliography

[1] Corma, A., Iborra, S., Velty, A., Chemical routes for the transformation of biomass to chemicals, Chem Rev **107** (2007) 2411–2502.
[2] Gallezot, P., Conversion of biomass to selected chemical products, Chem Soc Rev **41** (2012) 1538–1558.
[3] De Vyver, S. V., Geboers, J., Jacobs, P. A., Sels, B. F., Recent Advances in the catalytic conversion of cellulose, ChemCatChem **3** (2011) 82–94.
[4] Stöcker, M., Biofuels and biomass-to-liquid fuels in the biorefinery: catalytic conversion of lignocellulosic biomass using porous materials, Angew Chem Int Ed **47** (2008) 9200–9211.
[5] Mascal, M., Nikitin, E. B., Direct, high-yield conversion of cellulose into biofuel, Angew Chem Int Ed **47** (2008) 7924–7926.
[6] Tadesse, H., Luque, R., Advances on biomass pretreatment using ionic liquids: an overview, Energy Environ Sci **4** (2011) 3913–3929.
[7] Chrapava, S., Touraud, D., Rosenau, T., Potthast, A., Kunz, W., The investigation of the influence of water and temperature on the LiCl/DMAc/cellulose system, Phys ChemChemPhys **5** (2003) 1842–1847.
[8] Wada, M., Ike, M., Tokuyasu, K., Enzymatic hydrolysis of cellulose I is greatly accelerated via its conversion to the cellulose II hydrate form, Polym Degrad Stab **95** (2010) 543–548.
[9] Rinaldi, R., Palkovits, R., Schüth, F., Depolymerization of cellulose using solid catalysts in ionic liquids, Angew Chem Int Ed **47** (2008) 8047–8050.
[10] Onda, A., Ochi, T., Yanagisawa, K., Selective hydrolysis of cellulose into glucose over solid acid catalysts, Green Chem **10** (2008) 1033–1037.
[11] Zhang, Q., Jérôme, F., Mechanocatalytic deconstruction of cellulose: an emerging entry into biorefinery, ChemSusChem **6** (2013) 2042–2044.
[12] Young, R. A., Comparison of the properties of chemical cellulose pulps, Cellulose **1** (1994) 107–130.
[13] Kleinert, T. N., US Patent 3,585,104, 1971.
[14] eurobioref.org, accessed 10 September 2014.

[15] http://www.borregaard.com, accessed 10 September 2014.

[16] Holladay, J. E., Bozell, J. J., White, J. F., Johnson, D., Top Value-Added Chemicals from Biomass, Volume II – Results of Screening for Potential Candidates from Biorefinery, PNNL-16983, 2007.

[17] Zakzeski, J., Bruijnincx, P. C. A., Jongerius, A. L., Weckhuysen, B. M., The catalytic valorization of lignin for the production of renewable chemicals, Chem Rev **110** (2010) 3552–3599.

[18] Werpy, T., Petersen, G., Aden, A., Bozell, J., Holladay, J., White, J., Manheim, A., Top Value Added Chemicals from Biomass Volume I – Results of Screening for Potential Candidates from Sugars and Synthesis Gas, 2004. http://www1.eere.energy.gov/bioenergy/pdfs/35523.pdf.

[19] Bozell, J. J., Petersen, G. R., Technology development for the production of biobased products from biorefinery carbohydrates – the US Department of Energy's "Top 10" revisited, Green Chem **12** (2010) 525–728.

[20] Kosaric, N., Duvnjak, Z., Farkas, A., Sahm, H., Bringer-Meyer, S., Goebel, O., Mayer, D., Ulmann's Encyclopedia of Industrial Chemistry, Vol. 13, Ethanol Online version, 2009.

[21] Cavani, F., Centi, G., Perathoner, S., Trifiró, F., (Editors), Sustainable Industrial Chemistry, Wiley-VCH Verlag GmbH & Co. KGaA, Weinheim, Germany, 2009. doi: 10.1002/9783527629114. fmatter.

[22] Tret'yakov, V. F., Makarfi, Y. I., Tret'yakov, K. V., Frantsuzova, N. A., Talyshinskii, R. M., The catalytic conversion of bioethanol to hydrocarbon fuel: a review and study, Catal Ind **2** (2010) 402–420.

[23] http://www.braskem.com.br/site.aspx/green-products-USA, accessed 10 September 2014.

[24] http://www.axens.net/product/technology-licensing/20080/atol.html, accessed 10 September 2014.

[25] Tsuchida, T., Kubo, J., Yoshioka, T., Sakuma, S., Takeguchi, T., Ueda, W., Reaction of ethanol over hydroxyapatite affected by Ca/P ratio of catalyst, J Catal **259** (2008) 183–189.

[26] Lehmann, T., Seidel-Morgenstern, A., Thermodynamic appraisal of the gas phase conversion of ethylene or ethanol to propylene, J Chem Eng **242** (2014) 422–432.

[27] Makshina, E. V., Dusselier, M., Janssens, W., Degrève, J., Jacobs, P. A., Sels, B., Review of old chemistry and new catalytic advances in the on-purpose synthesis of butadiene, Chem Soc Rev (2014), DOI 10.1039/C4CS00105B.

[28] Vizcaíno, A. J., Carrero, A., Calles, J. A., Hydrogen Production: Prospects and Processes, Honnery, D., Moriarty, P., (Editors), Nova Science Publishers, New York, 2012.

[29] Pirez, C., Capron, M., Jobic, H., Dumeignil, F., Jalowiecki-Duhamel, L., Highly efficient and stable CeHzOy nano-oxyhydride catalyst for H2 production from ethanol at room temperature, Angew Chem Int Ed **50** (2011) 10193–10197.

[30] Sun, J., Wang, Y., Recent advances in catalytic conversion of ethanol to chemicals, ACS Catal **4** (2014) 1078–1090.

[31] http://www.solvaychemicals.com/EN/Sustainability/Issues_Challenges/EPICEROL.aspx, accessed 10 September 2014.

[32] Pagliaro, M., Rossi, M., The Future of Glycerol, 2nd edition, RSC, Cambridge, 2013.

[33] Katryniok, B., Paul, S., Dumeignil, F., Recent developments in the field of catalytic dehydration of glycerol to acrolein, ACS Catal **3** (2013) 1819–1834.

[34] Ochoa-Gómez, J. R., Gómez-Jiménez-Aberasturi, O., Ramírez-López, C., Belsué, M., A brief review on industrial alternatives for the manufacturing of glycerol carbonate, a green chemical, Org Process Res Dev **16** (2012) 389–399.

[35] Katryniok, B., Kimura, H., Skrzyńska, E., Girardon, J. S., Fongarland, P., Capron, M., Ducoulombier, R., Mimura, N., Paul, S., Dumeignil, F., Selectice catalytic oxidation of glycerol: perspectives for high value chemicals, Green Chem **13** (2011) 1960–1979

[36] Gaudin, P., Jacquot, R., Marion, P., Pouilloux, Y., Jérôme, F., Acid-catalyzed etherification of glycerol with long-alkyl-chain alcohols, ChemSusChem **4** (2011) 719–722.

[37] Gaudin, P., Jacquot, R., Marion, P., Pouilloux, Y., Jérôme, F., Homogeneously-catalzed etherification of glycerol with 1-dodecanol, Catal Sci Technol **1** (2011) 616–620.

[38] Gu, Y., Jérôme, F., Bio-based solvents: an emerging generation of fluids for the design of eco-efficient processes in catalysis and organic chemistry, Chem Soc Rev **42** (2013) 9550–9570.

[39] Gu, Y., Jérôme, F., Glycerol as a sustainable solvent for green chemistry, Green Chem **7** (2010) 1127–1138.

[40] Dusselier, M., Van Wouwe P, Dewaele A, Makshinaa E, Sels B, Lactic acid as a platform chemical in the biobased economy: the role of chemocatalysis, Energy Environ Sci **6** (2013) 1415–1442.

[41] Mäki-Arvela, P., Simakova, I. L., Salmi, T., Murzin, D. Y., Production of lactic acid/lactates from biomass and their catalytic transformations to commodities, Chem Rev **114** (2014) 1909–1971.

[42] Bechthold, I., Bretz, K., Kabasci, S., Kopitzki, R., Springer, A., Succinic acid: a new platform chemical for biobased polymers, Chem Eng Technol **31** (2008) 647–654.

[43] http://www.bio-amber.com/products/en/products/succinic_acid, accessed 10 September 2014.

[44] http://www.reverdia.com/products/biosuccinium/, accessed 10 September 2014.

[45] Delhomme, C., Weuster-Botz, D., Kühn, F. E., Succinic acid from renewable resources as a V4-building block chemical – a review of the catalytic possibilities in aqueous media, Green Chem **11** (2009) 13–26.

[46] http://www.global-bioenergies.com/index.php?option=com_content&view=article&id=60&-Itemid=157&lang=en, accessed 10 September 2014.

[47] Karinen, R., Vilonen, K., Niemelä, M., Biorefining: heterogeneously catalyzed reactions of carbohydrates for the production of furfural and hydroxymethylfurfural, ChemSusChem **4** (2011) 1002–1006.

[48] Dhepe, P. L., Sahu, R. A., Solid-acid-based process for the conversion of hemicellulose, Green Chem **12** (2010) 2153–2156.

[49] Nakagawa, Y., Tamura, M., Tomishige, K., Catalytic reduction of biomass-derived furanic compounds with hydrogen, ACS Catal **3** (2013) 2655–2668.

[50] Rackemann, D. W., Doherty, W. O., The conversion of lignocellulosics to levulinic acid, Biofuels Bioprod Bioref **5** (2011) 198–214.

[51] Alonso, D. M., Wettstein, S. G., Dumesic, J. A., Gamma-valerolactone, a sustainable platform molecule derived from lignocellulosic biomass, Green Chem **15** (2013) 584–595.

[52] http://www.amyris.com/News/123/Amyris-and-Michelin-Announce-Collaboration-to-Develop-and-Commercialize-Renewable-Isoprene, accessed 10 September 2014.

[53] Zakrzewska, M. E., Bogel-Lukasik, E., Bogel-Lukasik, R., Ionic liquid-mediated formation of 5-hydroxymethylfurfural – a promising biomass-derived building block, Chem Rev **111** (2011) 397–417.

[54] Hu, L., Zhao, G., Hao, W., Tang, X., Sun, Y., Lin, L., Liu, S., Catalytic conversion of biomass-derived carbohydrates into fuels and chemicals via furanic aldehydes, RSC Adv **2** (2012) 11184–11206.

[55] http://avantium.com/media/news.html, accessed 10 September 2014.

[56] http://www.roquette.com/polyols-sorbitol-maltitol-xylitol-isosorbide-mannitol/, accessed 10 September 2014.

[57] Wu, H., Lee, J., Karanjikar, M., San, K. Y., Efficient free fatty acid production from woody biomass hydrolysate using metabolically engineered *Escherichia coli*, Bioresour Technol **169** (2014) 119–125.

[58] Deuss, P. J., Barta, K., de Vries, J. G., Homogeneous catalysis of the conversion of biomass and biomass-derived platform chemicals, Catal Sci Technol **4** (2014) 1174–1196.

[59] Besson, M., Gallezot, P., Pinel, C., Conversion of biomass into chemicals over metal catalysts, Chem Rev **114** (2014) 1827–1870.

[60] Dumeignil, F., Propriétés et utilisation de l'huile de ricin, OCL **19** (2012) 10–5.

[61] http://www.arkema.com/en/products/product-finder/range-viewer/Rilsan-Arkema/, accessed 10 September 2014.

[62] http://www.rhodia.com/fr/binaries/BROCHURE%20EXTEN%20BD.pdf, accessed 10 September 2014.

[63] Vandesien, T., Hapiot, F., Monflier, E., Hydroformylation of vegetable oils and the potential use of hydroformylated fatty acids, Lipid Technol **25** (2013) 175–178.

[64] Dumeignil, F., Chemical catalysis and biotechnology: from a sequential engagement to a one-pot wedding, Chem Eng Technol **86** (2014) 1496.

Alfonso Grassi, Antonio Buonerba, and Sheila Ortega Sanchez

7 Bio-sourced polyolefins

Abstract: In the last 5 years (2010–2015) the worldwide annual production of plastics surpassed the figure of 300 Mt; more than 80% of the plastic market deals with the production, transformation and end use of polyolefins. Plastics are a fundamental component of everyday life providing cost effective, light and disposable tools which find application in different fields. The increased concerns about the depletion of fossil reserves, the greenhouse gas emissions and feedstock costs cause bio-derived polyolefins are emerging rapidly as valid alternative to the fossil fuel based counterpart. In this review the drop-in synthesis of conventional olefins is reviewed and the properties of some novel bioderived polyolefins are presented and discussed.

7.1 Introduction: why renewable polymers?

Synthetic and natural polymers have companioned humankind and radically changed the standard and quality of everyday life during history. Naturally occurring polymers such as cotton and starch (polysaccharides), wool and silk (polyamides), natural rubber, and gutta-percha (polyisoprene) are known from centuries and allowed facing and solving some issues of the daily life, providing threads for cloths, draperies, chords, padding, or other tools. Some of these applications are still in use and result cost-effective and environmentally friendly.

Cotton is known since the prehistoric age. Fragments of cotton fabric, dated since 5000 BC, were excavated in Mexico and the Indus Valley Civilization (Pakistan and India) [1]. This natural fiber consists of 95% pure cellulose, a linear polymer consisting of several hundreds to tens of thousands D-glucose units linked by β-1,4 glycosidic bonds. Current estimates for world production are of about 25 Mt, accounting for 2.5% of the world's arable land [2].

Starch grains from the rhizomes of *Typha* (cattails, bulrushes) as flour have been identified from grinding stones in Europe dating back to 30,000 years ago. Starch grains from sorghum were found on grindstones in caves in Ngalue, Mozambique, dating up to 100,000 years ago. In addition to human feeding, pure extracted wheat starch paste was used in ancient Egypt to glue papyrus. The extraction of starch is first described in the *Natural History of Pliny the Elder* around AD 77–79; Romans also used it in cosmetic creams, to powder the hair, and to thicken sauces [3].

Natural rubber, also called India rubber or *caoutchouc*, is recovered from tree secretions; the chemical composition consists of *cis*-1,4-poly(isoprene) as the prevalent ingredient and water with minor water-soluble organic contaminants. It can be considered the first polyolefin used by mankind. Ancient Mesoamericans in 1600 BC used latex from *Castilla elastica* for making rubber balls, ritual solid, and hollow figurines; they also used the liquid rubber for medicine, painting, and coating of paper [4]. The sticky white liquid was first processed to yield brittle solid, which could be shaped [5]. In 1751, La Condamine presented a paper, referred to as the first scientific report on rubber, in which he described many of the properties of natural rubber extracted from the secretion of *Hevea brasiliensis*, a tree commonly found in South America. La Condamine sent to Europe a package of natural rubber as well as a long memoir describing many aspects of its origins and production. In this report, he first included the words *Hévé* as the name of the tree from which the milk or *latex* flowed and the name given to the material by the Maninas Indians: *cahuchu* or *caoutchouc*. Moreover, he described the smoking procedure by which the natives made the rubber stable and the wide range of goods prepared with this natural product. Latex consists of *cis*-1,4-polyisoprene – with a molecular weight of 100,000 to 1,000,000 Da containing a small percentage (up to 5% of dry mass) of proteins, fatty acids, resins, and inorganic salts as minor components. The trade was well protected, and exporting seeds from Brazil was said to be a capital offense, although there was no law against it. Henry Wickham smuggled 70,000 rubber tree seeds from Brazil and delivered them to Kew Gardens, England. Only 2400 of these germinated, after which the seedlings were then sent to India, Ceylon (Sri Lanka), Indonesia, Singapore, and British Malaya. Malaya (now Peninsular Malaysia) was later to become the biggest producer of rubber.

Although 2000 rubber-producing species are known, only two have been exploited as commercial source of natural rubber: *Hevea brasiliensis* and *Parthenium argentatum* or guayule [6]. Although *Hevea* is the dominant rubber crop today, *Hevea* and guayule have had parallel histories of development. In the early 1900s, guayule was first considered as an alternative source of natural rubber in the USA because of the high price of *Hevea* rubber from the Amazon region. Economic forecasts later on suggested that guayule could become a crop, which can compete without subsidies, when rubber yields could be increased and/or commercial utilizations of processing co-products identified and developed. Demand for natural rubber continues to grow at a rate greater than new plantings of *Hevea*. Guayule should be able to fill this need, especially locally in arid and semi-arid environments. Under dryland conditions and using the same assumptions, annual rubber yields of 640 kg/ha (600 lb/acre) would be increased to be economically profitable, to mean an increase of 2.5 times the present dryland yields annual rubber yielding 1450 kg/ha (1,300 lb/acre). Several attempts were carried out all over the world to develop this new crop alternative to *Hevea*. In 1937, an agreement between Pirelli and Istituto per la Ricostruzione Industriale (IRI), favored by the Fascist government in Italy, created Società Agricola Italiana Gomma Autarchica (SAIGA), a new company for the development of natural rubber from guayule; the consultants of SAIGA were outstanding Italian scientists such as Giulio Natta and Francesco Giordani. In collaboration with the Intercontinental Rubber Company (San Diego, California,

USA), 25 million guayule trees were set and a production of 1000 kg/ha of rubber was expected. However, the unexciting results led to the closure the company in the following years, and the lands in Apulia were reconverted to more profitable cultivars such as cereals [7].

The natural polymers contributed to the sequestration of a huge amount of carbon dioxide from the atmosphere and have zero impact on the environment when the energy for their manufacturing/transformation and the corresponding CO_2 emission is omitted.

Although wood and coal pyrolysis are known from the ancient times to yield gas and liquid mixture of hydrocarbons used as sealing agent or for lighting, only recently has this approach been optimized to recover hydrocarbons, carbon monoxide, and hydrogen (syngas) from natural products or biomass (vide infra) [8].

The selective production of light olefins became accessible only after the discovery of oil and the intensive practice of oil distillation in the early 19th century. Ethylene is the organic compound produced in the largest amount, about 140 Mt/year, from steam cracking of gaseous and light hydrocarbons. Propene is the second most important starting material of the petrochemical industry after ethylene and results from a number of industrial processes comprising oil refining, cracking, olefin metathesis, and propane dehydrogenation. 1,3-Butadiene is isolated from the C4 fraction of steam cracking or by dehydrogenation of butane. An alternative approach comes from ethanol, a process known since 1934 [9]. Styrene results from dehydrogenation of ethyl benzene.

The easy and cheap availability of such olefins (Figure 7.1) at the beginning of the 20th century prompted several industrial and academic research teams to explore those chemicals as suitable monomers for the production of synthetic polymers. A breakthrough in this field was the discovery of metal-catalyzed ethylene polymerization, first reported by Karl Ziegler at Karlsruhe and the stereospecific polymerization of propene by Giulio Natta in Milano, acknowledged by the Nobel Prize in 1964 [10–12]. Parallel with the production of synthetic rubber (SBR; polyisoprene, polybutadiene) and thermoplastic elastomers (styrene-butadiene block copolymers) were successfully accomplished in Germany (1930) and USA (1950), respectively. These discoveries opened a new era in material science, permitting cheap and large-scale production of a number of tools, equipment, and items previously manufactured with metals, wood or cellulosic, and natural fibers [13]. This revolution significantly increased the quality of life, e.g., the hygienic manipulation and packaging of food using

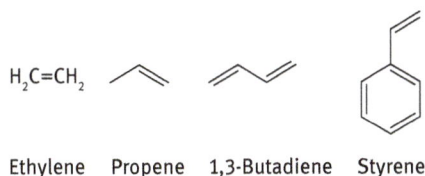

$H_2C=CH_2$

Ethylene   Propene   1,3-Butadiene   Styrene

Fig. 7.1: Common hydrocarbon olefins employed for the synthesis of commercial polyolefins.

disposable films, trays, and bottles, which also prolongs the shelf life of vegetables and meat, new medical devices, aseptic medical packaging, and the production of tools characterized by high durability, lightness, and mechanical resistance. Moreover, energy for metal extraction and shaping was saved as well as natural reservoir in favor of what, at that time, appeared as the most abundant and cheap natural resources, namely coal, oil, and natural gas.

This revolution also changed the mentality of the consumers that considered goods and items "not durable" but "disposable". After decades, several concerns of increasing interest for the future of the planet and new generations must be faced.

Natural resources are not unlimited, and the oil crisis, due to the depletion of the fossil sources or to an unacceptable quotation (e.g., 200 USD/barrel or higher) determined by the political control of the producers, could be expected in 50 years or, in the most optimistic case, in 70 years [14].

Huge amount of municipal wastes based on plastics must be withdrawn daily and treated in landfill or incinerators. The plastic debris accumulation in the open ocean, an estimated 6.6 to 35 ktons of floating plastic wastes, is the symbol of an ending era that cannot go on anymore [15].

Last but not least, the Intergovernmental Panel on Climate Change have indicated the risk of severe climate change impacts for a temperature increase of 2°C above pre-industrial levels (EPA, 2006). The current rate of global temperature increase is between 0.2°C and 0.3°C/decade (EPA, 2006). The global temperature increase will be limited to 2°C if CO_2 equivalent (CO_2e) concentrations will have to be stabilized at levels of between 400 and 450 ppm. The current level is 430 ppm and is rising by more than 2 ppm/annum: this requires a tight control of input and output of CO_2 in atmosphere as envisioned in the Kyoto Protocol.

The first and third issues require switching the synthesis of monomers toward renewable resources in order to preserve oil and provide a zero balance in the carbon footprint in polymer synthesis. To reach this goal successfully, attention must be paid to the life cycle assessment of the polymer, including all the steps (from the cradle to the grave) including energy employed for the monomer synthesis.

The second issue moves the researchers to search for biodegradable and compostable polymers. Alternatively, an economically feasible organization for the withdrawal of the exhaust polymeric material and monomer recovery must be set up.

7.2 Bioderived conventional polyolefins: the drop-in synthesis of olefins

Biopolymers are a variety of polymeric materials completely or in part synthesized by living organisms using partially or totally renewable chemicals available in nature [16]. There is no consensus on the amount of bio-sourced polymer in order to classify them as biopolymer; typically, the percentage of carbon atom from renewable source must be in the range of 25%–40% [17]. In most cases, biopolymers are also biodegradable, which means they can be depolymerized in the environment or in a reactor to give back the monomers; alternatively, they degrade aerobically or anaerobically to simple small molecules, e.g., water and carbon dioxide, in the presence of microorganisms (bacteria, yeast) or by simple chemical reactions, e.g., hydrolysis. Examples include starch, cellulose, polyhydroyalcanoates (PHAs) as polyhydroxybutyrate, polyamides. Bioderived polymers on the contrary are synthetic polymers obtained using monomers or bio-intermediates from renewable resources (biomass) and conventional chemical methodologies (polymerization processes). Examples include poly(lactic acid) (PLA) bio-polyethylene, bio-polyesters (via α,ω-diacids or α,ω-diols from biomass) bio-polyamides (via α,ω-diacids from biomass and α,ω-diamines) bio-polyurethanes incorporating polyols of vegetal origin.

It is noteworthy that plastic resins account for only 4%–5% of the world's fossil fuel, namely naphtha and natural gas, consumption. Thus, the need for bioderived polymers seems mainly related to environmental issues more than preservation of fossil fuel sources. Biomass for bioderived monomers is plant- or animal-based or consists of waste products whose chemical composition includes carbohydrates ($\approx$75%), lignin ($\approx$20%) triglycerides [18], fatty acids, proteins alkaloids, and terpenoids for the remaining about 5%.

At least 300 Mt of plastics are consumed annually worldwide [19] with an expected conservative annual growth rate of 5%. Hundreds of plastic materials are commercially available, but only a handful of these qualify as commodity thermoplastics in terms of their high volume and relatively low price [20], namely low-density polyethylene (LDPE), high-density polyethylene (HDPE), polypropylene (PP), polyvinyl chloride (PVC), and polystyrene (PS), which account for more than 80% of the total demand of plastics. In particular, PE (HDPE, LDPE, and linear low density polyethylene, or LLDPE) accounts for 56% [21] of the world plastics production (241 Mt in 2012) (http:www.plasticseurope. org), which testifies to the extreme versatility of this polymer. The chemistry and the industrial processes for the production of PE is well established and consolidated as well as its physical-chemical characteristics, which have made this polymer useful in a large number of technical and medical applications. Ethylene is produced via steam cracking of naphtha or heavy oil and then polymerized under pressure and temperature in the presence of a metal catalyst. The petrochemical industry invested large amounts of capital to reduce the energy consumption and the production cost of PE.

With the increase in oil price and the environmental concerns described above, bio-PE became increasingly interesting. Thus, efforts have been done to develop methodologies that allow to obtain ethylene from renewable and sustainable sources, the so-called drop-in approach.

If one looks at the relative price scale for a range of selected chemicals produced from either renewable or fossil resources, one can discover that bio-ethylene, resulting from dehydration of bio-ethanol, is one of the few compounds, as ethanol and acetic acid, whose biosynthetic route is favored compared to the fossil fuel one [9, 22].

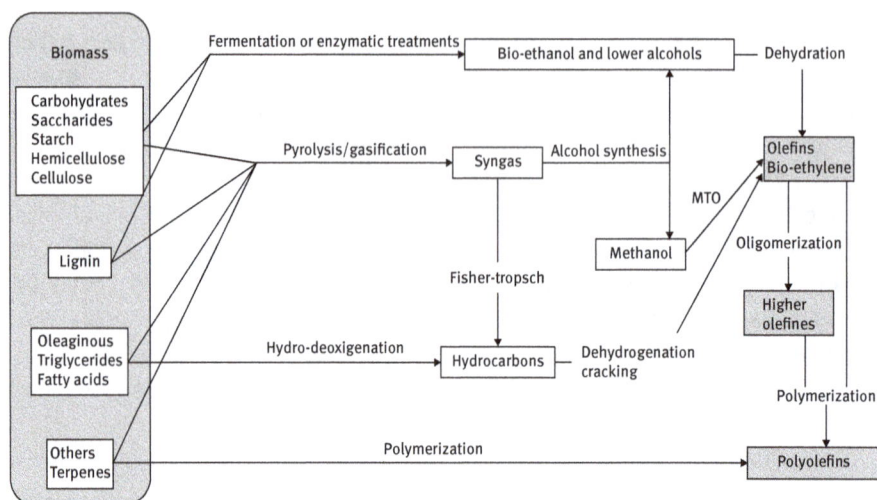

Fig. 7.2: Synthetic routes from biomass to olefins.

Biomass is currently exploited for the production of energy (by combustion) and fuels. Compared to this destination, the conversion of biomass into olefins is certainly a value-added route of relevant interest for chemical industry. Olefins can derive from four main kinds of biomass: (i) carbohydrates (sugars, starch, hemicellulose, and cellulose), (ii) lignin, (iii) triglycerides and fatty acids, and (iv) essential oils as source of terpenes and their derivatives (Figure 7.2). If we first focus our discussion on bio-ethylene, this compound results from dehydration of bio-ethanol, the key intermediate for the synthesis of olefins from biomass (Figure 7.2). The synthetic route involves saccharides from biorenewable feedstock, comprising sugar beet, starch crops such as corn, wood, wheat, and other plant wastes that, through microbial strain and biological fermentation processes, lead to bio-ethanol (the so-called first-generation bio-ethanol). The production of this compound reached 105 billion liters in 2011 and has significantly increased over the past years. The strong demand for bio-ethanol is mainly associated with its mandated use as a transportation fuel to replace non-renewable ones; this application accounts for around 67% of the total amount of bio-ethanol produced.

The use of bio-ethanol for added-value chemicals (Figure 7.3) would be more economically viable compared to its use in the transportation sector and could also lead to higher CO_2 savings. Currently, the world's largest producers of bio-ethanol are the USA, by fermenting corn-based glucose, and Brazil, by adapting sugarcane in roughly the same processes [23]. The two American countries contribute to 88% of the world production of bio-ethanol. As a result of the large increase in production volume, the price of bio-ethanol dropped substantially and can be estimated currently in 700 EUR/ton (November 2010, European market). Unfortunately, both the routes start from raw materials that compete with food reserves. Non-edible agricultural and forest wastes, mainly consisting of lignocellulosic fiber (the so-called second-generation bio-ethanol) [24] and aqueous plants as algae and seaweeds

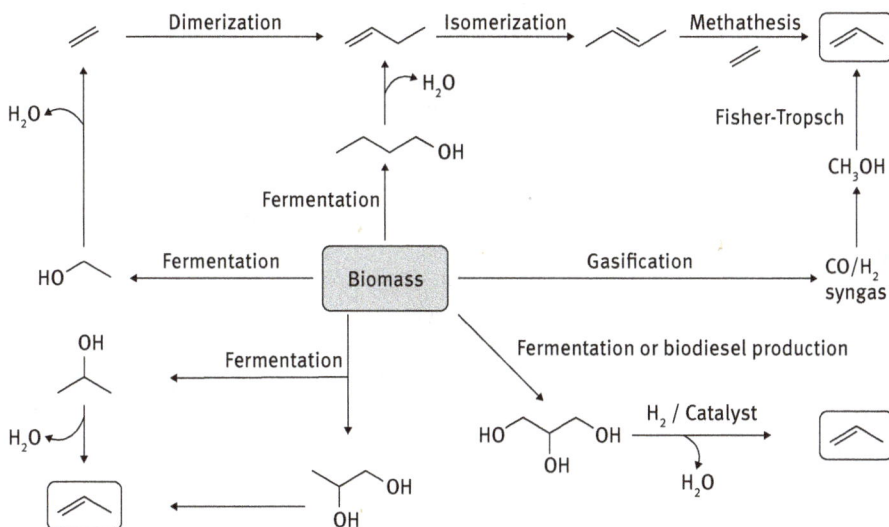

Fig. 7.3: Synthesis of propylene from biomass using fermentation and gasification strategies.

(the third-generation bio-ethanol) [25] are expected to be reasonable alternatives for avoiding the competition with food supply. Ethanol dehydration to ethylene has been known for a long time and only recently has become economically feasible. The most used catalysts are HZSM-5 and γ-alumina, which often serve as benchmark catalysts in comparative studies of new processes [26]. Based on the competitive production of ethanol from fermentation of sugarcane in Brazil, a large-scale plant has been built by the national petrochemical industry Braskem [27] and the completion of a new plant from Dow is expected in 2017 [9]. In sugarcane fermentation, the energy output/input ratio is 9:1, six times more than that of corn fermentation in the USA, which gives a ratio of 1.5:1 [26].

Bio-ethylene can be converted, via conventional metal-catalyzed oligomerization reactions, into higher olefins with even numbers of carbon atoms, such as 1-butene, 1-hexene, and other higher α-olefins that are important comonomers for the production of LDPE [28]. Noteworthy, olefins with odd number of carbon atoms can be also synthesized from bio-ethylene. Recently, Braskem claimed the synthesis of bio-propylene in pilot plant via metathesis reaction of bio-ethylene with 1-butene generated in situ. Braskem also announced a partnership with Novozymes in developing enzymatic strategies for the production of propylene from sugarcane [29]. An alternative approach is based on the selective fermentation of biomass leading to isopropyl alcohol, which is in turn readily converted to propylene through dehydration [23]. Fermentation of sugars via the well-known acetone-butanol-ethanol (ABE) process leads to n-butanol, an intermediate compound more attractive than ethanol because of its higher energy density and lower volatility. Dehydration of this alcohol followed by isomerization to 2-butene and metathesis with ethylene can provide a synthetic route to bio-propylene [23]. However, the cost-effective production of bio-propene for the synthesis of PP and ethylene-propylene copolymers (ethylene-propylene rubber (EPR) and linear low density polyethylene (LLDPE)) is still challenging [17, 30].

Fermentative pathways have been also developed for the production of isobutyl alcohol, an intermediate compound suitable for the synthesis of isobutylene, another interesting olefinic monomer.

1,3-Butadiene (Bd) is another important bulk chemical that finds industrial application in the synthesis of chloroprene (for the PVC production), adiponitrile (for nylon-6,6 production), and acrylonitrile-butadiene-styrene copolymer (ABS). More than 50% of Bd is consumed for SBR and tire production [9].

In addition, aromatic building blocks, such as styrene could also result from the consecutive dimerization/aromatization of Bd. As in the case of ethylene, Bd is obtained from paraffinic hydrocarbons in the steam cracker or by catalytic oxidative dehydrogenation of n-butane or n-butene. Bd is purified by repeated distillation of the C4 fractions, and this process is energy-consuming and expensive.

Two main processes were designed for a biomass-based synthetic pathway for Bd based on bio-ethanol. A single-step process, comprising the dehydration of two ethanol molecules followed by coupling and dehydrogenation, was discovered and patented by Lebedev in 1929 (Figure 7.4a) [31]. The reaction is catalyzed by a bifunctional catalyst, consisting of silicon and magnesium oxides, which exhibits both dehydrating and dehydrogenating properties.

Alternatively, several research groups in the USA designed a process consisting of two steps, in which ethanol is first dehydrogenated to acetaldehyde, which is in turn coupled with ethanol to yield Bd (Figure 7.4b) [32]. Several attempts are still in progress to improve yields and selectivity of the Lebedev process, which still appears as the most promising. The addition of a transition metal oxide, e.g., Ta_2O_5, Cr_2O_3, ZrO_2, N_2O_5, CuO, to the conventional catalyst improved Bd purity and the thermal stability of the catalyst. The life cycle assessment (LCA) for the bioproduction of Bd from bio-ethanol, based on five weighted parameters (economic constraints, environmental impact of raw materials, process costs, the EHS hazard index, and risk aspects) revealed that this process is promising because it is more cost-effective and potentially more sustainable than the naphtha-based process [33].

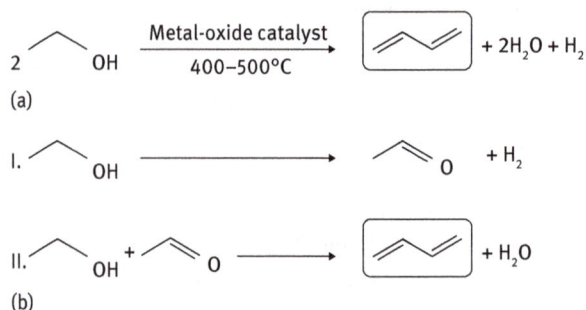

Fig. 7.4: Synthesis of 1,3-butadiene from ethanol: (a) Lebenev process (one step) and (b) two-step process.

Lignocellulosics are inexpensive polysaccharides, widely available in nature, that do not compete for human feeding as glucose, fructose, and saccharose; thus, they are a valid alternative to monosaccharides and disaccharides from vegetables for the synthesis of fuels and chemicals. Direct gasification [34] of lignocellulosics yields syngas, a mixture of CO and H_2 suitable for the synthesis of light olefins using conventional, well-established petrochemical processes [35] (Figure 7.2). Compared to the coal route, the biomass route provides a CO-richer syngas and lower feedstock cost, lower CO_2 contribution, and reduced concentration of sulfur and nitrogen contaminants, which require a thorough purification of the gas stream to avoid the poisoning of the catalyst. However, in the absence of tax incentives, the cost of collection and transportation of biomass to large-scale plants could become a severe limiting factor to this route, determining an increased price of the olefins. The synthetic routes can be divided in two main groups: direct and indirect processes.

The indirect methods include the methanol-to-olefin process (MTO) developed from several companies (Mobil, UOP) and catalyzed by zeolites to produce a mixture of C2-C4 with a selectivity value >95%. Higher selectivity in ethylene was found using the SAPO-34 catalyst, a silicoaluminophoshate [36], whereas the ZSM-5 catalyst allowed to produce up to 70% propylene from methanol via recycling of by-products.

Dimethyl ether from syngas is another possible entry to light olefins through a dedicated process, the dimethyl ether-to-olefins route, or DMTO, also known as syngas-via-dimethyl ether-to-olefins or SDTO route [37]. In principle, this process can be more efficient than MTO since dimethyl ether from syngas has more favorable thermodynamics. There are two reaction steps carried out in different reactors: in the first one, DME is produced from syngas using a metal-acid bifunctional catalyst, while DME is transformed in the second reactor into C2-C4 olefins with a selectivity of 90 wt% using the SAPO-34 catalyst. The selectivity in ethylene and propylene could be increased to 60% and 20%, respectively, when the Cu-Zn/ZSM-5 catalyst is used in the first step and metal-modified SAPO-34 in the second step.

Further synthetic pathways for the conversion of syngas to olefin are (1) the Texaco process, a two-step process in which the syngas is converted into carboxylic acid esters using a homogeneous ruthenium catalysts combined to a phosphonium salt, followed by pyrolysis of the esters that produces a mixture of alkenes in which ethylene or propylene could be obtained with a selectivity of 50% at variance of the experimental conditions [38]; (2) the Dow process, in which syngas is converted into a mixture of lower alcohols (C1-C5) using molybdenum sulfides, which are later dehydrated to give the corresponding olefins; (3) dehydrogenation of hydrocarbons deriving from the Fischer-Tropsch process [39]. All of the above-mentioned processes require multistep reactions that result in economic and technical drawbacks, e.g., the separation of the desired products from complex mixtures.

A direct process leading to olefins from syngas is the Fischer-Tropsch-to-olefins process (FTO) that has been subject of a number of patents and publications over the last decades and whose interest is surprisingly increasing in the last few years [40]. Although the reaction mechanism has been matter of investigation for several years, providing a vast amount of information concerning the fundamentals of the reaction, the industrial process, and the FT catalysts, a detailed description of the reaction mechanism is not yet available. Several comprehensive reviews cover this topic and the readers are referred to these papers for more detailed information [40]. The primary aim of FTO is to maximize the selectivity in lower olefins and to reduce methane and CO_2 production. The most efficient metal catalysts are based on iron compounds, which show lower conversion in methane, under the optimal conditions of high-temperature values necessary for obtaining high selectivity in light olefins. Noteworthy, the FTO process, compared to MTO or DMTO, displays lower selectivity in ethylene and propylene. Despite the efforts, there is no commercial application of this route because of the too low selectivity in C2-C4 olefins [35].

Lignin is the second most abundant renewable resource in nature, being the main constituent, along with cellulose and hemicellulose, of lignocellulosic biomass (15–30% by weight, 40% by energy). Lignin consists of an amorphous cross-linked polymer of molecular mass higher than 10,000 Da, in which phenyl propanoid monomer units (monolignols) are the repeated structural motif. This biopolymer has not been yet well characterized and its composition and structure can vary from species to species. Despite its diffusion and commercial availability, the existing market of lignin products remains limited and still unexplored. In fact, the structural complexity and heterogeneity and the presence of a variety of oxygenate functionalities make the selective synthesis of chemicals from lignin a challenging process [41]. Thus, lignin found main applications as dispersing and binding agents, nanofiller in rubber [42] and epoxy resins [43]. Despite the efforts for lignin valorization in the synthesis of chemicals, thermal breakdown is still the most common approach in degrading lignin, using pyrolysis and gasification routes, which are of interest for the synthesis of olefins and aromatic monomers [44].

Gasification of lignin gives syngas with different amounts of CO_2 and CH_4, depending on gasification temperature and pressure, the presence of steam or oxygen, and the heating rate. At variance of the fractionation procedure employed from biomass, sulfur by-product, e.g., H_2S, can be formed during lignin gasification. Theoretically, gasification with steam of 1 kg of isolated lignin approximately yields 62 mol of H_2 and 53 mol of CO, corresponding to a H_2/CO mole ratio of 1.2. This is an endothermic reaction and part of the lignin is thus combusted for producing the heat necessary for the process [45, 46]. Gasification of lignin is an alternative entry to the olefin synthesis via the syngas approach described above; to properly feed the plant for the MTO process, the $H_2/CO/CO_2$ mole ratio must be adjusted.

Catalytic fast pyrolysis of lignin consists of an anaerobic thermal treatment, which affords bio-oil, a mixture of organic compounds with an increased energy density that is still not sufficient for a convenient use as fuel or for the direct production of value-added chemicals. The hydrodeoxygenation (HDO) of bio-oil, comprising decarboxylation, hydrogenation, hydrogenolysis, hydrocracking, and dehydration reactions, requires the use of H_2 at high partial pressures (100–200 bar) and high temperatures (570–670 K) to produce water and hydrocarbons. This route is undeniably competitive to the corresponding pathway from fossil feedstock (naphtha and light diesel) [47, 48].

Biomass pyrolysis in the presence of zeolite catalysts is a promising technology for the direct conversion of solid biomass into olefins and gasoline range aromatics. Under specific conditions, good selectivity in ethylene or aromatic compounds such as benzene and toluene can be obtained, where the latter can give access to vinyl-aromatic monomers via conventional chemical transformations. There are significant advantages: (i) all the desired chemistry occurs in a single reactor using inexpensive silica-alumina catalyst; (ii) a range of different lignocellulosic feedstock can be processed; (iii) biomass pre-treatment is simple (drying and grinding) and the process occurs in fluidized bed reactors, which are currently commercially used in petroleum refinery. Furthermore, catalytic fast pyrolysis yields olefins and aromatic products that already fit into existing infrastructures.

Chemical depolymerization of lignin to aromatic hydrocarbons, such as phenolic derivatives and alkyl benzenes, is scarcely selective and typically gives a complex mixture of aromatic compounds from which the desired compound is separated with difficulty and uses high dispersion of energy. From alkylbenzenes to styrene, the synthetic approach follows conventional routes developed in the petrochemical industry [23]. Other hydrocarbons derived by pyrolysis and HDO of the lignin can afford conventional light olefins as ethylene, propylene, and butylene upon reforming [48].

7.3 New polyolefins from bioderived monomers

It is highly desirable, and actually expected, that bio-sourced polyolefins could cover in short time a significant quote of the plastics market. The forecast of a total capacity of 12 Mt by 2020 – a tripling of 2011 levels – suggests that bio-based polymers are definitely polymers for the future.

In different countries, the rules are becoming more and more binding, specifying that the polymers currently employed for the production of several commodities should be from renewable resources. European Union Lead Markets Initiative (LMI) supports actions to lower barriers to bringing new bio-based products to market. In contrast to biofuels, there is no European policy framework to support bio-based polymers, whereas biofuels receive strong and ongoing support during commercial production.

Japanese government's goal is that 20% of all plastics consumed in the country will be renewably sourced by 2020 [49]. In the US legislation introduced in March 2012 to foster a bio-based economy, the manufacturer has to declare the percentage of the combined weight of bio-based plastic materials, calculated as a percentage of total plastic (by weight) in each product.

The most dynamic development is foreseen for drop-in biopolymers, which are chemically identical to their petrochemical counterparts but partially derived from biomass.

Many efforts have been done to obtain conventional monomers as ethylene, butadiene, and isoprene from renewable sources in order to render the conventional thermoplastics such as LLDPE, HDPE, and PP available, as well as EPR and SBR, using bio-sourced monomers. This would help in the quest for new bio-sourced polymers.

Bioderived plastics are currently priced above their petroleum-based counterparts, and thus value-added applications have to be designed to make them competitive versus fossil fuel based polymeric materials.

This section deals with the synthesis of polyolefins from vinyl monomers derived by simple chemical transformations of bio-sourced platform molecules. For the sake of simplicity, polymers by bio-sourced monomer that underwent post-functionalization with, e.g., acrylic, maleimide, or other more reactive olefinic groups, will not be considered.

Furan-2-carbaldehyde (furfural) is a cheap biomass-derived platform molecule resulting from dehydration of the not edible fraction of vegetables (Figure 7.5) [50, 51]. Polysaccharides can be converted into furan derivatives by an acid-catalyzed dehydration [52]. 2-Vinylfuran (2VF) has been synthesized using a number of synthetic routes involving methylenation reaction of furfural.

Peterson olefination of furfural with (trimethylsilyl)methylmagnesium chloride in diethyl ether resulted to one of the most cost-effective routes that can be scaled up to mole scale without significant experimental concerns (Figure 7.6) [53].

The synthesis of random linear copolymer of styrene with 2VF (S-*co*-2VFs) was successfully accomplished in a wide range of composition via atom transfer radical

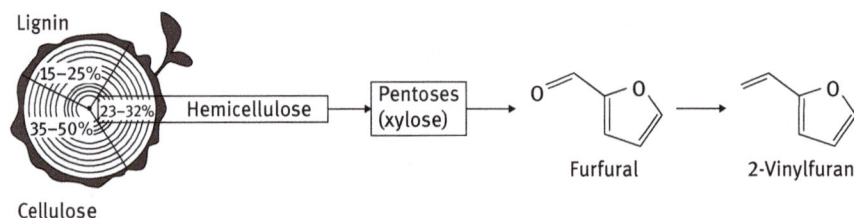

Fig. 7.5: Synthetic route for bio-derived 2VF.

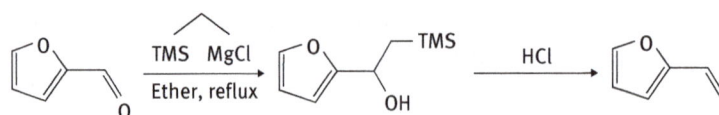

Fig. 7.6: Synthesis of 2VF using Peterson olefination reaction.

(ATR) Copolymerization using the CuBr/PMDETA catalyst and 1-PEBr as radical initiator [53]. Free radical polymerization (FRP) is a polymerization technique that is tolerant of polar functionalities and polar solvents and thus results are particularly useful in polymerization processes of bio-sourced monomer containing donor atoms. Conventional RP is employed to produce annually ca. 100 Mt of polymers, with thousands of different compositions [54]. However, the architectural control in these polymers is very limited. Free radical polymerization is initiated by addition of a radical species, the initiator, to the carbon-carbon double bond of an olefin, leading to the formation of a new radical species that is called the propagating species (Figure 7.7a). The initiator is typically a chemical compound containing a weak σ-bond, which undergoes homolytic cleavage by thermal heating or photo irradiation.

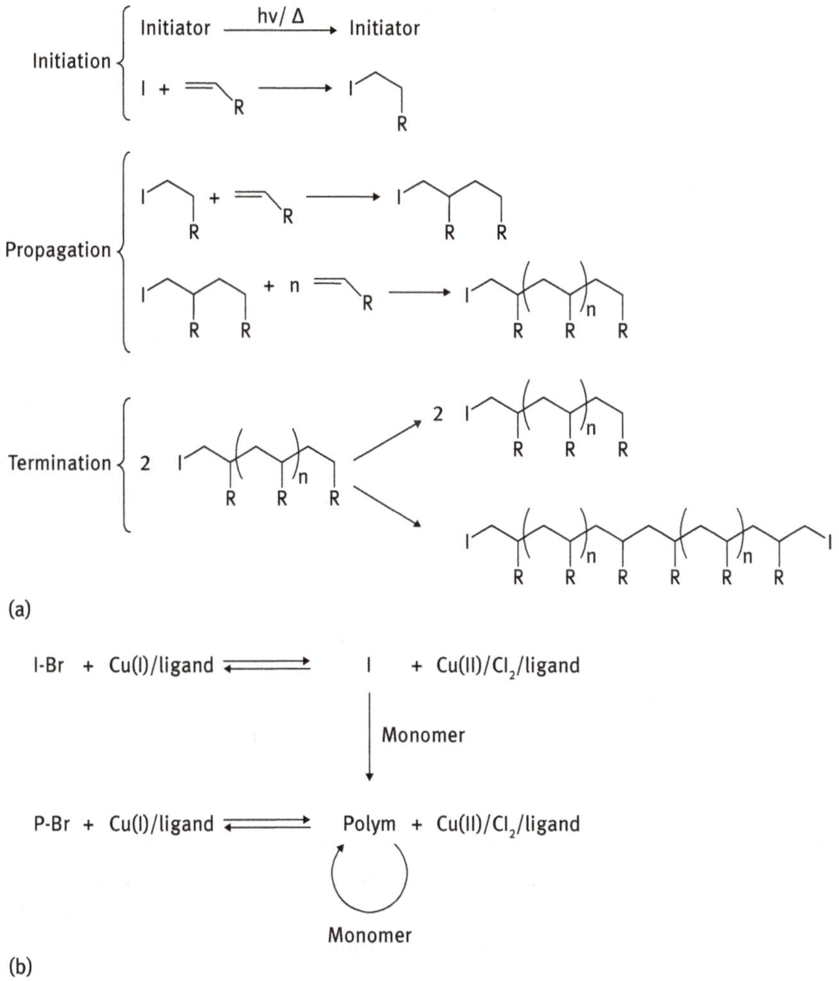

(a)

(b)

Fig. 7.7: (a) Free radical polymerization (FRP) and (b) atom transfer radical polymerization (ATRP).

Some examples are halogens, azo compounds (R–N=N–R') like azobisisobutyro-nitrile (AIBN), organic peroxides such as di-*tert*-butyl peroxide or benzoyl peroxide [55]. The propagating species reacts with an olefin, of the same or different kind, to yield the same or a new propagating radical species: this is called the propaga-tion step in which the polymer chain grows. The chain growth is terminated at some point by irreversible decomposition of the reactive radical center by an appropriate termination reaction (β-elimination, leading to a terminal olefinic unit and a new radical species, or bimolecular radical coupling). For a long time, control of molecu-lar architecture in RP was considered impossible since two radicals always terminate at a very fast, diffusion-controlled rate. The advent of ATRP revolutionized this field and gave access to polymers with precisely controlled molecular weight, relatively low disparities (M_w/M_n<1.1), and controlled molecular architecture in terms of chain topology (stars, cycles, combs, brushes, regular networks), composition (block, graft, alternate, gradient copolymers), and diverse functionality [54]. In ATRP, the radical initiator (R·) is produced through the reaction of an organic halide (RX) with a transi-tion metal halide (MLX_n) (Figure 7.7b), allowing the formation of a metal halide in higher oxidation state (MLX_{n+1}) and of the radical initiator (R·) via the so-called Khar-asch reaction [56]. At this stage, the propagation reaction follows the conventional pathway. However, the propagating radical is involved during propagation in an equi-librium with a dormant state, predominately in the form of initiating alkyl halides MLX_n and macromolecular species PX. The dormant species periodically react with the transition metal complexes in their lower oxidation state, MLX_n acting as acti-vators to intermittently form growing radicals (P·) and deactivators transition metal complexes in their higher oxidation state, coordinated with halide ligands MLX_{n+1}. A successful ATRP requires fast and quantitative initiation so that all propagating species begin the chain growth at the same time and fast and reversible deactivation of the propagating radicals. The latter equilibrium between the propagating species and the dormant state keeps the radical concentration low and minimizes the termi-nation reactions. As a result, a successful controlled mechanism is obtained, which ensures a narrow molecular weight distribution, tightly controls the average molecu-lar weight, and makes possible selective chain ending functionalization.

In the case of S-*co*-2VF, the ATR-copolymerization process is well controlled in neat monomers by the CuBr/PMDETA/1PEBr catalyst, leading to a narrow polydisper-sity index (M_w/M_n; PDI = 1.2–1.5) of the polymer products, linear increase of the M_n with both monomer conversion and polymerization time, particularly when polymer-ization time is longer than 5 h.

The chain endings of the S-*co*-2VFs exhibit bromine atoms in benzylic position as expected from an ATRP mechanism [53]. The distribution of the functional monomer 2VF in the copolymer chain is random as assessed by the evaluation of the reactiv-ity ratios determined by Fineman-Ross, Kelen-Tudos, extended Kelen-Tudos, and Meyer-Lowry methods and found r_s = 1.21–1.30 and r_{VF} = 0.98–1.10, corresponding to a reactivity ratio product value of ≈1.3–1.6. The S-*co*-2VFs are atactic in ^{1}H NMR and ^{13}C NMR analysis and amorphous at differential scanning calorimetry (DSC). The T_g values range from 105°C of PS to 54°C for S-*co*-2VF$_{60}$ (sample containing a mole fraction of 2VF, $x_{2VF} = 0.60$), where the latter T_g is very close to that of pure poly(2VF). The copolymer samples are thermally stable above T_g. Decomposition occurs in air in two steps: the first one observed at about 360°C and the second one at about 400°C. The weight loss is proportional to the 2VF concentration. The S-*co*-2VFs are thermally stable both in solid state and solution, without showing oxidation or covalent cross-linkings. Addition of the antioxidant Wingstay®K 1 wt%, significantly improves the thermal stability of the S-*co*-2VFs (Figure 7.8), which can be processed in air at temperature higher than T_g and lower than those typically used for PS. These copolymers were proposed as new entry to functional styrenic polymers since the furan moieties exhibit a rich chemistry. Actually, furans undergo to ring opening in acid conditions, electrophilic attack in positions 2 and 5, and aryl coupling by Suzuki reaction catalyzed by palladium complexes and thermoreversible Diels-Alder (DA) reaction. Each reaction can be used for the synthesis of functional olefin copolymers comprising this monomer.

Preliminary results showed that the treatment of a 10% w/w solution of S-*co*-2VF$_{44}$ in DMSO/tetrachloroethane (1:1) with 0.5 eq of bismaleimide (BMI) readily yields a polymer gel in few hours at room temperature (Figure 7.9) [53].

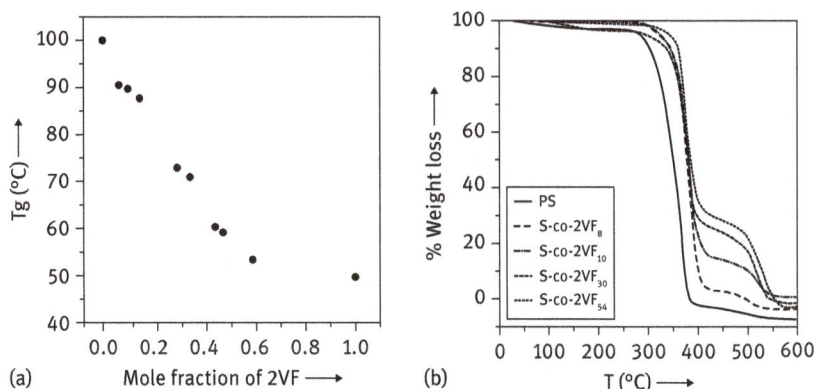

Fig. 7.8: (a) Glass transition temperature (T_g) versus molar fraction of 2VF in the S-*co*-2VF copolymer. (b) Thermogravimetric analysis of S-*co*-2VF copolymers under air.

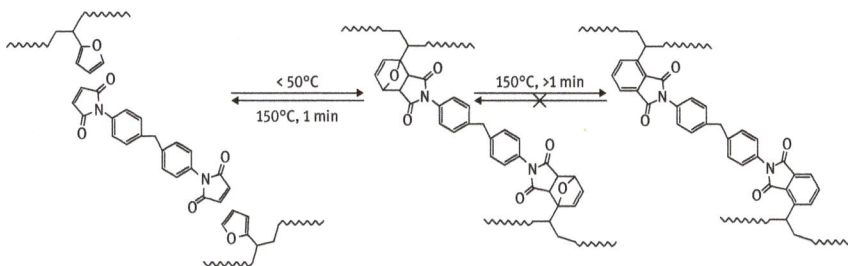

Fig. 7.9: DA equilibrium between furan moieties of S-*co*-2VF copolymer and BMI followed by aromatization of a furan DA adduct.

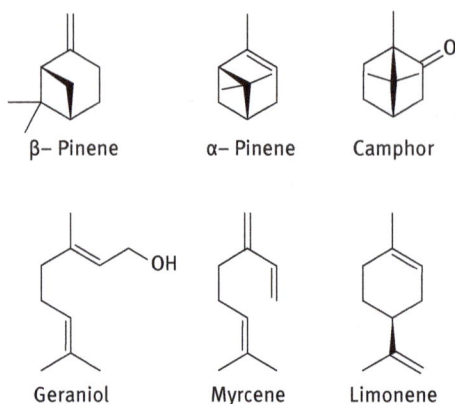

β– Pinene   α– Pinene   Camphor

Geraniol   Myrcene   Limonene

Fig. 7.10: Representative examples of commercial natural monoterpenes.

The pristine polymer solution was generated via retro DA reaction after treatment at 150°C for 1 min followed by quenching in ice water.

A variety of natural compounds comprising olefinic carbon-carbon double bonds are in principle reactive with Lewis acids, metal compounds, and radical initiators and could be polymerized using conventional carbo-cationic, metal-catalyzed addition polymerization or radical techniques leading to novel bio-sourced polymers and copolymers of general interest [57–60]. Terpenes, monoterpenes in particular, exhibit a chemistry similar to that of olefins and diolefins from oil and syngas. Naturally occurring terpenes are currently produced in megatons per year and are employed for the preparation of perfumes and fragrances, as diluting agents for dyes and varnishes, or to yield natural and synthetic resins [61]. Although the used amounts cannot certainly be compared with that of carbohydrates and fatty acids, terpenes can be considered among the promising renewable olefins for the future (Figure 7.10).

A representative example is myrcene, which is commercially available and relatively cheap [62]. Myrcene is an isoprene dimer resulting from the essential oil of several plants such as bay, cannabis, ylang-ylang, wild thyme, parsley, and hops; it is produced semi-synthetically from myrcia, the plant that gives the name to the compound, or by pyrolysis of β-pinene.

Despite the structural similarity with isoprene, there are few studies concerning the polymerization of this natural compound and the properties of the resulting polymers are thus unknown [63, 64]. The FRP of myrcene and copolymerization with styrene and methyl methacrylate (MMA) have been reported [65]. In FR copolymerization of myrcene with styrene, a reactivity ratio product of 1.20 was found (with $r_M = 1.36$ and $r_S = 0.27$, with M = myrcene and S = styrene) when AIBN was used as radical initiator at 65°C. Under the same conditions, copolymerization of myrcene with MMA led to reactivity ratio values $r_M = 0.44$ and $r_{MMA} = 0.27$, suggesting a preference for the cross-propagation reaction. The M_w/M_n values are of about 2 for all of the copolymers, regardless of monomer composition. Unfortunately, chemical physical properties of the resulting copolymers were not described.

Photolysis of myrcene in the presence of benzophenone as photosensitizer yields 5,5-dimethyl-1-vinylbicyclo-[2.1.1]hexane with a quantum yield of 0.05, a new bio-sourced olefin (Figure 7.11) [66, 67]. On the same line, co-dimerization of myrcene with other olefins and dimerization of myrcene provided a wide variety of linear unsaturated olefins of potential use in polymer synthesis [49]. However, no reports are available to date in scientific literature on this issue.

Ring closing metathesis (RCM) of myrcene yields isobutene and 3-methylene cyclo-pentene; while the former is a commercial monomer of interest for the production of the butyl rubber, the latter was considered attractive for the synthesis of new polyolefins [68, 69]. Hillmyer and Hoye et al. recently reported controlled cationic polymerization of 3-methylene cyclopentene using i-BuOCH(Cl)Me/Lewis acid/ Et$_2$O catalyst in toluene where suitable Lewis acids were SnCl$_4$ or ZnCl$_2$ (Figure 7.12a) [68]. The polymerization process catalyzed by the zinc complex is fast in the −4°C to 11°C temperature range and

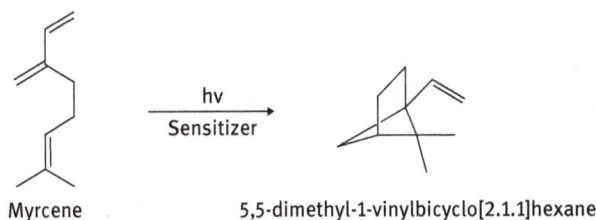

Myrcene   5,5-dimethyl-1-vinylbicyclo[2.1.1]hexane

Fig. 7.11: Cyclization of myrcene using a benzophenone-sensitized reaction.

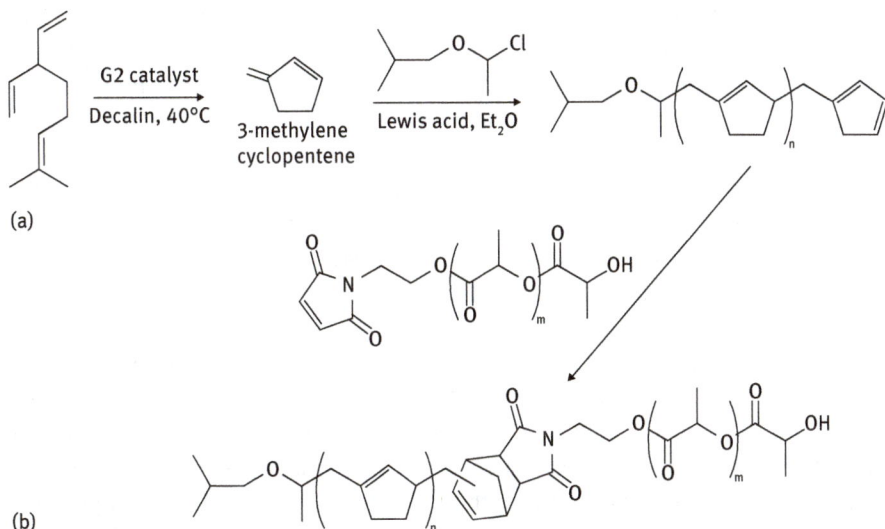

(a)

(b)

Fig. 7.12: (a) Cyclization of myrcene catalyzed by Grubbs second-generation initiator (G2) and its living cationic polymerization. (b) DA coupling of poly(3-methylene cyclopentene) with a maleimide end-functionalized polylactide.

gave polymer products with narrow PDI resulting from 1, 4-monomer addition. The M_n values increased with monomer conversion and were in agreement with the ones predicted, assuming a controlled mechanism [68]. Anionic polymerization of 3-methylene cyclopentene catalyzed by sec-BuLi/TMEDA provided high regioselective 3,4-monomer insertion. The T_g values of the 1,4-polymers are in the −17°C to 11°C temperature range, whereas that of the 4,3-polymer is 73°C.

The chain ending are of cyclopentadienyl type, providing functional groups suitable for DA reaction. Actually di-block copolymers with PLA were synthesized by click coupling through DA reaction of this end group with maleimide functionalized PLA (Figure 7.12b) [70, 71]. Recently, alternating copolymers of 3-methylene cyclopentene and N-substituted maleimides were obtained through controlled radical polymerization [72].

(+)-D-Limonene is an abundant chiral monoterpene with annual production estimated to be hundreds of tons per year; it is mainly used in the flavor and fragrance industry [58–61, 73, 74]. Cationic polymerization of limonene had been extensively investigated since the 1950s, with scarce success. On the contrary, both FRP- and the reversible-addition fragmentation transfer (RAFT)- [75] controlled copolymerization of limonene with maleimide were more successful [76]. Irrespective of the monomer feed ratios (1:1, 1:2, and 1:3), the AIBN-initiated FRP of limonene with N-phenylmaleimide or N-cyclohexylmaleimide in $PhC(CF_3)_2OH$ produced at 60°C a copolymer with AAB monomer sequence (A = maleimide = M_1; B = limonene = M_2). The copolymerization rate is rather low, yielding high monomer conversion (>80 mol%) in 120 h. The PDI of the copolymer products is about 2.0. The reactivity ratios were estimated using the penultimate model and assuming r_{11} and r_{21} are equal to zero because of the absence of M_1M_1 dyads in the copolymer chain. The calculated r_{12} and r_{22} values by the Kelen-Tudos method were 18.7×10^{-3} and 4.20×10^{-3}, respectively, which indicates that the M_1M_2 radical preferentially reacts with M_2, whereas the M_2M_2 chain ending exclusively reacts with M_1. The role of the fluorinated alcohol in determining the peculiar copolymer architecture was also discussed. The T_g values are in the range 226–243°C, as a result of the high maleimide incorporation and rigid alicyclic structure of the monomer. When chiral (+)-D-limonene was used, optical active polymers were obtained as assessed by circular dichroism measurements [77].

RAFT copolymerization of limonene with N-phenylmaleimide in $PhC(CF_3)_2OH$ in the presence of n-butyl cumyl trithiocarbonate or n-butyl 2-cyano 2-propyltrithiocarbonate using the AIBN initiator at 60°C led to copolymers with the same AAB macromolecular architecture previously observed (Figure 7.13) and slightly lower polydispersity index (1.5–1.8) [78]. Copolymers obtained by n-butyl cumyl trithiocarbonate exhibited at MALDI TOF a perfect main chain sequence and highly controlled chain ending consisting of trithiocarbonyl group. The authors claimed that this highly regular copolymer structure is promising in mimicking natural polymers, forming highly ordered structure that can possibly express biofunctions.

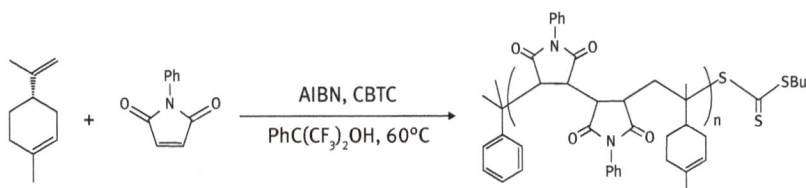

Fig. 7.13: AAB sequence radical copolymerization of D-limonene and phenylmaleimide.

The third-generation biomass, not competing with human feeding, is based on triglycerides or fats among which unsaturated fatty acids are becoming of increasing interest as testified by the large increase in their price in the market [73, 79–90]. Triglycerides are the major components in natural oils derived from plant and animals. Vegetable oils have been used for centuries in paints and coating [91], since when exposed to oxygen in air [92], they can polymerize or oligomerize via a free radical mechanism called lipid peroxidation. The primary products of the autoxidation reaction are peroxo or hydroperoxo species, which are frequently unstable and can decompose to aldehydes, ketones, and other reactive compounds.

Triglycerides are triesters of glycerol with a variety of fatty acids (Figure 7.14), typically differing in the number of carbon atoms, ranging from 14 to 22, and number of insaturations, namely carbon-carbon olefinic double bonds, ranging from 0 (saturated fatty acids of scarce value for human feeding due to their hepatotoxicity) to 3, e.g., in linoleic and eleostearic acids. Some fatty acids also include additional functionalities as epoxide in vernonia oil, hydroxyl in castor and lesquerella oil, and ketone in licania oil. The presence of these functional groups enlarges the array of polymerization techniques and the possibility of post-functionalization of the monomers or of the resulting polymers [90]. However, polymerization of post-functionalized monomers and the properties of the resulting polymer are not the subject of this review. Interest will be mainly focused on unmodified unsaturated acids that can undergo cationic and radical polymerization of the olefinic carbon-carbon double bond. Cationic polymerization is actually another polymerization technique that is compatible with polar functionalities and solvents and allows efficient polymerization of these monomers giving a wide variety of polyolefins carrying pendant polar functionalities (Figure 7.15). Thus, a variety of polymers with properties ranging from soft and rubbery to elastic have been obtained.

Triglyceride polymers from soybean, sunflower, cotton, and linseed oils are an important class of bioderived materials from renewable resources. Lewis acids such as $AlCl_3$, $TiCl_4$, $SnCl_4$, $ZnCl_2$, and BF_3-OEt_2 have been used in cationic polymerization of unsaturated fatty acids under mild conditions [83, 84], among which the boron halide is the most efficient [93]. Compared to ethylene and propylene, the carbon-carbon double bond of vegetable oils is slightly more nucleophilic, and the cationic

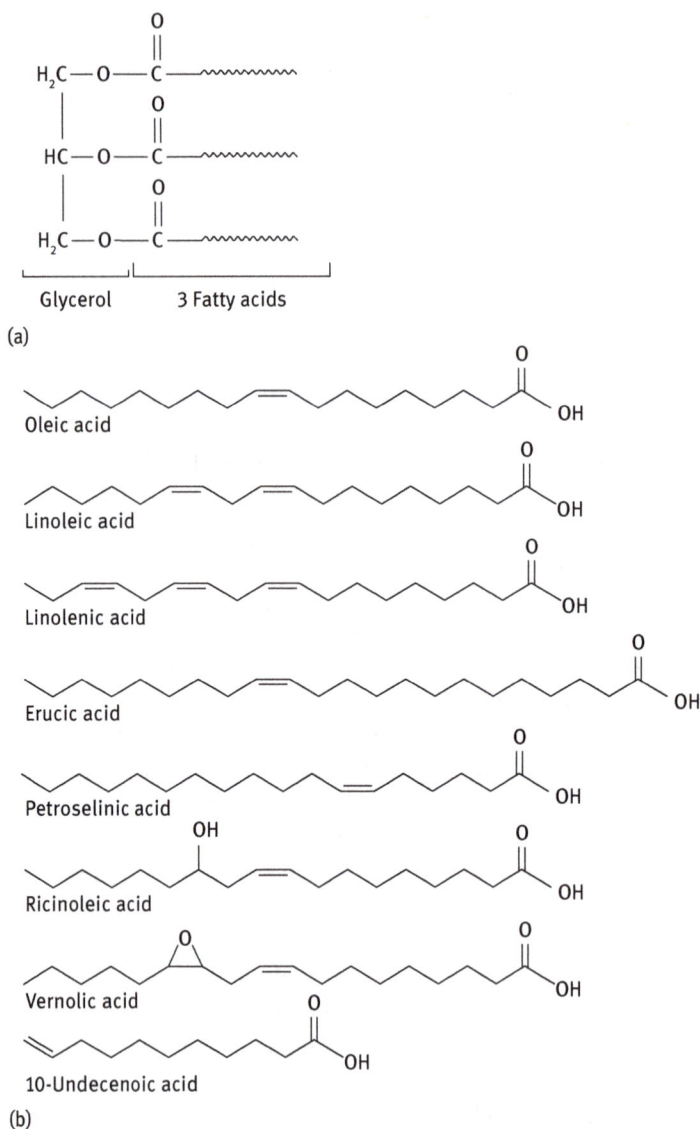

Fig. 7.14: (a) Triglyceride structure. (b) Most common unsaturated fatty acids used in polymerization.

intermediate species can be eventually stabilized by the presence of additional insa-turations along the main chain [83].

Tung oil, consisting of about 84 wt% α-eleostearic acid, a fatty acid possess-ing a triene functionality, is efficiently polymerized in the presence of BF_3-OEt_2 (Figure 7.16a). Each unsaturation of fatty acid units in the triglyceride is suitable for the polymerization process, leading to an extended cross-linked three-dimen-sional polymer network [83]. Cationic polymerization of various neat soybeans

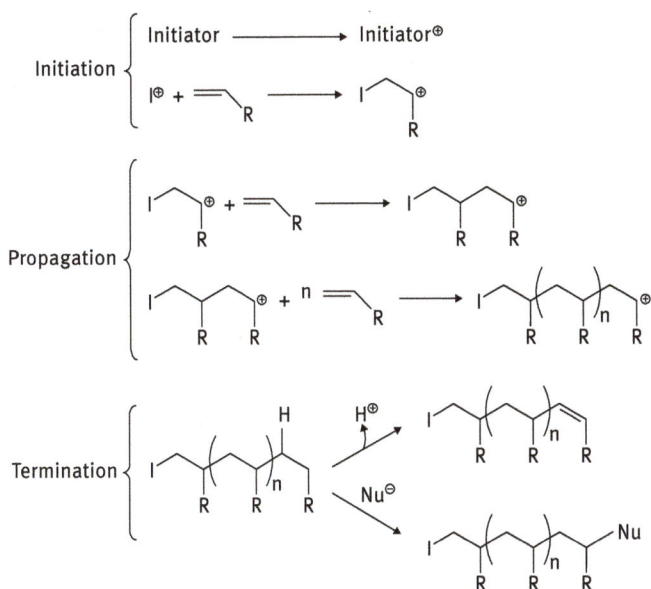

Fig. 7.15: Reaction scheme for cationic polymerization of olefins.

afforded a variety of products ranging from viscous oils to soft rubbery materials of limited utility [95]. On the contrary, cationic copolymerization of soybean oils with styrene, divinylbenzene, norbornadiene, dicyclopentadiene yielded thermosets of increased utility (Figure 7.16b) [96]. Copolymerization of soybean oil (50–60 wt%) with divinylbenzene initiated by BF_3-OEt_2 resulted in densely cross-linked polymer with a modulus ranging from 4×10^8 to 1×10^9 Pa at room temperature. Cationic polymerization of various neat soybeans afforded a variety of products ranging from viscous oils to soft rubbery materials of limited utility [95]. On the contrary, cationic copolymerization of soybean oils with styrene, divinylbenzene, norbornadiene, dicyclopentadiene yielded thermosets of increased utility [96]. Copolymerization of soybean oil (50–60 wt%) with divinylbenzene initiated by BF_3-OEt_2 resulted in densely cross-linked polymer with a modulus ranging from 4×10^8 to 1×10^9 Pa at room temperature. The mechanical properties of the polymers were significantly increased, as well as the amount of covalent cross-links significantly reduced, when the soybean oil is terpolymerized with divinylbenzene and styrene. Polymeric materials with moduli ranging from 6×10^6 to 2×10^9 Pa at room temperature and T_g from 0°C to 105°C were obtained; these terpolymers resulted thermally stable up to 200°C in air. A range of thermosets have been prepared by cationic polymerization of olive, peanut, sesame, canola, corn, soybean, grape seed, sunflower, low-saturation soy, safflower, walnut, and linseed oils with styrene using divinylbenzene as cross-linking agent under the same experimental conditions [86].

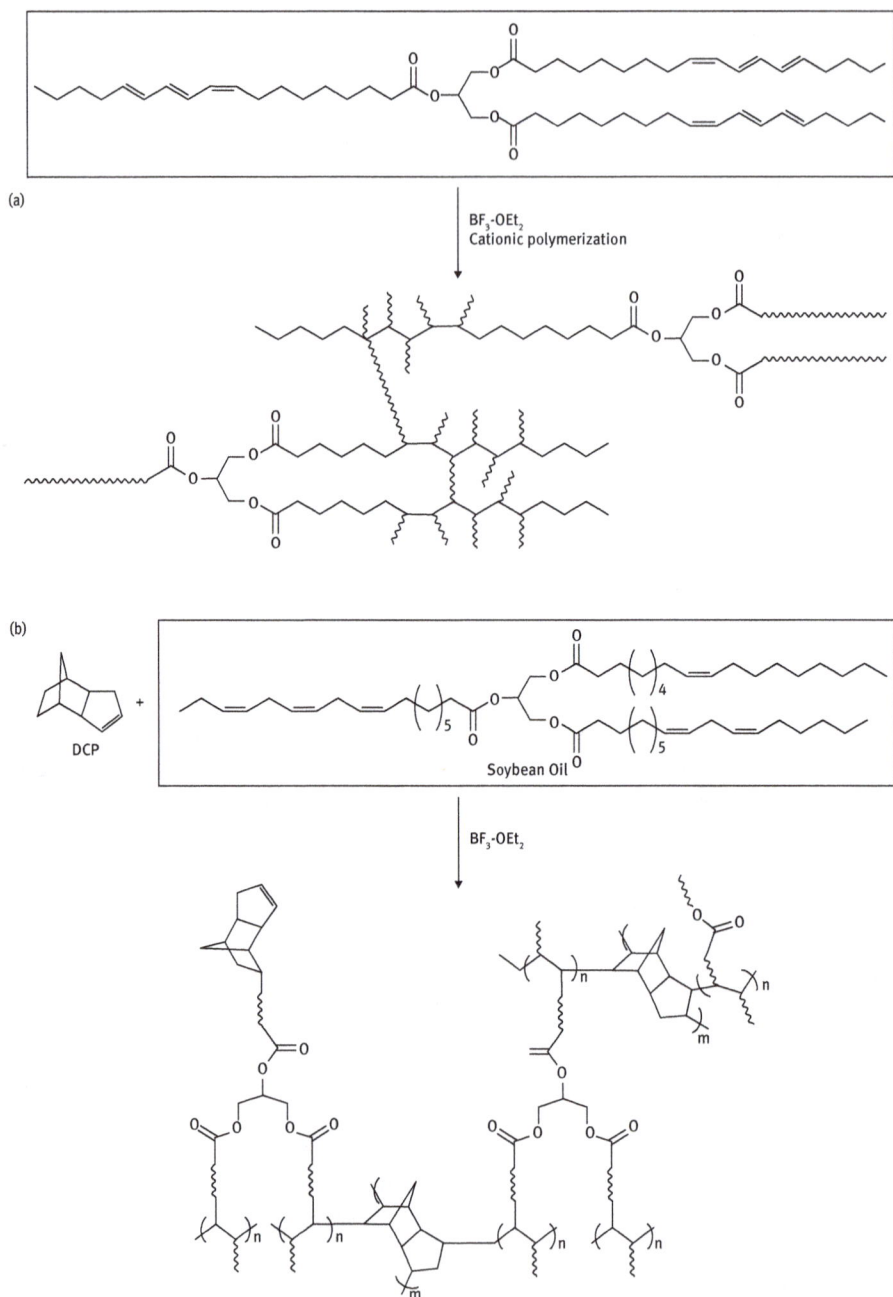

Fig. 7.16: (a) Tridimensional polymer network from cationic polymerization of tung oil. (b) Cationic copolymerization of soybean oil with DCP.

Replacement of dicyclopentadiene (0.29 USD/lb) for divinylbenzene (3 USD/lb) as cross-linking agent of soybean oil in cationic copolymerization initiated by BF_3-OEt_2 and modified with Norway Fish Oil produced copolymers with enhanced properties [97].

Fatty acids containing isolated double bonds are not readily reactive in radical polymerization as well as cationic polymerization; thus, they are conventionally modified to append polymerizable functionalities, such as acrylates [98] or norbornadienes [99], which are more reactive. Other chemical modifications include the isomerization of double bonds or the increase in the number of insaturations, leading to conjugated "diene-like" functionalities, which are more reactive in FRP [100].

Following this approach, linseed and soybean oils were first reacted with a rhodium catalyst to make fatty acids containing conjugated carbon-carbon double bonds; these chemically modified fatty acids were thus polymerized and copolymerized with styrene, acrylonitrile, dicyclopentadiene, and divinylbenzene under FRP conditions using the AIBN initiator. The copolymer include high concentration of oil and exhibit a wide range of thermal and mechanical properties [101–103]. Polymers resulting from FRP of soybean oil were melt blended with isotactic PLLA to increase toughness of the latter polymer. The PLLA/polySOY blends have a tensile toughness four times greater with the corresponding strain at break value six times bigger than PLLA [104].

The tung oil was terpolymerized with styrene and divinylbenzene using FRP techniques to yield rubbery materials to rigid plastics. Glass transition temperatures in the range of $-2°C$ to $116°C$, cross-link densities of 1.0×10^3 to 2.5×10^4 mol/m^3, coefficients of linear thermal expansion of 2.3×10^{-4} to $4.4 \times 10^{-4}/°C$, compressive strengths of 8 to 114 MPa, and compressive moduli of 0.02 to 1.12 GPa were reported. The addition

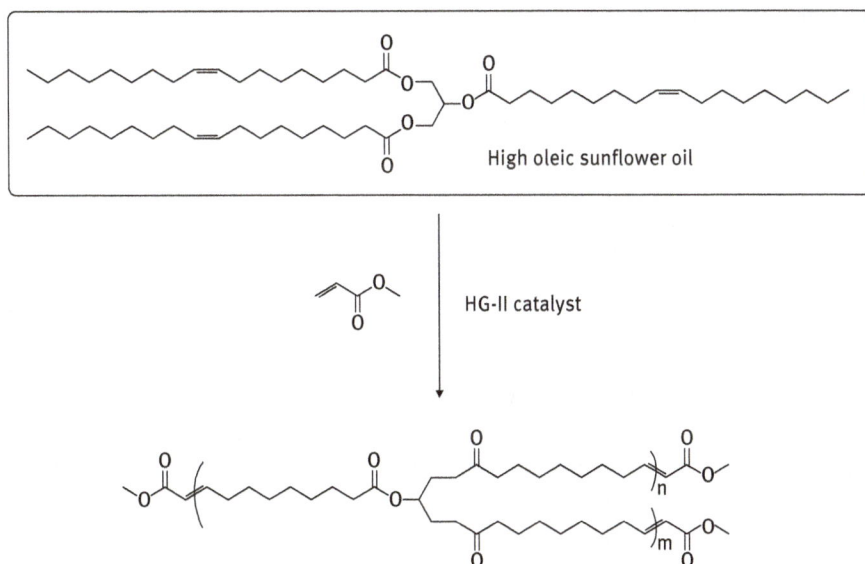

Fig. 7.17: ATMET polymerization of high oleic sunflower oil with methylacrylate as chain stopper.

of metallic salts of Co, Ca, and Zr as catalysts effectively accelerates the thermal copolymerization and increases the cross-link densities as well as the mechanical properties of the resulting copolymers [105].

The olefin metathesis reaction, catalyzed by W, Mo, and Re oxides supported main group metal oxides, developed by Schrock and Grubbs with Mo and Ru complexes, respectively, is tolerant of functional group as well as of oxygen and moisture and thus suitable in fatty acids polymerization via acyclic diene metathesis (ADMET) and acyclic triene metathesis (ATMET) approaches under sustainable conditions (Figure 7.17) [106]. Olefin metathesis of vegetal oils was investigated since 1972 [107]; however, post-functionalization of the fatty acids with strained olefins such as norbornene was conventionally applied to increase the reactivity of these monomers [108–110] Alternatively fatty acids are esterified with α,ω-alcohols and the olefinic group isomerized to yield α,ω diolefins. This kind of α,ω-unsaturated monomers are particularly reactive in ADMET polymerization using the first-generation Grubbs catalyst [84–89].

7.4 Outlook and perspective

Plastics are an important component of the range of materials used in the modern society; almost all aspects of daily life benefit of plastics or rubber in some form or the other. Traditional materials such as paper, metals, wood, ceramics, and glass have been systematically replaced over the course of time for producing clothing and footwear, packaging of food, building materials (windows and door frames, fixtures, and insulating materials), and high-performance equipment for sport. Plastics are also becoming smart, and special applications have been designed or expected in medicine (human tissues and organ transplants), high-performance and functional materials in photovoltaic technology, reusable graphic media, and flexible monitors. Moreover, plastics have contributed to a significant reduction of the CO_2 emission in public transportation. Actually, public and private transportation vehicles can contain from 20% up to 50% of plastic items, and this increases their lightness and reduce the amount of fuel consumed. When all these features are considered, it is difficult to imagine a future world without plastics.

Polyolefins are commodities that currently represent over the 60% of plastic market. The replacement of conventional polyolefins, namely PE, PP, PBd, and PS, with novel biopolymers or bio-based polymers is an expected trend, but this result cannot be successfully obtained in a short time. The approach of the manufacturer and, generally speaking, of industry toward novel plastics is typically conservative. The variety of the mechanical, physical, and chemical properties of new products, their cost, the search for new supplier, and, last but not least, the management of existing plants for the production and transformation of new polymers discourage their introduction and use [17]. Noteworthy is the example of poly(lactic acid) as thermoplastics substitute [20]. It is unthinkable that the polyolefins will be replaced in the short term by biopolymers or bioderived polymers. The only way for speeding up this process and to increase the demand for bioplastics is related to the decision of governments, at national and over national levels, e.g., European Community, to promote the adoption of legislation encompassing economic incentives for the use of bioplastics and at the same time enforce restriction to the trade of environmental unfriendly products. An example in Europe is the replacement of PE-based shoppers in favor of the cellulosic counterpart, based on, e.g., Mater-Bi. At the same time, real great challenges must be faced to penetrate the polymer market. Biopolymers should not only be biodegradable but also favorable to the environment, releasing only CO_2 and water after degradation or to fertilize the soil by composting, which means incorporating only renewable raw materials. In addition, the energetic demand for the monomer synthesis and polymer processing must be lower than for ordinary plastics.

Another important drawback for the future of the biopolymers is the competition with the food industry; the recent price increases of US corn triggered by the strong bio-fuel demand is a significant example. This perspective picture could radically change as soon as interesting breakthrough will be performed using new technologies based on genetically modified organisms, new processes of cellulose hydrolysis, or new catalytic processes for the production of bio-intermediates for polyesters and polyolefins synthesis in large scale.

For several countries that do not have enough oil to supply their olefin and petrochemical commodities demand but have wide areas with high solar insolation levels, farm lands and abundant water resources, the bio-derived approach is a big opportunity for designing a sustainable growth of industry. The well-known competition "oil for food" favors lignocellulosics or oleaginous fractions from non-edible biomass for the direct manufacturing of bioplastics or its bio-intermediates. Considering the state of the art in the bioplastic industry, most of the bioproducts are based on starch or PHAs, which cover a niche market of about 2% of global plastic market. However, the demand for bio-polyolefins, as well as that of bio-polyesters, bio-polyamides, and polyurethanes, is rapidly increasing and this could be a useful solution in the short and middle term. Actually, this approach allows the synthesis of the monomers from renewable resources and requires the use of conventional technologies for polymerization processes and polymer processing, leading to polymeric materials with well-established properties and applications.

The drop-in synthesis of olefins can follow different pathways. Pyrolysis of lignocellulosics or oleaginous fractions, or even of the entire biomass, affords the syngas from which olefins can be obtained using consolidated petrochemical processes. Hydrodeoxygenation processes, for the reduction of oxygen content and the contemporary energetic upgrading of biomass, have been developed for the direct drop-in synthesis of olefins. The process to be used depends not only on the operational costs but also on the kind biomass available.

Alternatively, fermentative or enzymatic processes for the conversion of the saccharide-based fractions (typically starch from corn and sugarcane) to bio-ethanol or the direct production of the desired olefins are currently an industrial reality. Bio-ethanol is a successful entry to bio-ethylene, bio-propylene, and bio-butadiene. However, it is worthy of note that, considering ethylene only, it is necessary to use more than three times of all bio-ethanol currently produced worldwide to face the current global demand of this monomer.

Finally, the production of specialty polyolefins can be achieved from a less abundant fraction of the biomass, as terpenes and fatty acids.

In a transition era such as what we are experiencing, the short-term approach should also aim to define a protocol for thermoplastic recycling through mechanical recovery and composting of wastes or a correct approach to energy production by polyolefin incineration.

Bibliography

[1] The Biology of Gosypium hirsutum, L., and Gossypium bardadense L (cotton), Department of Health and Aging, Office of the Gene Technology Regulator, Australian Government, 2008.

[2] The Deadly Chemicals in Cotton, Environmental Justice Foundation in collaboration with Pesticide Action Network UK, London, 2007.

[3] Revedin, A., Aranguren, B., Becattini, R., Longo, L., Marconi, E., Lippi, M. M., Skakun, N., Sinitsyn, A., Spiridonova, E., Svobodah, J., Thirty thousand-year-old evidence of plant food processing, Proc Natl Acad Sci USA **107** (2010) 18815–18819.

[4] Hosler, D., Burkett, S. L., Tarkanian, M. J., Prehistoric polymers: rubber processing in ancient Mesoamerica, Science **284** (1999) 1988–1991.

[5] Greve, H.-H., Ullmann's Encyclopedia of Industrial Chemistry, Kapitel "Rubber, 2, Natural", Wiley-VCH Verlag, Weinheim, 2000.

[6] Ray, D. T., Guayule: a source of natural rubber, in Janick, J., Simon, J. E., editors, New Crops, Wiley, New York, 1993, 338–343.

[7] Cianci, A., SAIGA Il progetto autarchico della gomma naturale, Terni, Thyrus, 2007.

[8] Mohan, D., Pittman, C. U., Steele, P. H., Pyrolysis of wood/biomass for bio-oil: a critical review, Energ Fuel **20** (2006) 848–889.

[9] Angelici, C., Weckhuysen, B. M., Bruijnincx, P. C. A., Chemocatalytic conversion of ethanol into butadiene and other bulk chemicals, ChemSusChem **6** (2013) 1595–1614.

[10] Hoff, R., Mathers, R. T., Handbook of Transition Metal Polymerization Catalysts, Wiley & Sons, 2010.

[11] Fink, G., Contributions to the Ziegler-Natta catalysis: an anthology, in Kaminsky W, editor, Polyolefins: 50 Years after Ziegler and Natta I, Adv Polym Sci **257**, Springer, Berlin, Heidelberg, 2013, 1–35.

[12] Busico, V., Giulio Natta and the development of stereoselective propene polymerization, in Kaminsky, W., editor, Polyolefins: 50 Years after Ziegler and Natta I. Adv Polym Sci **257**, Springer, Berlin, Heidelberg, 2013, 37–57.

[13] Andrady, A. L., Neal, M. A., Applications and societal benefits of plastics, Philos Trans R Soc. B **364** (2009) 1977–1984.

[14] Aguilera, R. F., Eggert, R. G., Lagos, G. C., Tilton, J. E., Depletion and the future availability of petroleum resources, Energy J **30** (2009) 141.

[15] Cózar, A., Echevarría, F., González-Gordillo, J. I., Irigoien, X., Úbeda, B., Hernández-León, S., et al., Plastic debris in the open ocean, Proc Natl Acad Sci USA **111** (2014) 10239–10344.

[16] Vert, M., Doi, Y., Hellwich, K. H., Hess, M., Hodge, P., Kubisa, P., Rinaudo, M., Schué, F., Terminology for biorelated polymers and applications (IUPAC Recommendations 2012), Pure Appl Chem **84** (2012) 377–410.

[17] Queiroz, A. U. B., Collares-Queiroz, F. P., Innovation and industrial trends in bioplastics, Polym Rev **49** (2009) 65–78.

[18] Koopmans, R. J., Polyolefin-Based Plastics from Biomass-Derived Monomers, Bio-Based Plastics, Wiley & Sons, 2013, 295–310.

[19] PlasticsEurope, http://www.plasticseurope.org/plastics-industry/market-and-economics.aspx, accessed 25 September 2014.

[20] Rudnik, E., editor, Compostable Polymer Materials, Amsterdam: Elsevier; 2008.

[21] Sagel, E., Forum Pemex, Poly(Ethylene) Global Overview, 2012.

[22] Rass-Hansen, J., Falsig, H., Jørgensen, B., Christensen, C. H., Bioethanol: fuel or feedstock?, J Chem Technol Biotechnol **82** (2007) 329–333.

[23] Mathers, R. T., How well can renewable resources mimic commodity monomers and polymers?, J Polym Sci A Polym Chem **50** (2012) 1–15.

[24] González-García, S., Luo, L., Moreira, M. T., Feijoo, G., Huppes, G., Life cycle assessment of hemp hurds use in second generation ethanol production, Biomass Bioenergy **36** (2012) 268–679.

[25] John, R. P., Anisha, G. S., Nampoothiri, K. M., Pandey, A., Micro and macroalgal biomass: a renewable source for bioethanol, Bioresour Technol **102** (2011) 186–193.

[26] Furumoto, Y., Tsunoji, N., Ide, Y., Sadakane, M., Sano, T., Conversion of ethanol to propylene over HZSM-5(Ga) co-modified with lanthanum and phosphorous, Appl Catal A **417–418** (2012) 137–144.

[27] Morschbacker, A., Bio-ethanol based ethylene, Polym Rev **49** (2009) 79–84.

[28] Speiser, F., Braunstein, P., Saussine, L., Catalytic ethylene dimerization and oligomerization: recent developments with nickel complexes containing P,N-chelating ligands, Acc Chem Res **38** (2005) 784–793.

[29] Straathof, A. J. J., Transformation of biomass into commodity chemicals using enzymes or cells, Chem Rev **114** (2013) 1871–1908.

[30] Lehmann, T., Seidel-Morgenstern, A., Thermodynamic appraisal of the gas phase conversion of ethylene or ethanol to propylene, Chem Eng J **242** (2014) 422–432.

[31] Lebedev, S. V., British Patent 331, 402, 1929.

[32] Toussaint, W. J., Dunn, J. T., US Patent 2, 357, 855, 1944.

[33] Patel, A. D., Meesters, K., den Uil, H., de Jong, E., Blok, K., Patel MK., Sustainability assessment of novel chemical processes at early stage: application to biobased processes, Energy Environ Sci **5** (2012) 8430–8444.

[34] Azadi, P., Inderwildi, O. R., Farnood, R., King, D. A., Liquid fuels, hydrogen and chemicals from lignin: a critical review, Renew Sust Energ Rev **21** (2013) 506–523.

[35] Torres Galvis, H. M., de Jong, K. P., Catalysts for production of lower olefins from synthesis gas: a Review, ACS Catal **3** (2013) 2130–2149.

[36] Boltz, M., Losch, P., Louis, B., A general overview on the methanol to olefins reaction: recent catalyst developments, Adv Chem Lett **1** (2013) 247–256.

[37] Al-Dughaither, A. S., de Lasa, H., Neat dimethyl ether conversion to olefins (DTO) over HZSM-5: effect of SiO_2/Al_2O_3 on porosity, surface chemistry, and reactivity, Fuel **138** (2014) 52–64.

[38] Knifton, J. F., Syngas reactions: V, Ethylene from synthesis gas, J Catal. **79** (1983) 147–155.

[39] Dupain, X., Krul, R. A., Schaverien, C. J., Makkee, M., Moulijn, J. A., Production of clean transportation fuels and lower olefins from Fischer-Tropsch synthesis waxes under fluid

catalytic cracking conditions: the potential of highly paraffinic feedstocks for FCC, Appl Catal B **63** (2006) 277–295.

[40] Jahangiri, H., Bennett, J., Mahjoubi, P., Wilson, K., Gu, S., A review of advanced catalyst development for Fischer-Tropsch synthesis of hydrocarbons from biomass derived syn-gas, Catal Sci Technol **4** (2014) 2210–3329.

[41] Zakzeski, J., Bruijnincx, P. C. A., Jongerius, A. L., Weckhuysen, B. M., The catalytic valorization of lignin for the production of renewable chemicals, Chem Rev **110** (2010) 3552–3599.

[42] Asrul, M., Othman, M., Zakaria, M., Fauzi, M. S., Process for producing reinforced natural rubber latex and the reinforced latex produced, Malaysian Patent Application No. PI 2011003945, 2011.

[43] Auvergne, R., Caillol, S., David, G., Boutevin, B., Pascault, J.-P., Biobased thermosetting epoxy: present and future, Chem Rev **114** (2013) 1082–1115.

[44] Rezaei, P. S., Shafaghat, H., Daud, W. M. A. W., Production of green aromatics and olefins by catalytic cracking of oxygenate compounds derived from biomass pyrolysis: a review, Appl Catal A **469** (2014) 490–511.

[45] Azadi, P., Inderwildi, O. R., Farnood, R., King, D. A., Liquid fuels, hydrogen and chemicals from lignin: a critical review, Renew Sust Energ Rev **21** (2013) 506–523.

[46] Saidi, M., Samimi, F., Karimipourfard, D., Nimmanwudipong, T., Gates, B. C., Rahimpour, M. R., Upgrading of lignin-derived bio-oils by catalytic hydrodeoxygenation, Energy Environ Sci **7** (2014) 103–129.

[47] Kang, S., Li, X., Fan, J., Chang, J., Hydrothermal conversion of lignin: a review, Renew Sust Energ Rev **27** (2013) 546–558.

[48] Sattler, J. J. H. B., Ruiz-Martinez, J., Santillan-Jimenez, E., Weckhuysen, B. M., Catalytic dehydrogenation of light alkanes on metals and metal oxides, Chem Rev **114** (2014) 10613–10653.

[49] Kuruppalil, Z., Green plastics: an emerging alternative for petroleum based plastics?, in Proceedings of the 2011 IAJC-ASEE International Conference, 2011, ISBN 978-1-60643-379-9.

[50] Resasco, D. E., Sitthisa, S., Faria, J., Prasomsri, T., Ruiz, M. P., Furfurals as chemical platform for biofuels production, in Kubička, D., Kubičková, I., editors, Heterogeneous Catalysis in Biomass to Chemicals and Fuels, Research Signpost, Kerala, India, 2011, chapter 5.

[51] Lange, J.-P., van der Heide, E., van Buijtenen, J., Price, R., Furfural – a promising platform for lignocellulosic biofuels, ChemSusChem **5** (2012) 150–166.

[52] Caratzoulas, S., Davis, M. E., Gorte, R. J., Gounder, R., Lobo, R. F., Nikolakis, V., et al., Challenges of and insights into acid-catalyzed transformations of sugars, J Phys Chem C **118** (2014) 22815–22833.

[53] Ortega Sánchez, S., Marra, F., Dibenedetto, A., Aresta, M., Grassi, A., ATR copolymerization of styrene with 2-vinylfuran: an entry to functional styrenic polymers, Macromolecules **47** (2014) 7129–7137.

[54] Matyjaszewski, K., Atom transfer radical polymerization (ATRP): current status and future perspectives, Macromolecules **45** (2012) 4015–4039.

[55] Denisov, E. T., Denisova, T. G., Pokidova, T. S., Frontmatter, Handbook of Free Radical Initiators, Wiley & Sons, 2005.

[56] Kharasch, M. S., Jensen, E. V., Urry, W. H., Reactions of atoms and free radicals in solution, X. The addition of polyhalomethanes to olefins, J Am Chem Soc **69** (1947) 1100–1105.

[57] Gandini, A., The irruption of polymers from renewable resources on the scene of macromolecular science and technology, Green Chem **13** (2011) 1061–1083.

[58] Zhao, J., Schlaad, H., Synthesis of terpene-based polymers, in Schlaad, H., editor, Bio-Synthetic Polymer Conjugates, Adv Polym Sci **253**, Springer, Berlin, Heidelberg, 2013, 151–190.

[59] Wilbon, P. A., Chu, F., Tang, C., Progress in renewable polymers from natural terpenes, terpenoids, and rosin, Macromol Rapid Commun **34** (2013) 8–37.

[60] Yao, K., Tang, C., Controlled polymerization of next-generation renewable monomers and beyond, Macromolecules **46** (2013) 1689–1712.

[61] Breitmaier, E., Terpenes: Importance, General Structure, and Biosynthesis, Wiley-VCH Verlag, Weinheim, 2006.

[62] Behr, A., Johnen, L., Myrcene as a natural base chemical in sustainable chemistry: a critical review, ChemSusChem **2** (2009) 1072–1095.

[63] Bolton, J. M., Hillmyer, M. A., Hoye, T. R., Sustainable thermoplastic elastomers from terpene-derived monomers, ACS Macro Lett **3** (2014) 717–720.

[64] Georges, S., Bria, M., Zinck, P., Visseaux, M., Polymyrcene microstructure revisited from precise high-field nuclear magnetic resonance analysis, Polymer **55** (2014) 3869–3778.

[65] Trumbo, D., Free radical copolymerization behavior of myrcene, Polym Bull **31** (1993) 629–636.

[66] Liu, R. S. H., Hammond, G. S., Photosensitization method of cyclization of myrcene, US Patent 3,380,903, 1968.

[67] Kaczor, A., Reva, I., Warszycki, D., Fausto, R., UV-induced cyclization in myrcene isolated in rigid argon environment: FT-IR and DFT study, J Photochem Photobiol A **222** (2011) 1–9.

[68] Kobayashi, S., Lu, C., Hoye, T. R., Hillmyer, M. A., Controlled polymerization of a cyclic diene prepared from the ring-closing metathesis of a naturally occurring monoterpene, J Am Chem Soc **131** (2009) 7960–7961.

[69] Delancey, J. M., Cavazza, M. D., Rendos, M. G., Ulisse, C. J., Palumbo, S. G., Mathers, R. T., Controlling crosslinking in thermosets via chain transfer with monoterpenes, J Polym Sci A Polym Chem **49** (2011) 3719–3727.

[70] Holmberg, A. L., Reno, K. H., Wool, R. P., Epps III, T. H., Biobased building blocks for the rational design of renewable block polymers, Soft Matter (2014).

[71] Tasdelen, M. A., Diels-Alder "click" reactions: recent applications in polymer and material science, Polym Chem **2** (2011) 2133–2145.

[72] Yamamoto, D., Matsumoto, A., Controlled radical polymerization of 3-methylenecyclopentene with N-substituted maleimides to yield highly alternating and regiospecific copolymers, Macromolecules **46** (2013) 9526–9536.

[73] Gandini, A., Monomers and Macromonomers from Renewable Resources, Biocatalysis in Polymer Chemistry, Wiley-VCH Verlag, Weinheim, 2010.

[74] Fahlbusch, K.-G., Hammerschmidt, F.-J., Panten, J., Pickenhagen, W., Schatkowski, D., Bauer, K., et al., Flavors and Fragrances, Ullmann's Encyclopedia of Industrial Chemistry, Wiley-VCH Verlag, Weinheim, 2000.

[75] Moad, G., Rizzardo, E., Thang, S. H., RAFT polymerization and some of its applications, Chem Asian J **8** (2013) 1634–144.

[76] Satoh, K., Matsuda, M., Nagai, K., Kamigaito, M., AAB-sequence living radical chain copolymerization of naturally occurring limonene with maleimide: an end-to-end sequence-regulated copolymer, J Am Chem Soc **132** (2010) 10003–10005.

[77] Singh, A., Kamal, M., Synthesis and characterization of polylimonene: polymer of an optically active terpene, J Appl Polym Sci **125** (2012) 1456–1459.

[78] Matsuda, M., Satoh, K., Kamigaito, M., 1:2-sequence-regulated radical copolymerization of naturally occurring terpenes with maleimide derivatives in fluorinated alcohol, J Polym Sci A Polym Chem **51** (2013) 1774–1785.

[79] Seniha Güner, F., Yağcı, Y., Tuncer Erciyes, A., Polymers from triglyceride oils, Prog Polym Sci **31** (2006) 633–670.

[80] Teramoto, N., Synthetic Green Polymers from Renewable Monomers, A Handbook of Applied Biopolymer Technology: Synthesis, Royal Society of Chemistry, London, 2011, 22–78.

[81] Meier MA. R., Metzger, J. O., Schubert, U. S., Plant oil renewable resources as green alternatives in polymer science, Chem Soc Rev **36** (2007) 1788–1802.

[82] Gandini, A., Polymers from renewable resources: a challenge for the future of macromolecular materials, Macromolecules **41** (2008) 9491–9504.

[83] Lu, Y., Larock, R. C., Novel polymeric materials from vegetable oils and vinyl monomers: preparation, properties, and applications, ChemSusChem **2** (2009) 136–147.

[84] Xia, Y., Larock, R. C., Vegetable oil-based polymeric materials: synthesis, properties, and applications, Green Chem **12** (2010) 1893–1909.

[85] Montero de Espinosa, L., Meier, M. A. R., Plant oils: the perfect renewable resource for polymer science?!, Eur Polym J **47** (2011) 837–852.

[86] Xia, Y., Quirino, R. L., Larock, R. C., Bio-based thermosetting polymers from vegetable oils, J Renew Mater **1** (2013) 3–27.

[87] Lligadas, G., Ronda, J. C., Galià, M., Cádiz, V., Renewable polymeric materials from vegetable oils: a perspective, Mater Today **16** (2013) 337–343.

[88] Mosiewicki, M. A., Aranguren, M. I., A short review on novel biocomposites based on plant oil precursors, Eur Polym J **49** (2013) 1243–1256.

[89] Quirino, R. L., Garrison, T. F., Kessler, M. R., Matrices from vegetable oils, cashew nut shell liquid, and other relevant systems for biocomposite applications, Green Chem **16** (2014) 1700–1715.

[90] Miao, S., Wang, P., Su, Z., Zhang, S., Vegetable-oil-based polymers as future polymeric biomaterials, Acta Biomater **10** (2014) 1692–1704.

[91] Alam, M., Akram, D., Sharmin, E., Zafar, F., Ahmad, S., Vegetable oil based eco-friendly coating materials: a review article, Arab J Chem **7** (2014) 469–479.

[92] Pratt, D. A., Tallman, K. A., Porter, N. A., Free radical oxidation of polyunsaturated lipids: new mechanistic insights and the development of peroxyl radical clocks, Acc Chem Res **44** (2011) 458–467.

[93] Liu, Z., Sharma BK., Erhan, S. Z., From oligomers to molecular giants of soybean oil in supercritical carbon dioxide medium: 1, preparation of polymers with lower molecular weight from soybean oil, Biomacromolecules **8** (2006) 233–239.

[94] Kennedy, J. P., Marechal, E., Carbocationic Polymerization, Wiley & Sons, New York, 1982, 31–55.

[95] Li, F., Larock, R., Novel polymeric materials from biological oils, J Polym Environ **10** (2002) 59–67.

[96] Li, F., Larock, R. C., New soybean oil-styrene-divinylbenzene thermosetting copolymers, I Synthesis and characterization, J Appl Polym Sci **80** (2001) 658–970.

[97] Andjelkovic, D. D., Larock, R. C., Novel rubbers from cationic copolymerization of soybean oils and dicyclopentadiene, 1. Synthesis and characterization, Biomacromolecules **7** (2006) 927–236.

[98] Lu, J., Khot, S., Wool, R. P., New sheet molding compound resins from soybean oil, I Synthesis and characterization, Polymer **46** (2005) 71–80.

[99] Xia, Y., Henna, P. H., Larock, R. C., Novel thermosets from the cationic copolymerization of modified linseed oils and dicyclopentadiene, Macromol Mater Eng **294** (2009) 590–598.

[100] Larock, R., Dong, X., Chung, S., Reddy CK., Ehlers, L., Preparation of conjugated soybean oil and other natural oils and fatty acids by homogeneous transition metal catalysis, J Am Oil Chem Soc **78** (2001) 447–453.

[101] Kundu, P. P., Larock, R. C., Novel conjugated linseed oil-styrene-divinylbenzene copolymers prepared by thermal polymerization, 1. Effect of monomer concentration on the structure and properties, Biomacromolecules **6** (2005) 797–806.

[102] Henna, P. H., Andjelkovic, D. D., Kundu, P. P., Larock, R. C., Biobased thermosets from the free-radical copolymerization of conjugated linseed oil, J Appl Polym Sci **104** (2007) 979–985.

[103] Valverde, M., Andjelkovic, D., Kundu, P. P., Larock, R. C., Conjugated low-saturation soybean oil thermosets: free-radical copolymerization with dicyclopentadiene and divinylbenzene, J Appl Polym Sci **107** (2008) 423–430.

[104] Robertson, M. L., Chang, K., Gramlich, W. M., Hillmyer, M. A., Toughening of polylactide with polymerized soybean oil, Macromolecules **43** (2010) 1807–1814.

[105] Li, F., Larock, R. C., Synthesis, structure and properties of new tung oil-styrene-divinylbenzene copolymers prepared by thermal polymerization, Biomacromolecules **4** (2003) 1018–1025.

[106] Mol, J. C., Application of olefin metathesis in oleochemistry: an example of green chemistry, Green Chem **4** (2002) 5–13.

[107] Van Dam, P. B., Mittelmeijer, M. C., Boelhouwer, C., Metathesis of unsaturated fatty acid esters by a homogeneous tungsten hexachloride-tetramethyltin catalyst, J Chem Soc Chem Commun (1972) 1221–1222.

[108] Henna, P., Larock, R. C., Novel thermosets obtained by the ring-opening metathesis polymerization of a functionalized vegetable oil and dicyclopentadiene, J Appl Polym Sci **112** (2009) 1788–1797.

[109] Henna, P. H., Kessler, M. R., Larock, R. C., Fabrication and properties of vegetable-oil-based glass fiber composites by ring-opening metathesis polymerization, Macromol Mater Eng **293** (2008) 979–990.

[110] Haman, K., Badrinarayanan, P., Kessler, M. R., Cure characterization of the ring-opening metathesis polymerization of linseed oil-based thermosetting resins, Polym Int **58** (2009) 738–844.

Eleni Heracleous, Efterpi S. Vasiliadou, Eleni F. Iliopoulou, Angelos A. Lappas, and Angeliki A. Lemonidou

8 Conversion of lignocellulosic biomass-derived intermediates to hydrocarbon fuels

Abstract: Biofuels – liquid and gaseous fuels derived from organic matter – can play an important role in reducing CO_2 emissions in the transport sector and enhancing energy security. Although lignocellulosic biomass presents the highest potential as feedstock, its conversion faces physical, chemical, and technical limitations due to the complex structure of biomass. Moreover, biofuels produced from primary processing of lignocelluloses require further upgrading to hydrocarbons, in order to be able to be used as drop-in fuels, i.e., fuels that are fully compatible with the existing fuel infrastructure and can run in today's engines without modifications. In this chapter, we present an overview on the secondary processing of lignocellulosic biomass-derived intermediates (namely pyrolysis bio-oil, sugars, and butanol) to hydrocarbon fuels.

8.1 Introduction

Biomass has been proposed as the only sustainable source of organic carbon currently available on earth, and thus, this resource is best suited as a potential substitute for petroleum for the production of fuels and chemicals [1]. Biofuels – liquid and gaseous fuels derived from organic matter – can play an important role in reducing CO_2 emissions in the transport sector, and enhancing energy security. By 2050, biofuels could provide up to 27% of total transport fuel and contribute in particular to the replacement of diesel, kerosene, and jet fuel. The projected use of biofuels could avoid around 2.1 Gt of CO_2 emissions per year when produced sustainably [2].

Biomass can be processed with several routes to obtain liquid and gaseous fuels and chemicals. The various conversion processes can be group into two main categories: chemical and thermochemical processes. These processes have been described in detail in the previous chapters of this book. The biochemical conversion of starch- or sugar-containing biomass to bioethanol is a well-established, mature technology for the production of biofuels that can be blended with existing petroleum fuels and can be utilized in the existing infrastructure. However, first-generation biofuels have several disadvantages, the most important of which are the competition with the food chain and the relatively low greenhouse gas (GHG) savings that can be achieved. The conversion of lignocellulosic biomass presents a much higher potential for large-scale biofuel production (due to its greater abundance), the higher yields that can be achieved (the whole biomass can be used as opposed to part of the crop in edible biomass), and the subsequent high GHG savings.

The conversion of lignocellulosic feedstock faces other physical, chemical, and technical limitations due to the complex structure of biomass. Depending on the process, the products obtained from the primary processing of lignocelluloses can have high water content, high oxygen concentration, are acidic and/or viscous, and in general have properties that do not allow their use as drop-in fuels, i.e., fuels that are fully compatible with the existing fuel infrastructure and can run in today's engines without modifications. The production of hydrocarbon fuels from biomass has many important advantages. First, "green" hydrocarbon fuels are essentially the same as those currently derived from petroleum, except that they are made from biomass. Therefore, it will not be necessary to modify existing infrastructure (e.g., pipelines, engines) and hydrocarbon biorefining processes can be tied into the fuel production systems of existing petroleum refineries. Second, biomass-based hydrocarbon fuels are energy equivalent to fuels derived from petroleum. Third, hydrocarbons produced from lignocellulosic biomass are immiscible in water; they self-separate, which eliminates the need for an expensive, energy-consuming distillation step. Fourth, biomass-based hydrocarbon fuels are produced at high temperatures, which allows for faster conversion reactions in smaller reactors. Thus, processing units can be placed close to the biomass source, thereby avoiding transport of biomass over long distances.

In this chapter, we look into secondary processing of lignocellulosic biomass-derived intermediates to hydrocarbon fuels. Focus is given on the upgrading processes of the most-common biomass-derived intermediates and specifically bio-oil obtained from thermochemical fast biomass pyrolysis, the aqueous-phase processing of sugars and various oxygenates derived from lignocellulosic biomass hydrolysis, and the oligomerization of butanol. Each section consists of a short description of the properties and quality of the initial biomass intermediate obtained from the primary biomass processing, followed by a description and a critical review of the most important upgrading processes in order to obtain fuel products with similar composition as conventional hydrocarbon fuels.

8.2 Upgrading of bio-oil from biomass pyrolysis

Fast pyrolysis is a high-temperature process in which the feedstock is rapidly heated in the absence of air, vaporizes, and condenses to a dark brown liquid that has a heating value of about half that of conventional fuel oil. The process is characterized by high heating rates and rapid quenching of the liquid products to terminate the secondary conversion of the products. Typical products yields in fast pyrolysis are 60–75 wt% of liquid, 15–25 wt% of solid char, and 10–20 wt% of non-condensable gases, depending on the feedstock used. The main advantages of the fast pyrolysis process are the high yields to liquid biofuel and the relatively low production cost of bio-oil due to lower cost of equipment and energy required for pyrolysis compared to other thermochemical processes [3]. The above renders biomass pyrolysis an excellent process for decentralized conversion of residual stranded biomass to optimized, high-energy-density carriers. The process has been demonstrated at a demo scale employing several different reactor types ranging from entrained flow to ablative and fluidized bed, with capacities between 1 and 50 t/h [4].

The main disadvantage of fast pyrolysis is the low quality of the attained liquid product. Pyrolysis oils are composed of a very complex mixture of oxygenated hydrocarbons (>300) [5], the main constituents being acids, aldehydes, ketones, alcohols, glycols, esters, ethers, phenols, and phenol derivatives, as well as carbohydrates and derivatives, and a large proportion (20–30 wt%) of lignin-derived oligomers [6]. The main properties of bio-oil are given in Table 8.1. Due to their oxygen-rich composition, they present a low heating value, immiscibility with hydrocarbon fuels, chemical instability, high viscosity, and corrosiveness [3, 7–10], restricting their direct use as fuels. The (partial) elimination of oxygen is thus necessary to transform the bio-oil into a liquid fuel that can compete with mineral oil refinery fuels.

Table 8.1: Typical properties of wood pyrolysis bio-oil (adapted from Zhang et al. [11], with permission of Elsevier).

Physical property	Bio-oil
Moisture content (wt%)	15–30
pH	2.5
Specific gravity	1.2
Elemental composition (wt%)	
C	54–58
H	5.5–7.0
O	35–40
N	0–0.2
Ash	0–0.2
HHV (MJ/kg)	16–19
Viscosity, at 773 K (cP)	40–100
Solids (wt%)	0.2–1.0
Distillation residue (wt%)	Up to 50

Catalytic pyrolysis of biomass, where the heat carrier in the pyrolysis reactor is substituted with a heterogeneous catalyst, constitutes a very attractive route for the in situ upgrading of the quality of bio-oil. In catalytic fast biomass pyrolysis, the heavy oxygenated volatiles from the decomposition of biomass are deoxygenated and converted to lighter fuels and chemicals by coming in contact with a suitable catalyst with an ultimate goal to produce a liquid with improved properties that could be either used directly as a liquid fuel or as a feedstock (or co-feedstock) in modern refineries, much like crude oil. These catalysts should be able to selectively favor the decarboxylation and decarbonylation reactions, producing high-quality bio-oil with low amounts of oxygen. They should also inhibit formation of undesirable oxygenated compounds, such as ketones, acids, and carbonyl compounds, which are known to be detrimental for the direct use or further co-processing of bio-oil. A plethora of catalytic materials such as zeolites, mesoporous materials with uniform pore size distribution (MCM-41, MSU, and SBA-15), microporous/mesoporous hybrid materials doped with noble and transition metals, and base catalysts have been investigated as candidate catalysts for biomass pyrolysis. A detailed review of catalysts for biomass pyrolysis has been recently published by our group [12].

As this chapter is devoted to the conversion of lignocellulosic biomass-derived intermediates to final fuels, we will give more emphasis on the downstream upgrading of pyrolytic oils, focusing on their deoxygenation, which is attempted either by a typical catalytic hydrotreatment with hydrogen under high pressure and/or in the presence of hydrogen donor solvents or utilizing zeolitic cracking catalysts. Integration of bio-oil upgrading in a conventional petrochemical refinery via co-processing bio-oil with conventional oil refinery feedstocks is also an approach that has received considerable interest in the last years. Steam reforming of either the whole bio-oil or the aqueous fraction of bio-oil to hydrogen is also a process that has been intensively investigated; however, we will only consider herein upgrading to liquid hydrocarbons. For those who are interested, an excellent overview of steam reforming of bio-oils to hydrogen has been published by some of the authors of this chapter [13].

8.2.1 Upgrading of bio-oil to hydrocarbons by hydrodeoxygenation

Hydrodeoxygenation (HDO) involves removal of most of the oxygen groups contained in the bio-oil via hydrogenation under high pressure (up to 200 bars) and temperatures in the range of 300–400°C. Typical products obtained from bio-oil HDO comprise of naphtha-like and diesel streams that could be blended in refineries with conventional transportation fuels. Typical products yields range between 0.3 and 0.5 L of product per liter of bio-oil depending on the extent of deoxygenation (partial or complete). The HDO process has been extensively investigated in literature and several reviews have been published [11, 14–19]. Catalysts for the reaction are traditional hydrodesulfurization catalysts, such as NiMo and CoMo catalysts on alumina or silica alumina supports, or metal catalysts, such as Pd/C. However, catalyst lifetimes of longer than 200 h have not been achieved with any current catalyst because of extensive carbon deposition that results in very short catalyst lifetimes. Carbon is principally formed through polymerization and polycondensation reactions on the catalytic surface, forming polyaromatic species that block the active sites on the catalysts. Studies on Co-MoS$_2$/Al$_2$O$_3$ catalysts showed that polyaromatic species strongly adsorb on the catalyst and fill up the pore volume during the start up of the system. Fonseca et al. [20, 21] reported that about one third of the total pore volume of a Co-MoS$_2$/Al$_2$O$_3$ catalyst was occupied with carbon during this initial carbon deposition stage, and hereafter, a steady state was observed where further carbon deposition was limited. An approach to reduce the chance of polymerization and thermal decomposition that forms coke and plugs the reactor is to perform the process in two stages: a low temperature (around 100–200°C) and a high temperature one (350–450°C) [14]. Initial thermal treatment of the bio-oil in a slurry reactor with the presence of hydrogen to attain 80% deoxygenation followed by complete deoxygenation in a fixed-bed reactor has also been suggested [6, 22]. Even then, the maximum period of uninterrupted operation reported in the open literature was 1 week. The large amounts of water in the feed also lead to catalyst deactivation, especially in the case of alumina support. Overall, due to the high oxygen content of bio-oil, which can reach up to over 40 wt% oxygen, full HDO involves high hydrogen consumption, making this process hardly competitive in economic terms with conventional fossil fuels.

The concept of using bio-oil as a petroleum refinery feedstock and co-process it in conventional refineries was extensively investigated by Veba Oil [6, 22, 23]. Due to the immiscible nature of bio-oil, they concluded that initial bio-oil pretreatment is necessary to process bio-oil together with petroleum products. Hydrotreating of bio-oil was investigated over sulfided CoMo and NiMo catalysts in a continuous-feed bench-scale reactor operated at 17.8 MPa. Over a temperature range of 350–370°C, deoxygenation rates of 88–99.9% were achieved with yields around 30–35% in oil, 50–55% water, and 15–20% gases. However, extensive coking was observed and catalyst deactivation took place within 8 days. CPERI/CERTH investigated in collaboration with Veba Oil both the thermal and the catalytic hydrotreating of bio-oil over conventional hydrotreating catalysts [6, 24]. In agreement with literature, it was found that catalytic hydrogenation of bio-oil has many operating problems due to plugging of the catalyst bed, although high deoxygenation (>85 wt%) is achieved. On the contrary, the thermal hydrogenation of bio-oil was feasible and without operational problems. The thermal hydrogenation achieved up to 85 wt% deoxygenation conversion, producing a bio-oil with an oxygen content of about 6.5 wt%. The hydrotreated bio-oil was further separated by in a light (LBFPL) and a heavy fraction (HBFPL), with the characteristics presented in Table 8.2. The light fraction comprised of components mainly in the gasoline and diesel range, and thus, it could be blended directly with the corresponding petroleum fractions, while the heavy fraction had characteristics similar to conventional vacuum gas oil (VGO). The total yield of the hydrotreated bio-oil was about 42 wt% (based on the non-hydrotreated bio-oil).

Table 8.2: Properties of the two fractions produced after thermally hydrotreating the bio-oil (reproduced from Lappas et al. [24], with permission of Elsevier).

Property	Light fraction	Heavy fraction
Elemental analysis (wt%)		
C	82.2	84.4
H	10.7	9.4
S	0.01	0.01
N	1.15	0.42
O	6.4	4.9
H_2O (wt%)	0.99	–
Density (g/cm³) (15°C)	0.942	1.036
Distillation (wt%)		
<200°C	27	
200–350°C	55.3	23
350–500°C	17.8	66
>500°C		11

8.2.2 Upgrading of bio-oil via C-C coupling reactions

A strategy to minimize hydrogen consumption in HDO, making it more attractive and a viable process option, is to deoxygenate bio-oil to hydrocarbons in a series of cascade catalytic transformations. This concept is currently being investigated in the frame of CASCATBEL (www.cascatbel.eu), a 4-year project funded by the EU (FP7) and aims to design, optimize, and scale up a novel multistep process for the production of second-generation liquid biofuels from lignocellulosic biomass in a cost-efficient way through the use of next-generation high-surface-area tailored nanocatalysts. The sequential coupling of catalytic steps is expected to achieve a progressive and controlled biomass deoxygenation, avoiding the previously highlighted problems that hinder bio-oil upgrading processes. The envisaged process consists of a cascade combination of up to three catalytic transformations: catalytic pyrolysis, intermediate deoxygenation, and HDO. In the first step, specially designed nanostructured catalysts, with mild acidity or basicity, based on hierarchical zeolites (ZSM-5 and Beta), 2D zeolites (ZSM-5 and MCM-22), and ordered mesoporous materials (Al-SBA-15) are investigated for the partial deoxygenation of bio-oil in order to avoid overcracking of the pyrolytic vapors and reduce the coke deposition. The bio-oil is then further deoxygenated in an intermediate step that targets in the condensation of the small molecules in bio-oil to increase the chain length and decrease its oxygen content. This intermediate step can occur via different types of oligomerization reactions, namely ketonization, aldol condensation, and esterification. In ketonization reaction, two carboxylic acid molecules couple to produce a larger ketone molecule, eliminating CO_2

and H_2O. Small carboxylic acids such as acetic or propanoic acid usually represent up to 10% of the bio-oils, and therefore, ketonization possesses a great potential for the catalytic upgrading of bio-oil. Vapor-phase ketonization of acetic acid on oxides has been widely studied in the past [25–27]. The reaction is typically catalyzed by inorganic oxides such as CeO_2, TiO_2, Al_2O_3, and ZrO_2 that possess high ketonization activity and selectivity to acetone at moderate temperatures (300–425°C) [28]. It was also found that the addition of catalytic metals with high hydrogen dissociation activity, such as Pd, Pt, Rh, and Co, significantly improves the reducibility of the metal oxides and consequently their ketonization activity [26, 29]. Vapor-phase ketonization is suitable for direct bio-oil upgrading coupled to catalytic pyrolysis, as it operates under vapor-phase conditions compatible with those of the outlet streams of the latter. Ketonization for bio-oil upgrading method in the vapor phase has been extensively investigated by the group of J. Dumesic [30–32], who showed that carboxylic acids can be ketonized with nearly 100% yield over a $CeZrO_x$ mixed-oxide catalyst at temperatures from 350°C to 400°C [30]. In the frame of the CASCATBEL project, the Laboratory of Environmental Fuels and Hydrocarbons in CPERI/CERTH investigates the vapor-phase ketonization of bio-oil model compounds over novel materials, such as metal-oxide-loaded activated carbons and gauge Fe and Mn oxide materials. Preliminary tests performed with acetic acid over commercial titania and zirconia showed that conversion yields higher than 60% can be achieved at temperatures over 350°C. Figure 8.1a and b presents the conversion of acetic acid and the yield to the different products as a function of temperature for TiO_2 and ZrO_2, respectively, obtained from experiments in a small-scale fixed-bed reactor at constant weight hourly space velocity (WHSV) (2/h). The main reaction products from acetic acid ketonization were the symmetric ketone (acetone), water, CO_2, and small amounts of coke. Selectivity to the respective ketone reached values close to the theoretical maximum at intermediate conversions, while it decreased at higher conversion levels at the expense of CO_2 formation.

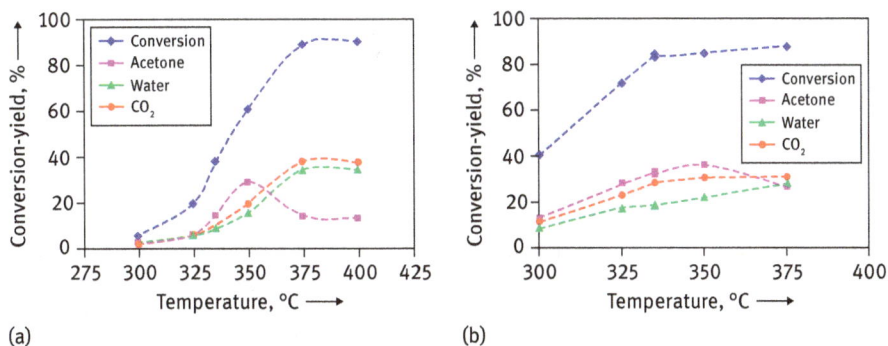

Fig. 8.1: Conversion and products yields as a function of temperature in the vapor-phase ketonization of acetic acid over TiO_2 (a) and ZrO_2 (b).

Liquid-phase ketonization at moderate temperatures has recently attracted attention as a promising approach for bio-oil upgrading [28, 33]. This approach claims to avoid undesirable issues with bio-oil upgrading such as excessive coking and thermal decomposition and repolymerization of other bio-oil components. However, studies on aqueous-phase ketonization using metal oxide catalysts are still very limited to date. The presence of water in high concentrations lowers the activity of the oxide due to competitive adsorption and active site blocking [28]. Moreover, the high temperatures necessary for ketonization to occur make this impractical. Therefore, a more active catalyst needs to be developed, allowing for lowering of the temperature at which significant conversion occurs [33].

Aldol condensation involves the reaction of two carbonyl compounds (aldehydes or ketones) to yield an aldol that after dehydration forms α,β-unsaturated ketones. The reaction has been mainly studied for the formation of C-C bonds in the context of cellulose hydrolysis but can also been applied for bio-oil upgrading due to the large content of aldehydes and ketones in bio-oil. Aldol condensation reactions are typically catalyzed by acid, base, and acid-base bifunctional catalysts. In the frame of CASCATBEL, a variety of hierarchical zeolites (X, Y, USY, A, L, ZSM-5, and Beta), with different levels of mesoporosity and Si/Al ratios are investigated. Esterification of bio-oil represents also an attractive intermediate deoxygenation reaction since the formation of esters would allow the concentration of carboxylic acids present in bio-oil to be significantly decreased. The reaction is usually performed with liquid acids, with the well-known dropbacks. In the CASCATBEL project, esterification is investigated as a potential intermediate bio-oil upgrading step in the presence of solid acid catalysts, such as nanostructured sulfated zirconia and low-silica zeolites functionalized with propylsulfonic or arylsulfonic groups. After the intermediate deoxygenation steps (ketonization, aldol condensation, esterification), bio-oil can be deoxygenated at moderate temperatures (200–350°C) and hydrogen pressures (20–150 bars). Since a substantial amount of oxygen would already be removed, hydrogen consumption is expected to be much lower compared to an initial dehydrogenation step of the bio-oil

8.2.3 Upgrading of bio-oil to hydrocarbons by fluid catalytic cracking (FCC)

Catalytic cracking of bio-oil to hydrocarbons in the presence of zeolites or aluminosilicates represents an attractive alternative process for the upgrading of bio-oil since it does not involve the consumption of large amounts of hydrogen as in the case of HDO presented in this previous section. Moreover, the product obtained is of higher value for the transportation sector, as it contains higher aromatics and therefore high octane number. On the down side, the yields are usually lower than those attained with deoxygenation, and a high coke formation rate is typically observed. The process of catalytic cracking for bio-oil upgrading has been studied by several research groups [34–37] mainly over zeolitic materials. Bulushev and Ross [19] have recently published a review on recent trends in the chemistry of biomass conversion to fuels via pyrolysis and gasification and the catalysts used. The review includes a comprehensive overview of catalytic materials for the upgrading of bio-oil via catalytic cracking. Among all investigated catalysts, ZSM-5 zeolites seem to be the most promising in the catalytic cracking of bio-oil, although they suffer under certain conditions from deactivation by coke deposition and dealumination [19].

The processing of bio-oil via catalytic cracking has been investigated with either model compounds, representative of bio-oil main components, or actual bio-oil fractions in pure form or in mixtures with VGO in fixed and fluid bed reactors. Gayubo et al. [35] studied the catalytic pyrolysis of various model compounds (1-propanol, 2-propanol, 1-butanol, 2-butanol, phenol, and 2-methoxyphenol) over an HZSM-5 catalyst in a fixed-bed reactor. At low temperatures ($\sim$250°C), the main reaction occurring with alcohols is dehydration to the respective alkenes, while at temperatures above 400°C, significant amounts of C_{5+} paraffins and aromatics are formed. The cracking of phenols was found to lead to high coke yields (up to 50%) on the acid sites. It was thus concluded that moderate acid sites are required for the dehydration of alcohols, while stronger acid sites are required for the reaction steps needed for heavier products. Moreover, it was shown that separation of phenolic products from bio-oil is highly recommended before any processing, since phenolics are the main precursors leading to thermal coking.

In order to reduce coking, Sharma and Bakhshi [38–40] studied the upgrading of wood-derived fast-pyrolysis bio-oil by catalytic cracking in a dual-reactor system. The dual-reactor system was particularly effective in producing higher amounts of organic liquid products. Thermal cracking of bio-oil in the first reactor, followed by catalytic upgrading (H-ZSM-5) in the second reactor, reduced catalyst coking in the second reactor, thus enhancing catalyst life.

As mentioned above, another approach to facilitate bio-oil upgrading in FCC is its co-processing with VGO, a conventional FCC refinery feed. Graca et al. [41] studied the cracking of VGO with high nitrogen content and mixtures of VGO with phenol, acetic acid, and hydroacetone in concentrations ranging from 6 to 10 wt% using either an FCC catalyst that has been equilibrated (Ecat) or a catalyst mixture of 90% Ecat with 10% steam-deactivated ZSM-5. Table 8.3 shows the product yields obtained at iso-conversion (63 wt%) for all the feeds, considering the two catalytic systems. Surprisingly, the coke yield with addition of oxygenated compounds was reduced, probably due to competitive adsorption of nitrogen compounds on the acid sites. With phenol in the co-feed, a 50–80% conversion of phenol was observed, being mainly dehydrated to benzene. Hence, co-processing of phenol might be critical due to benzene content in the gasoline. Acetic acid was converted to methane and CO_2, while hydroacetone mixed with VGO resulted in a slight increase in fuel gas, LPG, and gasoline yields.

Table 8.3: Product yields obtained at iso-conversion (63 wt%) for all the feeds, considering the two catalytic systems (reproduced from Graca et al. [41], with permission of Elsevier).

Yields (wt%)	E-CAT				E-CAT+ZSM-5			
	Pure gas oil	Gas oil/ acetic acid	Gas oil/ hydroxy-acetone	Gas oil/ phenol	Pure gas oil	Gas oil/ acetic acid	Gas oil/ hydroxy-acetone	Gas oil/ phenol
Coke	11.7	9.4	9.6	8.6	9.4	8.6	8.8	9.5
Fuel gas	2.7	3.7	3.0	2.5	3.3	3.7	3.4	2.8
CO	–	0.47	0.07	0.04	0.06	0.28	0.07	0.06
CO_2	0.25	0.54	0.35	0.17	0.38	0.62	0.44	0.34
LPG	12.4	15.7	14.0	13.2	20.4	20.7	19.7	16.7
Propane	1.42	1.63	1.56	1.39	1.98	1.89	2.02	1.70
Propylene	3.37	4.25	4.25	3.60	6.90	7.23	6.16	5.66
Butanes	4.20	5.13	4.40	4.22	5.61	5.29	5.85	4.67
Butylenes	3.39	4.73	3.82	3.91	5.83	6.28	5.60	4.67
Gasoline	35.8	33.4	35.9	38.1	29.6	29.3	30.7	33.3
H_2O	0.44	0.19	0.28	0.86	0.18	0.30	0.10	0.72
LCO	19.6	18.9	20.0	18.6	16.9	18.0	18.4	17.6
DO	17.2	17.7	16.8	18.0	19.9	18.6	18.4	19.0

LCO, light cycle oil; DO, decanted oil.

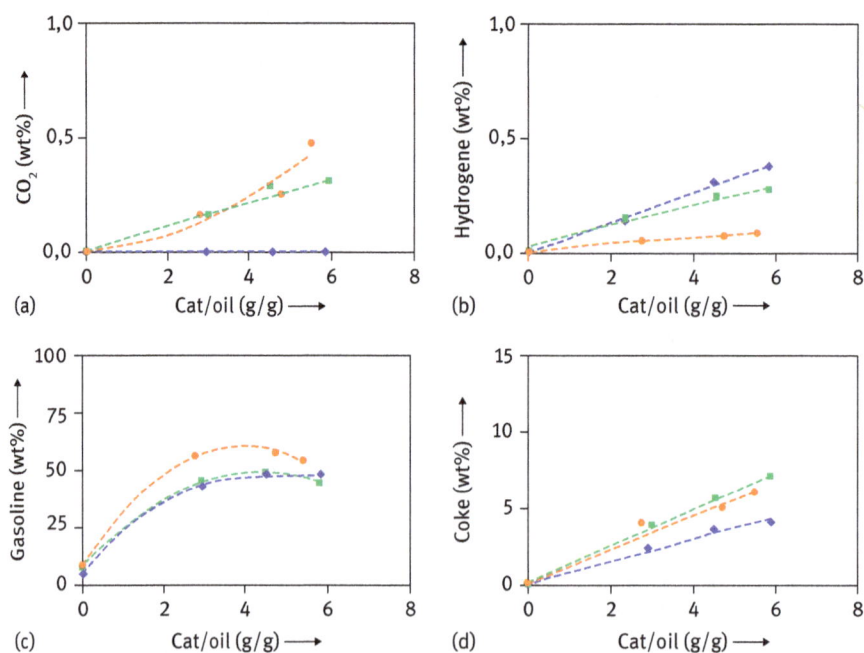

Fig. 8.2: Product distribution as a function of the catalyst/oil (cat/oil) ratio for VGO processing (♦), HDO/VGO co-processing (■), and CPO/VGO co-processing (•). (adapted from Thegarid et al. [42], with permission of Elsevier).

In a recent study [42], our group in collaboration with the group of Prof. Miroda-tos in CNRS investigated the co-processing of catalytic pyrolysis oil (CPO) and HDO-upgraded thermal pyrolysis oil with VGO in an FCC laboratory-scale fluidized-bed unit. Experiments included the following three cracking sequences run over an equilibra-ted FCC catalyst at different cat/oil ratios to vary the levels of conversion in a similar range: (i) pure VGO cracking at cat/oil ratios of 3.1, 6.0, and 8.9; (ii) 10 wt% HDO/VGO co-processing at cat/oil ratios of 3/4/6; (iii) 10 wt% CPO/VGO co-processing at cat/oil ratios of 2.9/3.2/5.5 and 8.3. The results of co-processing either HDO or CPO indicated only moderate differences on the overall performance, as shown in Figure 8.2, which presents the product distribution as a function of the cat/oil ratio for the three cases. Specific and significant effects were, however, noted on the product quality, such as a higher remaining fraction of phenolics in the gasoline or a higher content in aroma-tics. These differences could be reduced by further catalyst development both on the pyrolysis as on the FCC side. The main conclusion of the work was that pyrolysis oil could be upgraded directly by co-processing in FCC, thus eliminating a hydrogenation step by the addition of a suitable catalyst during the pyrolysis step. The organic yield of the catalytic pyrolysis route is estimated at approximately 30 wt% as compared to an overall yield for the thermal pyrolysis followed by an HDO step of 24 wt%. Bio-oils from catalytic pyrolysis thus offer an interesting potential for the production of bio-fuels via co-processing in a FCC unit both from an energetic as from a technical perspective.

The work in CERTH also showed that it is possible to successfully treat the heavy fraction of thermally hydrotreated bio-oil in FCC with conventional VGO without technical problems. In a work by Lappas et al. [24], thermally deoxygenated bio-oil with an oxygen content of about 6.5 wt% was separated in an LBFPL and an HBFPL (characteristics presented in Table 8.2). The light fraction could be blended directly with gasoline and diesel. The heavy fraction had characteristics similar to conven-tional VGO and its catalytic cracking with VGO was investigated in the FCC process. The bio-oil co-processing technology proposed by CPERI/CERTH is presented sche-matically in Figure 8.3.

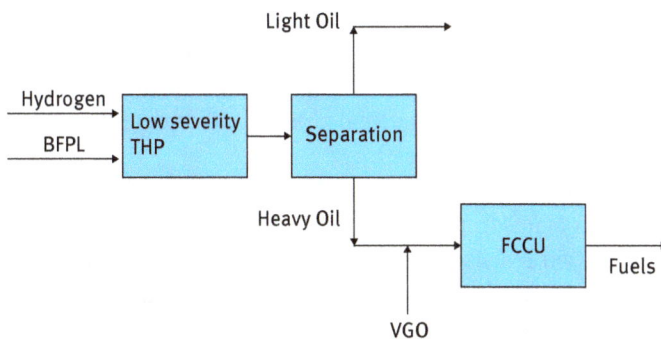

Fig. 8.3: Bio-oil co-processing technology investigated in CPERI. The heavy oil from the low-severity hydrotreating process of the bio-oil is introduced to the FCC after mixing with conventional VGO (reproduced from Lappas et al. [24], with permission of Elsevier).

Overall, the upgrading of bio-oil to hydrocarbon fuels via HDO, C-C coupling reactions, or catalytic cracking, although with challenges, is technically feasible. For deoxygenation, research is needed on catalyst development, for the design of novel catalysts with enhanced activity and selectivity, and especially better stability to deactivation. In catalytic cracking, the challenge is to design catalysts with less coke selectivity or to use bio-oil with less phenolic compounds. The concept of co-processing bio-oil fractions with conventional refinery streams is especially promising, as it provides a low-capital-cost route to high-quality biofuels.

8.3 Sugars to hydrocarbon fuels

As mentioned in the introduction, apart from the thermochemical route, lignocellulosic biomass can be deconstructed by enzymatic or acid hydrolysis to yield aqueous solutions of carbohydrates. The aqueous solution of carbohydrates containing mostly sugars can be used as raw materials for hydrocarbon fuels generation. Unlike the thermochemical route, the aqueous-phase processing of biomass-derived compounds is carried out at mild temperatures, allowing for better control of the catalytic chemistry and the possibility of achieving specific and well-defined liquid hydrocarbons at high yield. The catalytic upgrading pathway that involves conversion of these sugars (sorbitol, glucose, fructose, xylose) to hydrocarbon fuels has great potential in producing drop-in fuel blend stocks over the range of gasoline, jet fuel, and diesel (C_5-C_{12} for gasoline, C_9-C_{16} for jet fuel, and C_{10}-C_{20} for diesel applications). Research in this field is being conducted to develop new and improved processes that allow transformation of biomass-derived sugars to hydrocarbons via catalytic routes, at low temperatures (typically less than 600 K) and in liquid phase (at pressures close to 30 atm) [43]. The general scheme of the concept includes controlled conversion of sugars to platform compounds and subsequent transformation to hydrocarbons via reactions such as dehydration, reforming, hydrogenation, hydrogenolysis, aldol condensation, and oligomerization (Figure 8.4) [44]. Advantages of sugars to hydrocarbon fuels technology include high-energy-density transfer from the feed to the product and elimination of distillation steps due to the hydrophilic properties of the feed and the hydrophobic properties of the product stream. In addition, compared with oxygenated biofuels, the above strategy offers stability of the final product and compatibility with conventional fuels (gasoline, jet fuels, and diesel).

The research in this field is led by the group of Prof. Dumesic and coworkers at the University of Wisconsin-Madison [30, 43, 45–49]. The process that has been developed includes condensation reactions to form C-C bonds and oxygen removal via dehydration, hydrogenation, and hydrogenolysis reactions in aqueous-phase environment. The required hydrogen can be generated *in situ* via aqueous-phase reforming (APR) of a part of the feed or it can be supplied as molecular H_2 from an external source. A key challenge of these processes is, in any case, to achieve

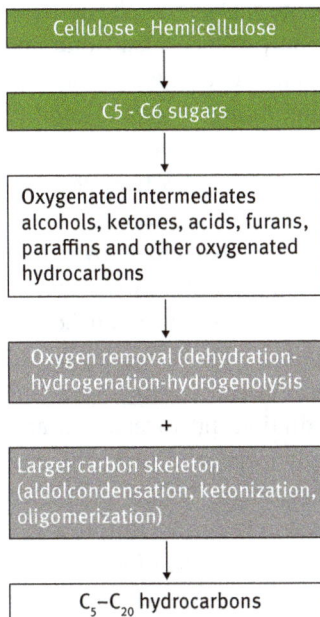

Fig. 8.4: Integrated transformation of sugars to hydrocarbon fuels.

complete oxygen removal with minimum hydrogen consumption, an issue that needs appropriate and careful catalyst design. As this process is highly hydrogen demanding, the option of in situ H_2 generation via APR is a very promising alternative allowing obtaining the required hydrogen and taking advantage of the favored water-gas shift reaction under the conditions applied. Depending on the feed used, the reaction pathways can be more or less complex; however, C-C and C-H bond scission as well as dehydration, hydrogenation, and dehydrogenation reactions are taking place. Based on the requirements dictated by the reaction pathways, efficient catalysts for this process include noble metal catalysts like Pd, Pt, Ru, Rh, and Ir as well as non-noble metal catalysts such as promoted (with Co, Fe, and Sn) nickel catalysts.

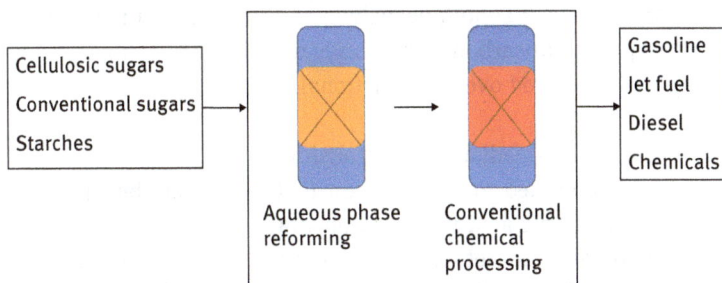

Fig. 8.5: BioForming Technology for sugars conversion to transportation fuels [50–54].

The most important feature of the concept described above is the parallel formation of reaction oxygenate intermediates like alcohols, ketones, acids, furans, paraffins, and other oxygenated hydrocarbons, with the mixture composition always depending on APR conditions (operating parameters and catalytic system) and the starting biomass-derived feed. These compounds undergo transformations via different reaction pathways (condensation, HDO, dehydration, oligomerization, etc.) targeted to longer-chain hydrocarbon formation, which can be used as fuels. The above process forms the basis of the BioForming Technology commercialized by Virent Company and is subject of a number of patents protecting the methods, systems, and processes under investigation [50–54]. Excellent reviews have been published by the above group concerning the process and are recommended for further reading to those who are interested [43, 55–57]. The Virent's BioForming process combines APR for the required hydrogen formation and several dehydration, hydrogenation, and base-catalyzed condensation reactions in order to finally form saturated hydrocarbons suitable for fuel applications. The above type of processing is flexible and utilize a wide range of biomass-derived compounds as illustrated in Figure 8.5 (http://www.virent.com/technology/bioforming/). Bifunctionality is a prerequisite of an efficient catalytic system, as not only hydrogenation but also C-C and C-O bond scission as well as dehydration reactions should occur. Moreover, concerning the economic evaluation of the process, preliminary analysis showed its competitiveness at crude oil prices greater than $60/bbl [44].

Conversion of highly functionalized molecules such as furfural, HMF, levulinic acid, etc. can be also realized under the above concept. This route involves dehydration pathways so as to remove hydroxyls, hydrogenation steps in order to reduce C=C bonds and aldol condensation reactions in order to form longer-chain alkanes (Figure 8.6) [47, 48]. These type of reactions can be self-aldol condensations or condensations with other carbonyl containing molecules (e.g., acetone). The latter reaction step is catalyzed by basic catalysis, e.g., basic solid Mg-Al oxides. An alternative option is the aldol condensation in biphasic systems so as to in situ extract the aldol adducts into the organic phase. A second condensation of the aldol adducts with the initial furfural molecules leads to longer chain compounds. Following is hydrogenation of C=C and C=O bonds of the aldol adducts catalyzed by a metallic phase (typically Pd). The preceding steps can also occur simultaneously over a bifunctional metal-basic catalyst [58]. The last step to hydrocarbon formation is dehydration/hydrogenation, where oxygen is completely removed over a bifunctional metal-acid catalyst such as $Pt/SiO_2-Al_2O_3$ [49].

Levulinic acid is another intermediate in the transformation of sugars into hydrocarbons, since it can be upgraded to hydrocarbon fuels (gasoline, jet fuel, and diesel). Transformation of levulinic acid involves dehydration/hydrogenation so as to minimize the oxygen content and ketonization in order to increase the molecular weight. Dumesic and coworkers reported [59] that a bifunctional (metal and acid sites) Pd/Nb_2O_5 catalyst allows this reaction sequence to be performed in a cascade

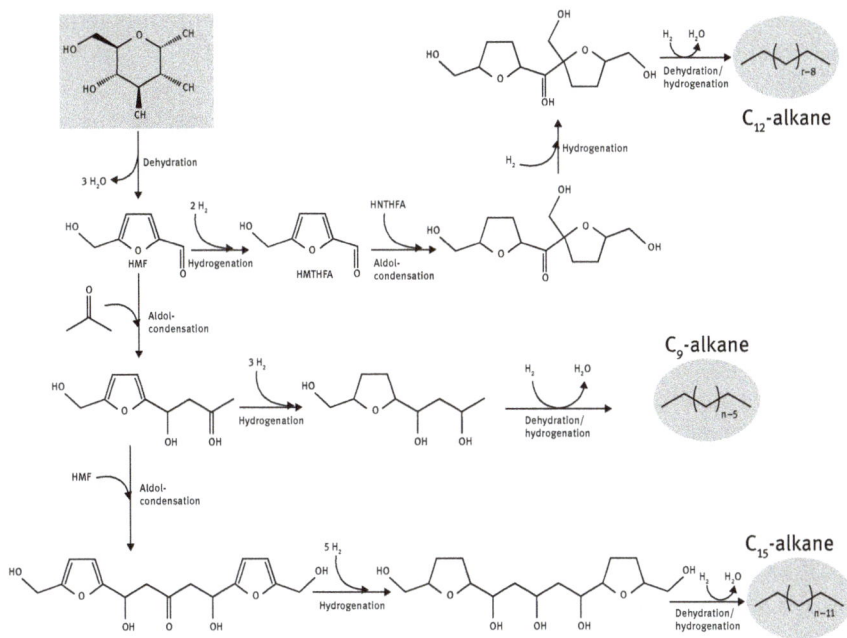

Fig. 8.6: Conversion of biomass-derived glucose acid to liquid hydrocarbons.

manner with minimum steps. The initial reaction step is levulinic acid hydrogenation to γ-valerolactone at low temperatures (423 K) over a Ru/C catalyst. Aqueous γ-valerolactone is subsequently converted to hydrophobic pentanoic acid over the Pd/Nb_2O_5 via an acid site-catalyzed ring opening reaction, which is in turn hydrogenated at moderate temperature and pressure conditions. Interestingly, pentanoic acid can be directly converted to 5-nonanone over the same Pd/Nb_2O_5 catalyst with 70% carbon yield at specific conditions (low space velocities). The latter process allows one-pot production of 5-nonanone from γ-valerolactone in a single reactor [60]. 5-Nonanone, which is obtained in an organic phase that spontaneously separates from the aqueous phase, serves as a platform molecule for the production of liquid hydrocarbon fuels. 5-Nonanone transformation pathway includes hydrogenation-dehydration cycles leading to linear n-nonane, which has excellent cetane number and lubricity that makes it a diesel blender agent [48]. Alternatively, 5-nonanol, formed by hydrogenation of 5-nonanone can be dehydrated and isomerized over a USY zeolite catalyst to produce a mixture of branched C9 alkenes, which, after hydrogenation to the corresponding alkanes, can be used as gasoline blend stocks.

Another process that converts aqueous mixtures of γ-valerolactone into liquid hydrocarbon fuels without the need of an external source of hydrogen has been developed by Prof. Dumesic's group [59]. According to this method, the feed undergoes decarboxylation at elevated pressures (36 bars) in the presence of a $SiO_2-Al_2O_3$ catalyst forming butane isomers and CO_2. This gaseous stream is afterward oligomerized, a solid acid catalyst (H-ZSM5, Amberlyst) yielding alkenes with molecular weights suitable for gasoline and jet fuel applications.

The discussion presented above clearly points out to the potential of sugars and sugar-derived platform molecules conversion to hydrocarbons via energy- and atom-efficient processes.

8.4 Upgrading of butanol to fuels

Another class of compounds that recently attracted a lot of interest as additives and/ or after further processing as substitutes of jet fuel is heavier alcohols and most importantly butanols. Butanols, derived from fossil sources, are well-known gasoline additives for over 40 years. Recently, efforts are directed toward the development of renewable processes to produce butanols using biomass-derived sugars. This strategy aims at converting existing cornstarch ethanol plants into isobutanol plants (http:// www.gevo.com/our-markets/isobutanol/). The latter is due to lower ethanol energy content and difficulties in its handling. The above process has been realized by Gevo (http://www.gevo.com/our-markets/isobutanol/). By analogy to ethanol, isobutanol can be shipped in pipelines, both inbound to and outbound from a refining/blending plant. Isobutanol's properties (low Reid vapor pressure, above average octane, good energy content, low water solubility, and low oxygen content) allow blending with gasoline and transformation to other valuable products, thus providing great flexibility in the industry.

Focusing on hydrocarbon fuels, isobutanol serves as a platform molecule, as it can be converted to isobutylene, a precursor for a variety of transportation fuel products such as iso-octene (gasoline blend stock), iso-octane (alkylate-high-quality gasoline blend stock and/or aviation gasoline blend stock), iso-paraffinic kerosene (jet fuel), and diesel. The strategy to convert isobutanol to jet fuel includes dehydration followed by oligomerization, hydrogenation, and distillation steps (Figure 8.7). Gevo (Englewood, CO, USA) claims the following properties of the renewable jet fuel: high blend rate, very low freeze point (-80°C), high thermal oxidation stability, and ASTM distillation curve requirements.

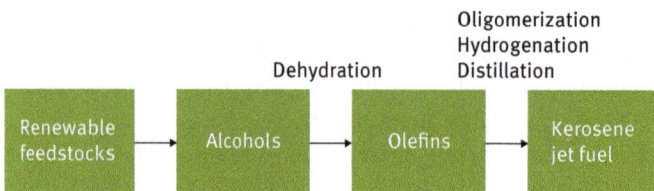

Fig. 8.7: Isobutanol conversion to jet fuel.

Table 8.4: Conventional and bio-jet fuel properties (www.cobalttech.com).

	Cobalt/Navy jet fuel profile	
Physical properties	**ATJ-5 DLA requirements**	**Cobalt/Navy bio-jet fuel**
Freeze point (D5972)	<−46°C	<−82°C (LoWax at −82°C)
Distillation (D86)	T90−T10>25°C	49.6°C
Hydrocarbon type (D6379)	<30% cycloparaffins	0%
Heat of combustion (D4809)	>42.8 MJ/kg	44.1 MJ/kg
Density at 15°C (D4052)	0.76−0.845	0.777
Flash point (D93)	>60°C	67°C
Aromatics (D2425)	<0.5 vol%	0.1% (ASTM D6379)
Elemental composition	Report	85% C, 15% H

In 2010, Cobalt Technologies (www.cobalttech.com), a company that commercialized biobutanol production as a renewable chemical and fuel, announced the signature of a Cooperative Research and Development Agreement (CRADA) with US Navy designed to develop a process for the conversion of biobutanol into jet and diesel fuels. Specific research goals include optimization of the dehydration chemistry for the conversion of bio-*n*-butanol to 1-butene, followed by oligomerization of the latter into jet fuel. Additional focus will be paid on converting the biobutanol into butyl ether, which will be mixed with *n*-butanol and other compounds targeted to a viable drop-in diesel fuel replacement. Table 8.4 shows the physical properties of the Cobalt/Navy bio-jet fuel, verifying all quality requirements are met.

Conversion of butanol to gasoline-range hydrocarbons has been studied by various research groups [35, 61–63]; however, research on this issue seems to be limited. The reaction sequence includes dehydration of butanol as the first step of alcohol-gasoline conversion over solid acids. Moreover, the formed butenes convert into gasoline hydrocarbons via dimerization, isomerization, aromatization, and alkylation reactions, which take place in the zeolite pores. Early studies reported by Le Van Mao and McLaughlin [62] revealed that among various light alcohols used as feedstock, the higher yields in liquid hydrocarbons are obtained with butanol. The most recent study on this issue has been reported by Varvarin et al. [63] over a series of zeolites as catalysts. In the presence a H-ZSM-5, a gasoline yield of 50–55 wt% at 300–350°C and liquid hourly space velocity (LHSV) = 0.3/h (Figure 8.8) was obtained. The liquid product stream formed composed of aliphatic, olefinic, and aromatic hydrocarbons, while it was observed that product distribution strongly depends on temperature and different zeolites evaluated. Similar results reported by the other research groups cited above.

Another approach has been quite recently reported by Nahreen and Gupta from the Auburn University [64]. An acetone-butanol-ethanol mixture, containing 62.9 wt% *n*-butanol, 29.3 wt% acetone, and 7.8 wt% ethanol, was subjected to dehydration

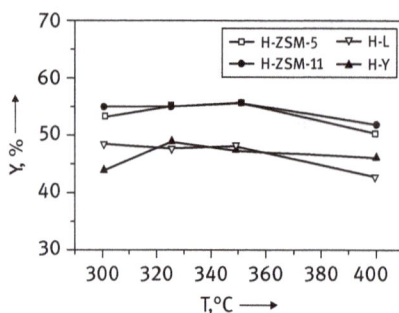

Fig. 8.8: Yield of liquid hydrocarbons over studied H-zeolites at different temperatures (LHSV =0.3/h) (reproduced from Varvarin et al. [63], with permission).

reactions in order to deoxygenate the mixture. The process was realized with two different catalysts: alumina (γ-Al$_2$O$_3$) and zeolite (ZSM-5). The dehydration products formed from the acetone-butanol-ethanol mixture are mostly unsaturated hydrocarbons in the range of C$_2$-C$_{16}$. Three phases were formed consisting of gas products (light hydrocarbons and carbon dioxide), organic liquid phase (heavy hydrocarbons), and an aqueous phase (dissolved oxygenated hydrocarbons). Interestingly, the catalytic dehydration of the mixture was different from the individual compounds, revealing a synergy of the reacting components and resulting in a product with high heating value. Best performance in dehydration routes was shown by γ-Al$_2$O$_3$ at 400°C, which produced the highest amount of useful products (butane) and high heating value liquid. The above discussed processes show that butanol can be regarded as a versatile bio-based chemical that could be effectively transformed to renewable hydrocarbons.

An alternative strategy for the production of fuels and fuel additives based on alcohols has been explored within the framework of Eurobioref project funded by FP7 EU research framework (www.eurobioref.org). The simplified flow scheme of the main process routes and products is depicted below (Fig. 8.9). The main characteristics of the proposed value chain, which is based on alcohols, is the production of heavy alcohols/branched paraffins to be blended as components of aviation gasoline and jet fuel. Integrated tests at demo scale proved the feasibility of this approach, while complete technology development showed the potential for industrial implementation.

8.5 Challenges – outlook

Lignocellulose is the cheapest and most abundant form of biomass and, on an energy basis, is significantly cheaper than crude oil. However, the recalcitrance and the

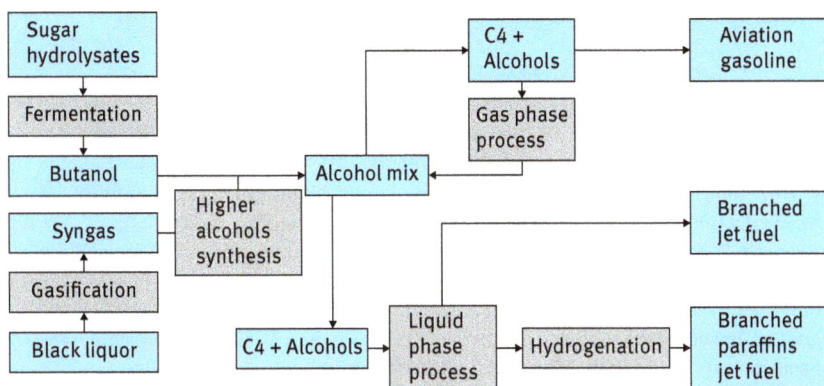

Fig. 8.9: Simplified flow scheme of the main process routes and products based on heavy alcohols (www.eurobioref.org).

complexity of non-edible lignocellulosic biomass are the main hurdles for the large-scale utilization of lignocellulose as a source of liquid hydrocarbon fuels. Fuels are a commodity produced on a large scale, and therefore, development of processes economically feasible requires large investments. The technology level of all the processes described in the previous sections is not fully developed, and cost analysis shows that the hydrocarbon biofuels are not competitive with the conventional fossil-based fuels unless a subsidy is provided. As most of the processes are catalytic, problems related with low activity, poor selectivity to the target product, and most importantly unsatisfactory stability are common in all the routes described. The fundamental chemistry of most of the reactions is not well understood, and it is likely that further scientific understanding will lead to improved processes. It is vital that new and more efficient catalysts, taking into account the nature of the renewable feedstock, be developed . It is likely that heterogeneous catalysts that have been the backbone of the chemical and petrochemical industry will continue to play a key role in the biomass to fuels sector. It should be also underlined that the engineering of the processes like the reactor design and the separation steps has not been adequately addressed. The importance of both is high because with new reactor designs, the overall product yield could be further improved, while with the detailed investigation and design of the separation steps, issues like the presence of impurities and their effect on the final performance of the fuels will be identified and eventually resolved.

The production of hydrocarbon fuels from lignocellulose is technically feasible, with good perspectives for further improvements process-wise and could be economically feasible only within a biorefinery highly integrated with other process routes and products.

Bibliography

[1] Ragauskas, A. J., Williams, C. K., Davison, B. H., Britovsek, G., Cairney, J., Eckert, C. A., Frederick, WJ, Jr., Hallett, J. P., Leak, D. J., Liotta, C. L., Mielenz, J. R., Murphy, R., Templer, R., Tschaplinski, T., The path forward for biofuels and biomaterials, Science **311** (2006) 484–489.

[2] Technology roadmap: biofuels for transport, IEA report, 2011.

[3] Mohan, D., Pittman, CU, Jr., Steele, P. H., Pyrolysis of wood/biomass for bio-oil: a critical review, Energy Fuel **20** (2006) 848–889.

[4] Venderbosch, R. H., Prins, W., Fast pyrolysis technology development, Biofuel Bioprod Bioref **4** (2010) 178–208.

[5] Marsman, J. H., Wildschut, J., Mahfud, F., Heeres, H. J., Identification of components in fast pyrolysis oil and upgraded products by comprehensive two-dimensional gas chromatography and flame ionisation detection, J Chromatogr A **1150** (2006) 21–27.

[6] Samolada, M. C., Baldauf, W., Vasalos, I. A., Production of a bio-gasoline by upgrading biomass flash pyrolysis liquids via hydrogen processing and catalytic cracking, Fuel **7** (1998) 1667–1675.

[7] Oasmaa, A., Czernik, S., Fuel oil quality of biomass pyrolysis oils – state of the art for the end users, Energy Fuel **13** (1999) 914–921.

[8] Czernik, S., Bridgwater, A. V., Overview of applications of biomass fast pyrolysis oil, Energy Fuel **18** (2004) 590–598.

[9] Yaman, S., Pyrolysis of biomass to produce fuels and chemical feedstocks, Energy Convers Manage **45** (2004) 651–671.

[10] Adjaye, J. D., Sharma, R. K., Bakhshi, N. N., Characterization and stability analysis of wood-derived bio-oil, Fuel Process Technol **31** (1992) 241–256.

[11] Zhang, Q., Chang, J., Wang, T., Xu, Y., Review of biomass pyrolysis oil properties and upgrading research, Energy Convers Manage **48** (2007) 87–92.

[12] Lappas, A. A., Kalogiannis, K. G., Iliopoulou, E. F., Triantafyllidis, K. S., Stefanidis, S. D., Catalytic pyrolysis of biomass for transportation fuels, WIREs Energy Environ **1** (2012) 285–297.

[13] Lemonidou, A. A., Kechagiopoulos, P. N., Heracleous, E., Voutetakis, S. S., Catalytic Steam reforming of bio-oils to hydrogen, in Triantafyllidis, K., Lappas, A. A., Stöcker, M., editors, The Role of Catalysis for the Sustainable Production of Bio-fuels and Bio-chemicals, Elsevier, Oxford, (2013) 469–495.

[14] Huber, G. W., Iborra, S., Corma, A., Synthesis of transportation fuels from biomass: chemistry, catalysts, and engineering, Chem Rev **106** (2006) 4044–4498.

[15] Elliott, D., Historical developments in hydroprocessing bio-oils, Energy Fuel **21** (2007) 1792–1815.

[16] Furimsky, E., Catalytic hydrodeoxygenation, Appl Catal A Gen **199** (2000) 147–190.

[17] Demirbas, A., Progress and recent trends in biofuels, Prog Energy Combust Sci **33** (2007) 1–18.

[18] Demirbas, M. F., Biorefineries for biofuel upgrading: a critical review, Appl Energy **86** (2009) S151–S161.

[19] Bulushev, D. A., Ross, J. R. H., Catalysis for conversion of biomass to fuels via pyrolysis and gasification, Catal Today **171** (2011) 1–13.

[20] Fonseca, A., Zeuthen, P., Nagy, J., ^{13}C n.m.r, quantitative analysis of catalyst carbon deposits, Fuel **75** (1996) 1363–1376.

[21] Fonseca, A., Zeuthen, P., Nagy, J., Assignment of an average chemical structure to catalyst carbon deposits on the basis of quantitative ^{13}C n.m.r. spectra, Fuel **75** (1996) 1413–1423.

[22] Baldauf, W., Balfanz, U., Rupp, M., Upgrading of flash pyrolysis oil and utilization in refineries, Biomass Bioenergy **7** (1994) 237–244.

[23] Baldauf, W., Balfanz, U., Upgrading of pyrolysis oils from biomass in existing refinery structures; final report JOUB-0015, Veba Oel, Gelsenkirchen, 1992.

[24] Lappas, A. A., Bezergianni, S., Vasalos, I. A., Production of biofuels via co-processing in conventional refining processes, Catal Today **145** (2009) 55–62.

[25] Martinez, R., Huff, M. C., Barteau, M. A., Ketonization of acetic acid on titania-functionalized silica monoliths, J Catal **222** (2004) 404–409.

[26] Dooley, K. M., Bhat, A. K., Plaisance, C. P., Roy, A. D., Ketones from acid condensation using supported CeO$_2$ catalysts: effect of additives, Appl Catal A Gen **320** (2007) 122–133.

[27] Nagashima, O., Sato, S., Takahashi, R., Sodesawa, T., Ketonization of carboxylic acids over CeO$_2$-based composite oxides, J Mol Catal A Chem **227** (2005) 231–239.

[28] Pham, T. N., Shi, D., Resasco, D. E., Evaluating strategies for catalytic upgrading of pyrolysis oil in liquid phase, Appl Catal B Environ **145** (2014) 10–23.

[29] Idriss, H., Diagne, C., Hindermann, J. P., Kiennemann, A., Barteau, M. A., Reactions of acetaldehyde on CeO$_2$ and CeO$_2$-supported catalysts, J Catal **155** (1995) 219–237.

[30] Kunkes, E. L., Simonetti, D. A., West, R. M., Serrano-Ruiz, J. C., Gaertner, C. A., Dumesic, J. A., Catalytic conversion of biomass to monofunctional hydrocarbons and targeted liquid-fuel classes, Science **322** (2008) 417–421.

[31] Gaertner, C. A., Serrano-Ruiz, J. C., Braden, D. J., Dumesic, J. A., Catalytic coupling of carboxylic acids by ketonization as a processing step in biomass conversion, J Catal **266** (2009) 71–718.

[32] Gurbuz, E., Kunkes, E. L., Dumesic, J. A., Integration of C-C coupling reactions of biomass-derived oxygenates to fuel-grade compounds, Appl Catal B Environ **94** (2010) 134–141.

[33] Snell, R. W., Shanks, B. H., Ceria calcination temperature influence on acetic acid ketonization: mechanistic insights, Appl Catal A Gen **451** (2013) 86–93.

[34] Srinivas, S. T., Dalai, A. K., Bakhshi, N. N., Thermal and catalytic upgrading of a biomass-derived oil in a dual reaction system, Can J Chem Eng **78** (2000) 343–354.

[35] Gayubo, A. G., Aguayo, A. T., Atutxa, A., Aguado, A., Bilbao, J., Transformation of oxygenate components of biomass pyrolysis oil on H-ZSM-5 zeolite, alcohols and phenols, Ind Eng Chem Res **43** (2004) 2610–2618.

[36] Gayubo, A. G., Aguayo, A. T., Atutxa, A., Aguado, R., Olazar, M., Bilbao, J., Transformation of oxygenate components of biomass pyrolysis oil on a HZSM-5 zeolite, II. Aldehydes, ketones, and acids, Ind Eng Chem Res **43** (2004) 2619–2626.

[37] Gayubo, A. G., Aguayo, A. T., Atutxa, A., Valle, B., Bilbao, J., Undesired components in the transformation of biomass pyrolysis oil into hydrocarbons on an HZSM-5 zeolite catalyst, J Chem Technol Biotechnol **80** (2005) 1244–1251.

[38] Sharma, R. K., Bakhshi, N. N., Catalytic conversion of fast pyrolysis oil to hydrocarbon fuels over HZSM-5 in a dual reactor system, Biomass Bioenerg **5** (1993) 445–455.

[39] Sharma, R. K., Bakhshi, N. N., Conversion of non-phenolic fraction of biomass-derived pyrolysis oil to hydrocarbon fuels over HZSM-5 using a dual reactor system, Bioresource Technol **45** (1993) 195–203.

[40] Sharma, R. K., Bakhshi, N. N., Upgrading of pyrolytic lignin fraction of fast pyrolysis oil to hydrocarbon fuels over HZSM-5 in a dual reactor system, Fuel Proc Technol **35** (1993) 201–218.

[41] Graca, I., Ribeiro, F. R., Cerqueira, H. S., Lam, Y. L., de Almeida, M. B. B., Catalytic cracking of mixtures of model bio-oil compounds and gasoil, Appl Catal B Environ **90** (2009) 556–563.

[42] Thegarid, N., Fogassy, G., Schuurman, Y., Mirodatos, C., Stefanidis, C., Iliopoulou, E. F., Kalogiannis, K., Lappas, A. A., Second-generation biofuels by co-processing catalytic pyrolysis oil in FCC units, Appl Catal B Environ **145** (2014) 161–166.

[43] Chheda, J., Huber, G. W., Dumesic, J. A., Liquid-phase catalytic processing of biomass derived oxygenated hydrocarbon to fuels and chemicals, Angew Chem Int Ed **46** (2007) 7164–7183.

[44] Melero, J. A., Iglesias, J., Garcia, A., Biomass as renewable feedstock in standard refinery units, feasibility, opportunities and challenges Energy Environ Sci **5** (2012) 7393–7420.

[45] Kim, Y. T., Dumesic, J. A., Huber, G. W., Aqueous-phase hydrodeoxygenation of sorbitol: a comparative study of Pt/Zr phosphate and PtReOx/C, J Catal **304** (2013) 72–85.

[46] Serrano-Ruiz, J. C., Wang, D., Dumesic, J. A., Catalytic upgrading of levulinic acid to 5-nonanone, Green Chem **12** (2010) 574–577.

[47] Huber, G. W., Chheda, J. N., Barrett, C. J., Dumesic, J. A., Production of liquid alkanes by aqueous-phase processing of biomass-derived carbohydrates, Science **308** (2005) 1446–1450.

[48] West, R. M., Liu, Z. L., Peter, M., Dumesic, J. A., Liquid alkanes with targeted molecular weights from biomass-derived carbohydrates, ChemSusChem **1** (2008) 417–424.

[49] Huber, G. W., Cortright, R. D., Dumesic, J. A., Renewable alkanes by aqueous-phase reforming of biomass-derived oxygenates, Angew Chem Int Ed **43** (2004) 1549–1551.

[50] Cortright, R. D., Blommel, P. G., Synthesis of liquid fuels and chemicals from oxygenated hydrocarbons, US Patent Application 13/157247, October 20, 2011.

[51] Cortright, R. D., Blommel, P. G., Synthesis of Liquid fuels and chemicals from oxygenated hydrocarbons, US Patent Application 12/044908, December 4, 2008.

[52] Beck, T., Blank, B., Jones, C., Woods, E, Cortright, R., Production of aromatics from di- and polyoxygenates, US Patent Application 14/210925, September 18, 2014.

[53] Held, A., Woods, E., Cortright, R., Gray, M., Processes for converting biomass-derived feedstocks to chemicals and liquid fuels, US Patent Application 14/215283, September 18, 2014.

[54] Cortright, R. D., Blommel, P., Synthesis of liquid fuels and chemicals from oxygenated hydrocarbons, US Patent Application 13/793068, July 25, 2013.

[55] Simonetti, D. A., Dumesic, J. A., Catalytic production of liquid fuels from biomass-derived oxygenated hydrocarbons: catalytic coupling at multiple length scales, Catal Rev **51** (2009) 441–484.

[56] Alonso, D. M., Wettsteina, S. G., Dumesic, J. A., Bimetallic catalysts for upgrading of biomass to fuels and chemicals, Chem Soc Rev **41** (2012) 8075–8098.

[57] Davda, R. R., Shabaker, J. W., Huber, G. W., Cortright, R. D., Dumesic, J. A., A review of catalytic issues and process conditions for renewable hydrogen and alkanes by aqueous-phase reforming of oxygenated hydrocarbons over supported metal catalysts, Appl Catal B Environ **56** (2005) 171–186.

[58] Barret, C., Chheda, J., Huber, G. W., Dumesic, J. A., Single-reactor process for sequential aldol-condensation and hydrogenation of biomass-derived compounds in water, Appl Catal B Environ **66** (2006) 111–118.

[59] Bond, J. Q., Martin-Alonso, D., Wang, D., West, R. M., Dumesic, J. A., Integrated catalytic conversion of g-valerolactone to liquid alkenes for transportation fuels, Science **327** (2010) 1110–1114.

[60] Serrano-Ruiz, J. C., West, R. M., Dumesic, J. A., Catalytic conversion of renewable biomass resources to fuels and chemicals, Annu Rev Chem Biomol Eng **1** (2010) 79–101.

[61] Fuhse, J., Bandermann, F., Conversion of organic oxygen compounds and their mixtures on H-ZSM-5, Chem Eng Technol **10** (1987) 323–329.

[62] Le Van Mao, R., McLaughlin, G. P., Conversion of light alcohols to hydrocarbons over ZSM-5 zeolite and asbestos-derived zeolite catalysts, Energy Fuel **3** (1989) 620–624.

[63] Varvarin, A. M., Khomenko, K. M., Brei, V. V., Conversion of n-butanol to hydrocarbons over H-ZSM-5, H-ZSM-11, H-L and H-Y zeolites, Fuel **106** (2013) 617–620.

[64] Nahreen, S., Gupta, R. B., Conversion of the acetone-butanol-ethanol (ABE) mixture to hydrocarbons by catalytic dehydration, Energy Fuel **27** (2013) 2116–2125.

Marco Ricci and Carlo Perego

9 Use of bio-sourced syngas

Abstract: Syngas is one of the most important feedstock of the chemical industry, largely used to produce several major chemicals including hydrogen, ammonia (and then urea), methanol, and its derivatives, oxo-chemicals (aldehydes and their derivatives), and even fuels obtained by the Fischer-Tropsch process. In the last few years, syngas has been also used as feedstock for fermentations, mostly to produce ethanol. Since syngas can be obtained from a variety of raw materials including biomass (bio-syngas), it is likely that it will play a significant role in the development of bio-refineries. Chemical and, to a lesser extent, energy use of syngas and bio-syngas will be shortly reviewed, and some perspective view will be provided.

9.1 Introduction

Synthesis gas, most often referred to as syngas, is basically a mixture of hydrogen and carbon monoxide, along with several impurities, largely used as feedstock for the production of many important chemicals. In the next few years, the market of syngas and its derivatives is expected to grow at a compound annual growth rate higher than 8%. One of the main drivers for this increase is the feedstock flexibility of the syngas production process: in fact, it can be produced from a number of different sources including coal, natural gas, oil, and biomass [1].

Historically, syngas availability played a very important role in the development of the chemical industry, originally based on coal. Syngas production by reaction of coal with steam

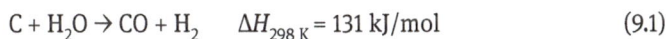

$$C + H_2O \rightarrow CO + H_2 \qquad \Delta H_{298\,K} = 131 \text{ kJ/mol} \tag{9.1}$$

is an endothermic process and the heat is provided by burning part of the feed. Since different applications usually require a different CO to H_2 ratio, the latter can be adjusted by exploiting the water gas shift (WGS) reaction, for the discovery of which (in 1780!) the Italian physicist Felice Fontana is credited [2]:

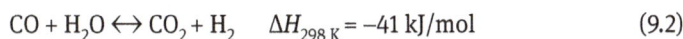

$$CO + H_2O \leftrightarrow CO_2 + H_2 \qquad \Delta H_{298\,K} = -41 \text{ kJ/mol} \tag{9.2}$$

Well into the 20th century, oil-based materials became the main feedstock for the chemical industry. Nevertheless, syngas retained, and still retains, its importance even if most of it is no longer prepared from coal but rather through the steam reforming of natural gas:

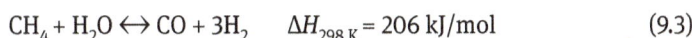

$$CH_4 + H_2O \leftrightarrow CO + 3H_2 \qquad \Delta H_{298\,K} = 206 \text{ kJ/mol} \tag{9.3}$$

Finally, syngas can be also prepared by gasification of biomass, including organic wastes. Syngas produced this way is often referred to as bio-syngas. It can be prepared by exploiting the least valuable parts of a biomass (i.e., not sugars or lipids). Since it is likely that industrial implementation of the biorefinery concept will only succeed if it will be possible to use as much of the biomass as possible, the use of bio-syngas prepared from poor biomass residues assumes an obvious interest. Along with hydrogen and carbon monoxide, bio-syngas also contains carbon dioxide, water, some methane, and several minor, characteristic impurities deriving from biomass heteroatoms, mainly HCN, NH_3, H_2S, COS, and HCl.

In the following, chemical, and energy uses of syngas will be shortly reviewed, paying particular attention to bio-syngas.

9.2 Uses of syngas

Current uses of syngas can be arranged into three main classes:
- as a fuel (and as a biofuel if bio-syngas is concerned), mainly for power generation
- as a chemical feedstock for producing a number of chemical intermediates
- as an intermediate for the production of transportation fuels or biofuels.

9.2.1 Syngas as a fuel

Syngas has approximately half of the energy density of natural gas and can be used in steam cycles, gas engines, fuel cells, or turbines for power generation with co-production of heat.

Syngas, particularly bio-syngas, was also used as a transportation fuel in Italy, during the embargo that followed the Ethiopia invasion (1935–1936). At that time, many cars were equipped with the *gasogeno*, a cumbersome device in which wood (or coal) was burned in an O_2-poor atmosphere to provide a mixture of CO, CO_2, N_2, and H_2. This mixture was a cheap fuel that, despite its low energy content, could be used instead of gasoline. Both power and mileage of modified cars were very poor, and frequent stops were needed to recharge the wood, since 2.5 kg of it provided the energy of just 1 L of gasoline.

9.2.2 Syngas as a chemical feedstock

Many syngas derivatives have a large number of chemical applications that range from chemical intermediates to fertilizers, monomers, solvents, fine chemicals, etc. The main current processes for the production of both chemicals and fuels starting from syngas are summarized in Figure 9.1.

Hydrogen, ammonia, and methanol are the most important syngas derivatives.

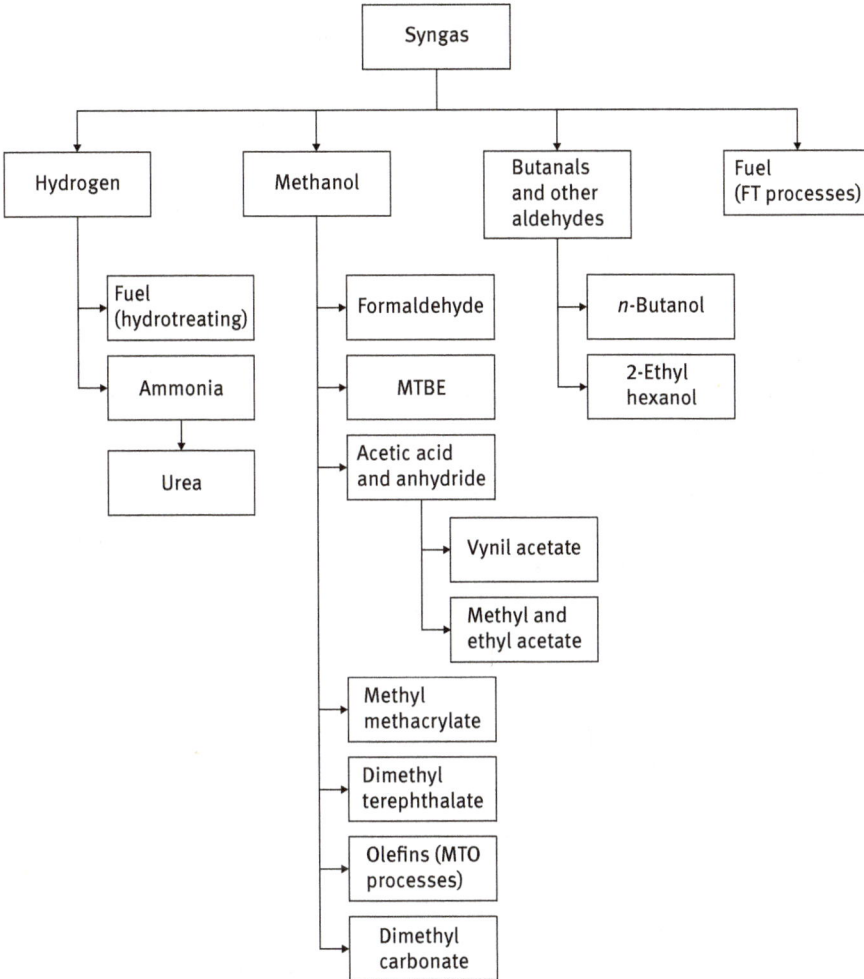

Fig. 9.1: Main chemical and fuel productions based on syngas.

9.2.2.1 Hydrogen production and uses

With more than 50 Mt (millions of metric tons) produced globally each year, hydrogen is a critical feedstock for both the chemical industry and the refinery [3].

More than 95% of hydrogen production occurs via the steam reforming of methane [Eq. (9.3)], oil, or coal [Eq. (9.1)], all affording syngas, which then undergoes WGS [Eq. (9.2)]. The use of methane, oil, and coal as raw materials account, respectively, for almost 50%, 30%, and 18% of the global hydrogen production, with a further 4% being provided by electrolysis of aqueous solutions.

According to the geographical location (e.g., USA or Europe), the main use of hydrogen occurs either in the refinery or for the ammonia synthesis. Particularly, in the refinery, hydrogen is needed (and will be needed in future biorefineries as well) for the hydrotreatment processes in the production of high-quality fuels with low environmental impact (e.g., for hydrodesulfurization and hydrodearomatization) [4] and for the conversion of heavy crude oil fractions into middle distillates (kero and diesel fuel) by hydrocracking [5].

9.2.2.2 Ammonia production and uses

Ammonia production accounts for about 1.6% of the world consumption of fossil energy and amounted in 2012 to ca. 137 Mt [6]. Due to its nitrogen content, ammonia is the most important source of nutrients for plant growth and its most relevant use is in the production of fertilizers. In addition, every single nitrogen atom of the industrially produced chemicals comes, directly or indirectly, from ammonia, which is a fundamental building block for the production of intermediates, plastics, fibers, and explosives.

By far, the most important method for manufacturing ammonia is the synthesis from the elements, which accounts for over 90% of the global production and which is still basically run according to the Haber-Bosch process, first industrialized in 1913. Hydrogen and nitrogen in stoichiometric 3:1 molar ratio are reacted at high pressure (100–250 bar) and temperature between 350°C and 550°C, usually over iron-based catalysts:

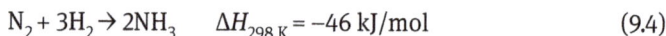

$$N_2 + 3H_2 \rightarrow 2NH_3 \qquad \Delta H_{298\,K} = -46\ \text{kJ/mol} \qquad (9.4)$$

The once-through conversion is low (20–30%), and a substantial part of the unconverted gas is recirculated to enhance the total conversion.

9.2.2.3 Urea production and use

As already stated, ammonia is largely consumed for producing fertilizers, particularly urea. Urea is manufactured via a two-step reaction of carbon dioxide and ammonia. In the first step, the reagents combine to give ammonium carbamate,

$$2NH_3 + CO_2 \leftrightarrow H_2N-COO^-\,NH_4^+ \qquad (9.5)$$

which, in the second step, affords urea upon dehydration:

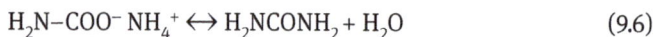

$$H_2N-COO^-\,NH_4^+ \leftrightarrow H_2NCONH_2 + H_2O \qquad (9.6)$$

The importance of urea can hardly be overestimated. The growth of any food crop is usually limited by the availability of nitrogen, which is much needed for the synthesis of proteins. Natural nitrogen fixation probably allows the production of roughly 50% of the global protein demand or so. Thus, in order to satisfy such demand, we need nitrogen fertilizers, and urea is, by far, the most important among them: almost two fifths of the world's population rely, particularly in less developed countries, on urea for food supply. As a consequence, urea is one of the world's most produced chemicals with a global supply that, in 2012, surpassed 161 Mt.

9.2.2.4 Methanol production and uses

Methanol is another major product of chemical industry with a global production (2013) around 70 Mt/year. It is produced from syngas by catalytic hydrogenation of carbon monoxide over zinc/copper-based catalyst:

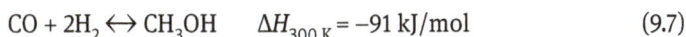

$$CO + 2H_2 \leftrightarrow CH_3OH \qquad \Delta H_{300\,K} = -91 \text{ kJ/mol} \qquad (9.7)$$

Methanol synthesis is exothermic and equilibrium limited, favored by low temperature and high pressure. Commercially, fixed-bed reactors are used, operating at 230–270°C and 50–100 atm.

Stoichiometry requires 2 mol of hydrogen per 1 mol of CO. Thus, when syngas is prepared, as usual, by steam reforming of natural gas [Eq. (9.3)], some excess of hydrogen is available with respect to Eq. (9.7). In this case, it can be useful to co-feed some CO_2, the reduction of which is still an exothermic reaction, although to a lesser extent compared to that of carbon monoxide:

$$CO_2 + 3H_2 \leftrightarrow CH_3OH + H_2O \qquad \Delta H_{300\,K} = -49 \text{ kJ/mol} \qquad (9.8)$$

Interestingly, from a mechanistic point of view, it is increasingly accepted that methanol forms almost exclusively by hydrogenation of the CO_2 contained in the syngas or formed upon the WGS. In fact, it has been shown that the reaction on a conventional methanol catalyst of a CO/H_2 mixture carefully purified from both CO_2 and water affords no or very little methanol [7].

Methanol is largely used in refrigeration systems or as antifreeze, inhibitor of hydrates formation in natural gas pipelines, absorption agent in gas scrubbers, and, to a lesser extent, as a solvent. However, about 70% of its production is actually used as a raw material for the synthesis of other intermediates and products. The most relevant of them, in order of decreasing importance, are

- Formaldehyde. It accounts for ca. 30% of the world methanol demand and is mainly used in the construction industry to produce adhesives for the manufacture of various construction board.
- Acetic acid. Mainly obtained by methanol carbonylation, i.e., by the reaction of methanol with CO. The reaction is catalyzed by salts or carbonyl complexes of cobalt, rhodium, or iridium. Acetic acid production consumes ca. 10% of the world methanol market. Acetic acid is mainly required for the manufacturing of vinyl acetate monomer or as a solvent for terephthalic acid production.

- Methyl *ter*-butyl ether (MTBE). Produced by acid-catalyzed addition of methanol to isobutene. Heterogeneous catalysts are used, e.g., acid resins. MTBE production accounts for ca. 7% of the global methanol demand. MTBE is largely used as a gasoline additive.
- Methyl methacrylate.
- Dimethyl terephthalate.

A relatively minor consumption of methanol occurs in the production of dimethylcarbonate (DMC). Its 2011 production was 429 kt, mostly for captive use in the phosgene-free synthesis of polycarbonates via a double transesterification process: of DMC with phenol to get diphenylcarbonate, and then of the latter with bisphenol A to afford polycarbonate. Potential for further DMC development, however, is still very high both as a poorly toxic oxygenated solvent and as a friendly methylating or carbonylating reagent [8].

Finally, should methanol be available in sufficient amounts from biomass, methanol-to-olefin (MTO) processes, currently under development, may in principle allow to refound on renewable raw materials most of the current chemical industry, which is largely based on olefins (particularly ethylene and propylene) obtained by the cracking of naphtha, liquefied petroleum gas, ethane, propane, or butane.

9.2.2.5 Oxo chemicals production and uses

Olefins react with syngas in the presence of homogeneous catalysts to afford aldehydes with one more carbon atom. The reaction, discovered in 1938 by the German chemist Otto Roelen at Ruhrchemie, is usually referred to as hydroformylation or oxosynthesis. α-Olefins can afford two isomeric aldehydes, linear and branched:

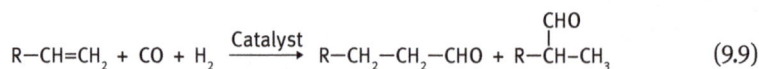

$$
R-CH=CH_2 + CO + H_2 \xrightarrow{\text{Catalyst}} R-CH_2-CH_2-CHO + R-\overset{\overset{\displaystyle CHO}{|}}{CH}-CH_3 \tag{9.9}
$$

Cobalt carbonyl or rhodium complexes are most used as catalysts. Rhodium-based catalysts usually show high selectivity toward linear aldehydes.

The most important oxo products are butyraldehydes (butanals), manufactured by propene hydroformylation. Linear butyraldehyde is mostly hydrogenated to *n*-butanol or subjected to aldol condensation on the way to prepare 2-ethylhexanol. Alcohols prepared by reduction of C6-C13 aldehydes are widely employed as plasticizers, whereas mixtures of aldehydes with 12–15 carbon atoms are used as intermediates in the production of surfactants for detergency. In 2012, the total worldwide production of oxo chemicals exceeded 50 Mt, 60% of which was accounted for by *n*-butyraldehyde.

9.2.3 Transportation fuels from syngas: the Fischer-Tropsch process [9]

As already seen, methanol is obtained by the heterogeneously catalyzed hydrogenation of CO. Different catalysts and conditions dramatically change the reaction output: substantial formation of carbon-carbon bond occurs and linear alkenes and alkanes are the main products, even if it is likely that the alkenes are the only primary products,

$$nCO + 2nH_2 \rightarrow C_nH_{2n} + nH_2O \qquad \Delta H_{298} = ca. -150 \text{ kJ/mol}_{CO} \qquad (9.10)$$

and the alkanes only arise upon their hydrogenation.

The reaction is usually referred to as Fischer-Tropsch (FT), after the names of Franz Fischer and Hans Tropsch who discovered it in 1923 at the Kaiser Wilhelm Institute für Kohlenforschung in Mülheim an der Ruhr (Germany) [10, 11].

9.2.3.1 Fischer-Tropsch to fuels: biomass-to-liquids processes

FT technologies allow the production of fuels with excellent properties, basically not different from those of diesel and gasoline obtained by oil refining. So far, however, only particular geopolitical situations favored the realization of industrial plants to produce synthetic fuels starting from syngas obtained, in turn, from coal (CtL processes). This was the case in Germany during World War II and in South Africa during the period of the embargo. More recently, however, there was a renewed interest for the FT technology for two reasons. One of the reasons is the availability, often in very remote locations, of huge amounts of natural gas difficult to transport. Indeed, transporting a liquid fuel is much easier, and several initiatives flourished to get liquids from natural gas via steam reforming [Eq. (9.3)] followed by FT [gas-to-liquid (GtL) processes] [12]. Another reason is the impetus on biofuels, which prompted research on biomass gasification followed by FT. This latter approach provides routes to transform biomass into liquid fuels (BtL processes [13]) with typical energy efficiency (i.e., FT fuel heat value/biomass heat value) of ca. 30%.

Eq. (9.10) is basically the reverse of a steam reforming reaction and is definitely exothermic. Although linear alkenes and alkanes are the main products, several other reactions also occur, including methane formation by CO hydrogenation:

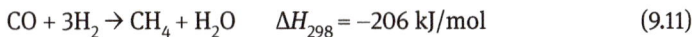

$$CO + 3H_2 \rightarrow CH_4 + H_2O \qquad \Delta H_{298} = -206 \text{ kJ/mol} \qquad (9.11)$$

Apart from methane (which usually forms in amounts higher than expected), the distribution of the products obtained from an FT process arises from a chain growth polymerization mechanism usually described by a model developed by Anderson, Schultz, and Flory [14]. According to this model, the product distribution is ruled by the equation $W_n/n = (1-\alpha)^2\alpha^{n-1}$, where n is the number of carbon atoms in the product, W_n is the product weight fraction, and α is a chain growth probability factor that can assume values ranging from 0 to 1. As a consequence, the FT output is always a complex mixture of products ranging from methane to waxes formed by high-molecular-weight linear paraffins. However, the α factor is a function of the chain propagation and termination rates ($\alpha = r_p/(r_p + r_t)$, where r_p and r_t are the propagation and termination rates, respectively), and is characteristic of the reaction conditions and catalyst. Thus, a proper choice of catalyst and reaction conditions allows to tune, to some extent, the composition of the final mixture, even if it remains impossible to force the process to produce selectively a well-defined range of products. Figure 9.2 shows the distribution of the products of the FT reaction according to different α values.

The core of any FT process is its catalyst. Only few metals show catalytic activity in the FT synthesis. The FT reaction starts with the adsorption of the reagents on the catalyst surface, followed by a chain initiation, chain propagation, and finally chain growth termination. After CO adsorption on the catalyst, two main classes of mechanisms have been proposed for the next steps: those in which the C-O bond of carbon monoxide is first cleaved and those in which some hydrogenation by adsorbed hydrogen atoms precedes the C-O cleavage [15] (Figure 9.3).

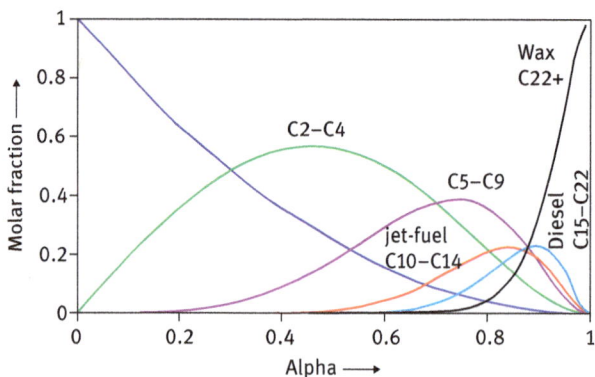

Fig. 9.2: Hydrocarbons selectivity as a function of the chain growth probability factor α.

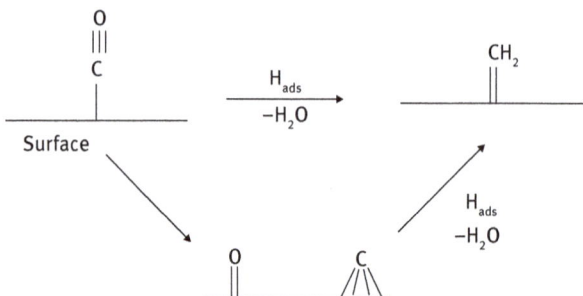

Fig. 9.3: Possible initial steps of the FT reaction.

Most mechanisms, however, converge on the formation of a surface-bound methylene species, which would be responsible for the chain growth. Thus, a good catalyst should adsorb both CO, possibly in a dissociative way, and H_2. Furthermore, since metal oxide formation is always possible under FT conditions either by dissociative CO absorption or by metal reaction with co-produced water, the metal oxide should be easily reduced under the reactions conditions.

Early transition metals are able to adsorb CO in a dissociative way, but they show poor or no capability to adsorb H_2, and their oxides are not reduced under usual FT conditions; accordingly, these metals are not active FT catalysts. Late transition metals, as well as group 12 elements (Zn, Cd, and Hg), either show poor or no CO adsorption (and, consequently, no FT activity) or, in few cases (Pd, Ir, Pt), have good H_2 adsorption capability and reducible oxides but adsorb CO in a non-dissociative way and, as a matter of fact, are poor FT catalysts. So, all the best FT catalysts are actually based on just three elements: iron, cobalt, and ruthenium. Nickel, rhodium, osmium, and possibly rhenium behave in the middle [13].

FT industrial plants, however, require huge amounts of catalyst, and ruthenium is too rare and expensive to be used on this scale; thus, cobalt and iron are the only metals suitable for industrial applications. Iron is obviously cheaper then cobalt but, to select between them, a key issue is the carbon feedstock. Iron is a good WGS catalyst and, for this reason, is particularly suitable to enrich hydrogen-poor syngas, such as those obtained from coal or biomasses. Cobalt performs better with an almost stoichiometric ratio between hydrogen and carbon monoxide, so it is preferred when the carbon feedstock is natural gas. Alternatively, cobalt can also be used with hydrogen-poor syngas provided that the H_2/CO molar is adjusted by a WGS unit between the gasification and the FT reactors.

The choice of the catalyst also has important consequences on the feed purity. Cleaning up the syngas is a key aspect of any FT process, but, as already stated, bio-syngas has, in addition, a number of peculiar impurities including hydrogen sulfide (usually, 150–350 ppm), COS (20–40 ppm), nitrogen compounds (mainly ammonia and hydrogen cyanide, 2000–3000 ppm overall), and hydrogen chloride (100–250 ppm). Concentrations of several of these compounds (particularly, of sulfur compounds) must be greatly reduced since they can cause catalyst poisoning and/ or reactor corrosion. Usually, iron catalysts are more robust than cobalt ones. Gas cleaning, however, is very expensive, and there is some tradeoff between the catalyst cost and the investment and operating cost of the gas-cleaning facility. As a matter of fact, a good feed has <1 ppm nitrogen (NH_3, NO_x, HCN), <1 ppm of sulfur for iron catalyst (<4 ppb for cobalt ones), and <10 ppb of halides.

An important concern for any BtL plant is the biomass availability. FT plants are quite expensive, and integrated BtL plants will be even more expensive, since they also require biomass gasification step and bio-syngas cleanup. So far, significant savings can be only envisaged with rather huge plants: commonly used estimates agree on figures of 15–30,000 bpd (barrels per day) as the best choice for a BtL plant or, in more traditional units, 750–1500 kt/year. To feed a 750-kt/year BtL plant (energy efficiency 0.3; fuel heat value of 37.8 GJ/t) with giant reed (*Arundo donax*; dry biomass productivity: 40 t/ha *per* year with a heat value of 17.4 GJ/t) harvested with an efficiency of 0.8, a circular area with a radius of ca. 23 km is needed. Such a huge surface is not only difficult to find, at least in densely populated countries such as most of the European ones, but is also a source of significant costs for biomass transportation. Thus, the viability of oil-scale refineries fed with biomass is still an open question.

New catalysts and compact reactors design, however, may offer a route to small, but still profitable, BtL plants, possibly able to treat different biomass including agricultural waste, paper, wood chips, food scraps, and even the organic part of municipal waste. In this way, transportation costs would be greatly reduced by bringing the reactors to the biomass, instead of vice versa [16].

A significant pilot experience of a BtL plant has been done in Germany by Choren-Shell. According to press releases and presentations by the company, the Alpha plant built up in Freiberg by Choren produced 200 t/year of top-quality diesel fuel (SunDiesel™) with typical yields, based on the dry biomass, around 20%. The company, however, shut down the plant and discontinued its activities in July 2011. Other companies that have been developing commercial FT technologies, mainly devoted to GtL applications (e.g., Sasol, Shell, BP, COP, eni-IFP/Axens, ExxonMobil, Statoil, Rentech, Syntroleum), are now in a favorable position for BtL projects. Among the more innovative projects, based on cheap feed (mainly wastes) and compact reactors, it can be mentioned the ambitious attempt to fuel, by the end of 2015, all British Airways flights out of London City Airport with 60 kt/year of jet fuel produced (along with a similar amount of a mixture diesel fuel and naphtha and with 40 MW of power) by GreenSky London starting from 500 kt/year of municipal waste. The project will rely upon an expensive but versatile plasma technology by Solena Fuels for waste gasification and compact reactors by Velocys for the FT reaction [16].

9.2.3.2 Fischer-Tropsch to olefins

As already stated, olefins are probably the only primary FT products and alkanes only form upon olefin hydrogenation. Thus, in the few last years, FT chemistry has been also investigated as a possible source of light olefins, which are currently the base building blocks for petrochemistry. Significant improvements are still necessary in order to get CO conversion and selectivities to C2-C4 olefins of industrial interest, but interesting results have been already obtained [17], and it is reasonable to expect that the industrial targets can be reached in few years of further research. The availability of an FT route to light olefins would represent an alternative to the already mentioned MTO processes in order to refound on renewable feedstock the current chemical industry.

9.2.3.3 Feeding Fischer-Tropsch reactions with CO_2

Carbon dioxide can be a significant component of the gas fed to FT plants, particularly when the syngas is obtained by biomass gasification. Any FT-like reaction fed with CO_2 instead of CO [Eq. (9.12)] may provide a route to recycle the CO_2 produced in a number of anthropogenic processes [1 and references therein]:

$$nCO_2 + 3nH_2 \rightarrow C_nH_{2n} + 2nH_2O \quad \Delta H_{298} = \text{ca.} -100 \text{ kJ/mol}_{CO_2} \quad (9.12)$$

Due to the increasing concern about the role of CO_2 in environment pollution and global warming, within the general frame of green chemistry, considerable effort has been devoted to the study of these FT-like reactions of CO_2.

From the thermodynamic point of view, a CO_2 FT is less favorable than the classical FT process but nevertheless is still favorable since additional water is formed, thus providing the chemical energy for the conversion of the very stable CO_2 molecule.

CO_2 FT reactions have been studied on both cobalt- and iron-based catalysts. Cobalt catalysts are not satisfactory, since they mostly catalyzed the CO_2 hydrogenation (Sabatier reaction) and methane accounts for up to 95% of the organic products:

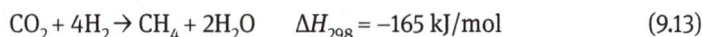

$$CO_2 + 4H_2 \rightarrow CH_4 + 2H_2O \quad \Delta H_{298} = -165 \text{ kJ/mol} \quad (9.13)$$

The output of the CO_2 FT reaction is completely different over iron catalysts. At 250°C, the products distribution is basically unaffected with respect to that of a classical FT process. Particularly, the selectivity to methane remains relatively low (less than 15%). Activity tests on different iron catalysts showed that, for CO_2 transformation, Al_2O_3 is a better support than TiO_2 or SiO_2 and that alkali (potassium) are essential promoters to speed up both direct and reverse WGS reactions (i.e., CO formation if the reactor is fed with CO_2: vide infra) and greatly inhibit the methane formation.

From the mechanism point of view, CO_2 conversion under FT conditions can be, in principle, achieved either through its direct hydrogenation or through a reverse WGS [Eq. (9.14)] followed by conventional FT conversion of CO.

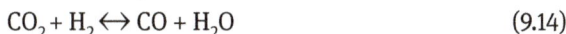

$$CO_2 + H_2 \leftrightarrow CO + H_2O \quad (9.14)$$

The latter pathway is supported by increasing evidence, provided, for instance, by elegant isotopic experiments, indicating that CO_2 is much less reactive than CO in chain initiation and propagation, except when close to the WGS equilibrium, where CO and CO_2 interconvert at a much higher rate than that of the FT process, thus becoming kinetically indistinguishable from each other [18].

From the process point of view, Eq. (9.12) obviously requires more hydrogen than the classical FT process and is therefore less attractive, unless (i) a cheap source of hydrogen will be available, possibly not involving CO_2 co-production, e.g., by water splitting promoted by sun light or by hydroelectric, wind, or nuclear power, or (ii) the reverse WGS [Eq. (9.13)] is substituted for a thermal dissociation of CO_2 to CO at high temperatures (for instance, by a thermochemical solar approach), followed by a standard FT process.

Meanwhile, compared to the classical FT process, Eq. (9.12) is less exothermic, thus making the temperature control of the reactor easier, even if CO_2 transformation will likely require high temperatures, which favor the reverse WGS [Eq. (9.14)].

A challenging possibility to improve the performances of CO_2 FT reactions is the use of membranes that can selectively remove water from the reaction medium, thus forcing the reverse WGS [Eq. (9.14)] and, consequently, the whole process.

Thus, an industrial FT-type process fed with CO_2 appears technically feasible. Some iron-based catalyst will probably be the catalyst of choice for such a process. Much optimization, however, is still to be done on issues such as reaction conditions, catalyst composition, and reactor configuration.

9.3 Syngas fermentation

Although it may appear surprising, syngas can be exploited by few microorganisms able to grow autotrophically on CO or on mixtures of H_2 and CO_2 rather than on sugars, as in more traditional fermentations. Microbes' capability to grow on gases is probably ancient, predating the appearance of photosynthesis, and probably developed to take advantage from gas emissions from hydrothermal vents. It is even possible that these gas emissions were the main, or perhaps the only, carbon and energy source for the first life forms. Today, gas from both hydrothermal vents and from several industrial emissions (e.g., by steel manufacturing) have quite similar compositions including carbon monoxide and dioxide and some hydrogen, hydrogen sulfide, and methane. Thus, some flue gases can be exploited as both nutrient and energy source to feed fermentations and to produce a number of chemicals, providing a new route to carbon capture and reuse. Work on gas fermentation started in the late 1980s at the University of Kansas. The microorganisms are obligate anaerobes, and the best known among them are Clostridia, e.g., *Clostridium ljungdahlii* or *C. autoethanogenum*. They are able to exploit carbon monoxide, with or without hydrogen, or carbon dioxide/hydrogen mixtures. The best developed fermentations mostly afford ethanol, along with some acetic acid or acetate anion, according to the fermentation conditions. The overall transformations are

Ethanol production:

$$6CO + 3H_2O \rightarrow C_2H_5OH + 4CO_2 \tag{9.15}$$

$$2CO_2 + 6H_2 \rightarrow C_2H_5OH + 3H_2O \tag{9.16}$$

Acetic acid production:

$$4CO + 2H_2O \rightarrow CH_3COOH + 2CO_2 \tag{9.17}$$

$$2CO_2 + 4H_2 \rightarrow CH_3COOH + 2H_2O \tag{9.18}$$

Carbon monoxide is the preferred substrate with respect to the CO_2/H_2 mixture: typical CO conversions for laboratory-scale fermentations are about 90%, while hydrogen conversions are around 70%.

The ratio of ethanol to acetic acid depends upon the strain and the fermentation conditions. The microorganisms are inhibited by low pH and high concentrations of acetate ion. When acetic acid is formed, the pH drops and the acetate concentration rises. So, the microorganisms switch to ethanol production to alleviate further stress. Typically, pH is kept around 4.5 in ethanol production.

Many of the microorganisms are mesophiles or even thermophiles, with the optimum temperature in the range between room temperature and 90°C. A fairly rich medium is required, with possible contamination problems. However, contamination risks are greatly reduced by the harsh fermentation conditions: high temperatures, low nutrients levels, and low pH. Furthermore, the high level of carbon monoxide inhibits the growth of methanogenic bacteria.

Syngas purity could not be critical, since some tolerance is expected to sulfur compounds, tars, and other impurities. Published work on this issue, however, is still too scanty for any conclusion to be drawn.

Simple gas-sparged tank reactors, operating either in batch or in continuous mode, can be used. Ethanol can be recovered using procedures close to those already in use in the corn ethanol industry: e.g., the ethanol/water azeotropic mixture can be distilled overhead, and then water can be removed by an adsorption unit.

The energy efficiency (heat of combustion of products/heat of combustion of feed) is 0.80–0.81 for Eqs. (9.15) and (9.16) and 0.77 for Eqs. (9.17) and (9.18). These figures are rather low for an anaerobic fermentation: by comparison, the ratio for the glucose fermentation affording ethanol is 0.98.

Current leading players in the gas fermentation field are LanzaTech and Coskata companies. At the laboratory scale, methodologies have been developed for producing a number of organic chemicals including propanols, butanols, 2,3-butanediol, *iso*-butene, acetic acid, etc. At the same time, ethanol production has been scaled up to pre-commercial stage. Particularly, in 2012 and 2013, LanzaTech started operations at two plants built in partnership with two different Chinese steel producers, each plant producing 300 t/year (100.000 gallon/year) of ethanol.

9.4 Perspectives

Many technologies for producing several of the most important chemicals, and even fuels, starting from syngas, including bio-syngas, are already available or mature. In few cases, such technologies may find application in biorefineries to exploit the whole biomass value. However, in most cases, the viability, from both the energetic and the economic point of view, of using biomass to feed large-scale plants is still an open question. Thus, eventually, it could be worthy to pay most attention to waste-to-fuel and waste-to-chemicals routes, thanks to the feedstock versatility of the gasification technology and to the development of chemical and biotechnological routes for chemicals production either from syngas or directly from industrial flue gases. Such approaches would afford valuable products starting from raw materials with nearly zero (or even negative) cost, at the same time providing routes for carbon recycling and, more generally, to mitigate the environmental impact of several human activities.

Bibliography

[1] Zennaro, R., Ricci, M., Bua, L., Querci, C., Carnelli, L., d'Arminio Monforte, A., Syngas: the basis of Fischer-Tropsch, in Maitlis, P. M., de Klerk, A., editors, Greener Fischer-Tropsch Processes for Fuels and Feedstocks. Weinheim, Germany, Wiley, 2013, 19–51.
[2] http://en.wikipedia.org/wiki/Felice_Fontana, accessed 30 June 2014.
[3] US Department of Energy, Report of the Hydrogen Production Expert Panel: A Subcommittee of the Hydrogen and Fuel Cell Technical Advisory Committee, http://www.hydrogen.energy.gov/pdfs/hpep_report_2013.pdf, accessed July 4, 2014.
[4] Song, C., Ma, X., Ultra-clean diesel fuels by deep desulfurization and deep dearomatization of middle distillates, in Hsu, C. S., Robinson, P. R., editors, Practical Advances in Petroleum Processing, Springer, New York, 2006, 317–312.
[5] Rana, M. S., Samano, V., Ancheyta, J., Diaz, J. A. I., A review of recent advances on process technologies for upgrading of heavy oils and residua, Fuel **86** (2007) 1216–1231.
[6] US Geological Survey. Mineral commodity summaries, nitrogen (fixed) ammonia, January 2013, http://minerals.usgs.gov/minerals/pubs/commodity/nitrogen/mcs-2013-nitro.pdf, accessed 5 July 2014.
[7] Olah, G. A., Goeppert, A., Prakash, G. K. S., Chemical recycling of carbon dioxide to methanol and dimethylether: from greenhouse gas to renewable, environmentally carbon neutral fuels and synthetic hydrocarbons, J Org Chem **74** (2009) 487–498.
[8] Tundo, P., Selva, M., The chemistry of dimethyl carbonate, Acc Chem Res **35** (2002) 706–716.
[9] Maitlis, P. M., de Klerk, A., editors, Greener Fischer-Tropsch Processes for Fuels and Feedstocks, Weinheim, Germany, Wiley, 2013.
[10] Fischer, F., Tropsch, H., Über die direkte Synthese von Erdöl-Kohlenwasserstoffen bei gewöhnlichem Druck, Erste Mitteilung Chem Ber **59** (1926) 830–831.
[11] Fischer, F., Tropsch, H., Über die direkte Synthese von Erdöl-Kohlenwasserstoffen bei gewöhnlichem Druck, Zweite Mitteilung Chem Ber **59** (1926) 832–837.
[12] Perego, C., Bortolo, R., Zennaro, R., Gas to liquids technologies for natural gas reserves valorization: the Eni experience, Catal Today **142** (2009) 9–16.

[13] Perego, C., Ricci, M., Diesel fuel from biomass, Catal Sci Technol **2** (2012) 1776–1786.

[14] Flory, P. J., Molecular size distribution in linear condensation polymers, J Am Chem Soc **58** (1936) 1877–1885.

[15] Maitlis, P. M., Zanotti, V., The role of electrophilic species in the Fischer-Tropsch reaction, Chem Commun (2009) 1619–1634.

[16] Krieger, K., Biofuels heat up, Nature **508** (2014) 448–449.

[17] Lanzafame, P., Centi, G., Perathoner, S., Evolving scenarios for biorefineries and the impact of catalysis, Catal Today **234** (2014) 2–12, and references therein.

[18] Krishnamoorthy, S., Li, A., Iglesia, E., Pathways for CO_2 formation and conversion during Fischer-Tropsch synthesis on iron-based catalysts, Catal Lett **80** (2002) 77–86.

Jean-Luc Couturier and Jean-Luc Dubois

10 Oil chemistry: chemicals, polymers, and fuels

Abstract: Vegetable oil chemistry is a powerful source of valuable chemicals, polymers, and fuels. Industrial applications have been developed for a long time, but recent advances with homogeneous catalytic transformations such as metathesis, oxidative cleavage, and hydroformylation open new opportunities. In the EuroBioRef project, new concepts of chemical-driven biorefineries based on non-edible vegetable oils (castor, crambe, safflower, etc.) have been developed to give access to a broad range of renewable long-chain polyamides. The new value chains defined have been completely assessed in terms of risks, competition, life cycle assessment, costs, and job creation.

10.1 Introduction

Vegetable oils are used worldwide not only for food and feed applications but also industrial uses. World production of major oils and fats in 2011 is estimated to be around 170 million tons, with a strong growth in recent years [1, 2] (Figure 10.1).

The main producing countries are located in Asia (Indonesia, Malaysia, China, India), with about 50% of the market, while Europe represents less than 20%. The most widely used oils are edible oils such as palm, soybean, canola, and sunflower oils, and together account for about 70% of total production (Figure 10.2). Non-edible or industrial oils such as castor, linseed, or tall oils are limited to about 1% of the global oil production.

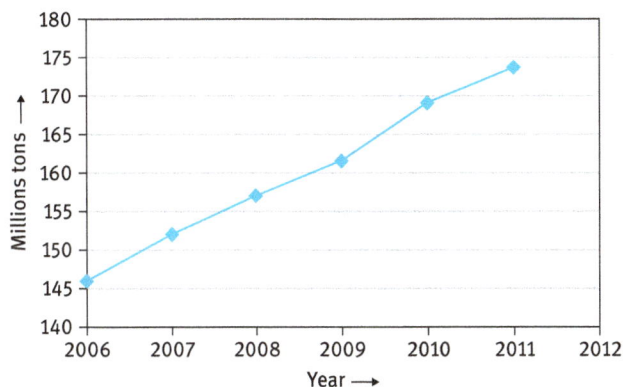

Fig. 10.1: Historical worldwide oils supply.

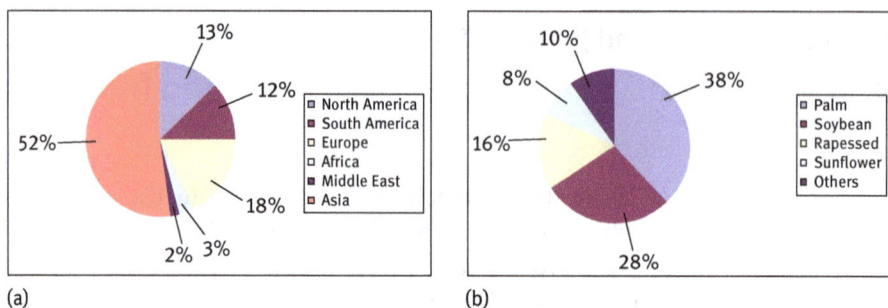

(a)   (b)

Fig. 10.2: Oils supply (2011) by areas (left) and by type (right).

In the chemical industry, vegetable oils are historically, and are still currently, one of the most important renewable feedstock. This feedstock is abundant and cheap and offers numerous advantages such as low toxicity and inherent biodegradability. Today, about 15% of oil production is consumed for chemicals, polymers, and biofuels [3]. The chemical industry paid especially much attention to non-edible oils to avoid the food competition. Cultivation of the respective crops for the production of oils is a way to increase the agricultural biodiversity, an important aspect for the sustainability of renewable feedstocks. It is especially interesting to grow non-edible crops in currently uncultivated lands, the so-called set-aside lands, in a crop rotation strategy with a food crop. The crop rotation strategy has proven to be profitable for

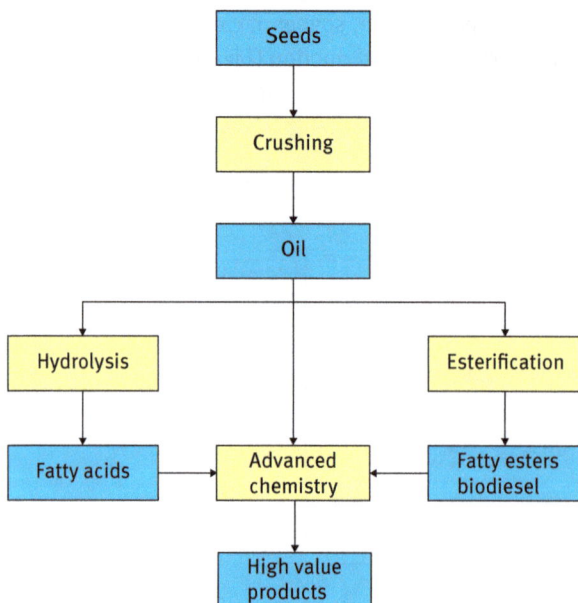

Fig. 10.3: General oil biorefinery scheme to fuels and/or high-value products.

both the food and non-food crops with the advantages of saving fertilizers, avoiding diseases and saving pesticides, and globally improving the crops yields [4]. There are also opportunities for industrial crops in marginal lands that could be understood as "too far from the village to get there by bicycle" and could mean good soil but no irrigation and low inputs resulting in low cost but low yield.

Vegetable oils are extracted from seeds and can afford several kinds of chemistries to lead to fatty acids and fatty esters or derivatives for biofuels or higher-value products (Figure 10.3).

Vegetable oils are composed mainly of triglycerides. The fatty acid composition depends on the oil nature and includes different chain lengths (typically C12 to C24), different unsaturation numbers (typically 0 to 3), and, in the specific cases of castor and lesquerella, the presence of hydroxyl group on the chain [5] (Figure 10.4 and Table 10.1).

Chemical reactions on oils may occur on the ester function, the double bond, or the hydroxyl group in the case of castor and lesquerella oils [6, 7] (Table 10.2).

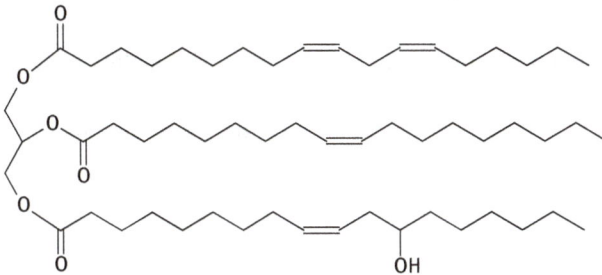

Fig. 10.4: General oil structure.

Table 10.1: Typical compositions of common food and industrial oils.

Oil	C16:0	C18:0	C18:1	C18:2	C18:3	C22:1	C18:1-OH
Fatty acid (%)	Palmitic	Stearic	Oleic	Linoleic	Linolenic	Erucic	Ricinoleic
Palm	45	5	39	9	–		–
Soybean	7	5	19	68	1		–
Rapeseed	3	1	22	14	7	35	–
Sunflower	7	5	19	68	1		–
Sunflower (high oleic)	3	5	82	8	<1	–	–
Linseed	7	4	39	15	35	–	–
Castor	1	1	3	4	–		89
Safflower	6	3	11	78	–		–
Safflower (high oleic)	5	1	87	9	1		–
Crambe	2	1	17	9	5	62	–

Table 10.2: Typical chemical reactions on oils and oils derivatives.

Ester/acid reactions	Double-bond reactions	Hydroxyl group reaction (castor or lesquerella)
– Hydrolysis	– Oxidation	– Dehydration
– Esterification	– Epoxidation	– Caustic fusion
– Saponification	– Polymerization	– Pyrolysis
– Reduction/hydrogenation	– Hydrogenation	– Alkoxylation
– Amidation	– Halogenation	– Esterification
– Halogenation	– Addition reaction	– Halogenation
– Nitrilation/ammoniation	– Sulfonation/sulfurization	– Urethane formation
	– Metathesis	– Sulfation
	– Oxidative cleavage	
	– Hydroformylation	

These different reactions can advantageously be used in a cascade way to lead to high-value products or building blocks. Important advances have been made in the recent years with selective reactions on the double bonds of vegetable oils or vegetable oil derivatives through homogeneous catalytic techniques with metal catalysts such as metathesis [3], oxidative cleavage [8] and hydroformylation [9]. This is a powerful way to go to α,ω-bifunctional products useful for the monomers and polymers industries.

10.2 Industrial applications of oil chemistry

The most classical oleochemical transformations involve the ester functionality such as hydrolysis to free fatty acids and glycerol and transesterification to fatty acid esters. Fatty acids are the main outlet for vegetable oils in chemistry with a worldwide production of 6.1 million tons in 2011 [2]. These basic oleochemicals or derivatives such as fatty amines, fatty nitriles, fatty amides, or fatty alcohols are used for versatile industrial products in soaps, surfactants, lubricants, polymers, or coatings applications.

Epoxidized vegetable oils especially from soybean and linseed are used for a long time in the chemical industry for applications as plastic additives or plasticizers [10]. Some specific vegetable oils are the direct source of valuable materials [11] such as linseed oil, which is converted to linoleum for floor-covering applications through an oxidation process, or castor oil, which acts as a polyol for polyurethane foams for automotive applications [12]. Vegetable oils or fatty acids are also key components for the alkyd resins for paints applications [13].

Vegetable oils are the source of powerful building blocks for chemistry. New opportunities for the vegetable oil conversions have recently emerged with the metathesis reaction [3]. Olefin metathesis was used in industry at large scale about

50 years ago for the transformation of petrochemical commodities such as ethylene and propylene. It was originally possible to practice the reaction without understanding the catalyst's role. The mechanism discovery by Chauvin in 1971 was a great step for the further development of this technology. Another breakthrough came in 1990 when Schrock and coworkers synthesized a group of very active well-defined molybdenum carbene catalysts. The latest breakthrough came in 1992 when Grubbs and coworkers disclosed ruthenium-based catalysts that efficiently perform the metathesis reaction with functionalized olefins including acid and ester functions. Because of this, the Nobel Prize in Chemistry was awarded to Grubbs, Schrock, and Chauvin in 2005 for the development of the metathesis reaction in organic synthesis, which enabled the reaction to take place on vegetable oils. Elevance Renewable Science (http://www.elevance.com) developed a new concept of biorefinery based on the metathesis reaction to make biodiesel and specialty chemicals such as performance waxes, olefins, esters, or dicarboxylic acids [14]. Among the fatty esters, one example is the methyl 9-decenoate, which combines the functionalities of olefins and oleochemicals in one product and which is a valuable building block for lubricants, surfactants, and polymers. Elevance Renewable Science and Wilmar International Limited announced in 2013 the startup of a plant located in Gresik in Indonesia with a capacity of 180,000 t of products [15] (Figure 10.5).

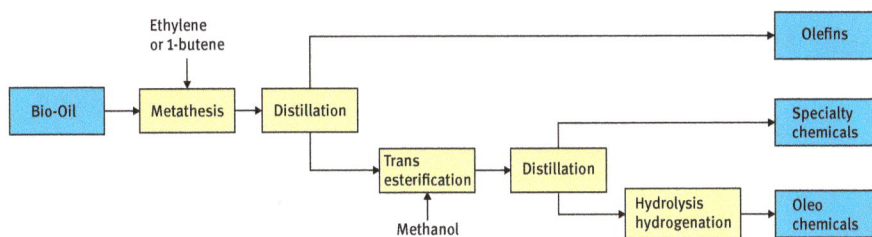

Fig. 10.5: Elevance biorefinery concept.

Another technology for the transformation of vegetable oils and vegetable oil derivatives is oxidative cleavage. Matrica, a 50:50 joint venture between Versalis and Novamont, announced in 2014 the startup of a bio-monomer and bio-lubricant plant with global capacity about 50,000 t/year [16]. The purpose is to convert the Porto Torres petrochemical site into a biorefinery. The main products are azelaic acid and pelargonic acid. Oxidative cleavage chemistry can be performed with hydrogen peroxide in a two-phase system starting either from fatty acids and esters or directly from the oil [17, 18]. Azelaic acid finds applications in polyamides, polyesters, pharmaceuticals, plasticizers, lubricant, or hydraulic fluids. Azelaic acid is also industrially produced via the ozonolysis technology from oleic acid [19, 20].

Castor oil is a very remarkable renewable resource for the chemical industry [12]. It is derived from the bean of the castor plant *Ricinus communis* (Euphorbiaceae family), native of Africa. Castor oil is known for a long time for medicinal applications. It is a non-edible oil whose seeds are toxic due to the presence of the toxic ricin protein. Castor oil is used in the manufacturing of a number of valuable organic derivatives. Ricinoleic acid, a specific acid present at about 85% in castor oil, can be cleaved in different ways. Alkali pyrolysis at about 280°C yields sebacic acid and 2-octanol (Oleris® by Arkema). These products are manufactured by Arkema and are precursors for industrially important bioplastics (polyamides, polyesters), plasticizers, surface coatings, and perfumery chemicals. The mechanism proposed by Dynthan and Weedon is given in Figure 10.6 [21].

Non-catalyzed pyrolysis of methyl ricinoleate at 400–575°C gives a different cleavage chemistry and yields methyl-10-undecenoate and heptanal, which have versatile applications in the polymer, lubricant, and fragrance industries [22] (Figure 10.7).

This unique chemistry is practiced by Arkema in its biorefinery in Marseille (France) to manufacture especially the 11-amino-undecanoic acid, which is the monomer for polyamide (PA) 11 (Figure 10.8). The performance of Rilsan® PA11 has been trusted in highly demanding applications for nearly 60 years. It affords unrivaled balance of technical and economic benefits that are valuable in markets such as automotive, energy, consumer goods, and electronics. Undecylenic acid obtained through this chemical process is also a valuable and physiologically active building block [23].

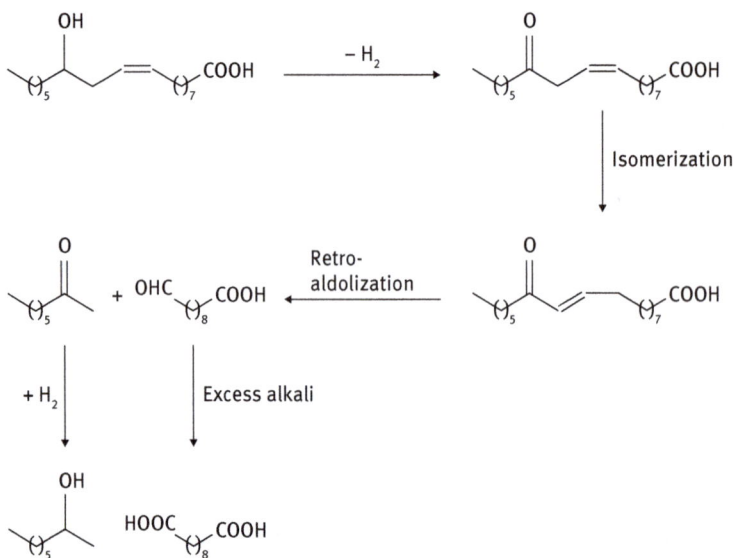

Fig. 10.6: Alkali cleavage of ricinoleic acid to sebacic acid and 2-octanol.

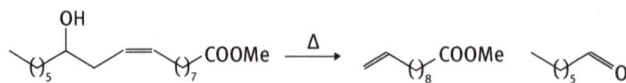

Fig. 10.7: Thermal cleavage of methyl ricinoleate to methyl 11-undecenoate and heptanal.

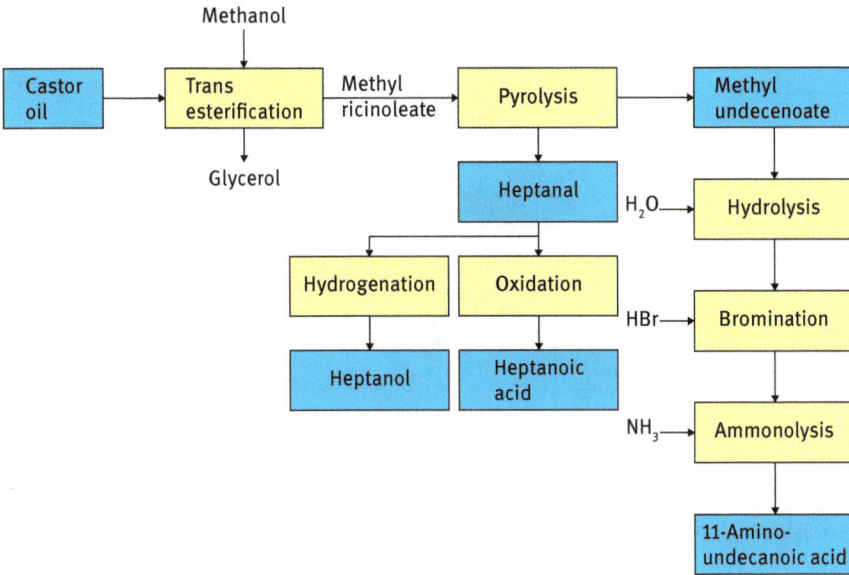

Fig. 10.8: Principal steps of Arkema's castor biorefinery in Marseille.

10.3 EuroBioRef case studies

In the EuroBioRef project (funding from the European Union Seventh Framework Programme FP7/2007 2013 under grant agreement N°241718), some of the partners [Arkema, BKW, CECA, The Center for Research and Technology Hellas (CERTH), Center for Renewable Energy Sources and Savings (CRES), Centre National de la Recherche Scientifique (CNRS), Danish Technological Institute (DTI), Imperial College, Process Design Center, Quantis, Soabe, Technische Universität Dortmund (TUDO), Technische Universität Hamburg-Harburg (TUHH), Umicore, University of Warmia and Mazury (UWM)] designed two complete value chains going from vegetable oils to high-value monomers. The methodology used for the technical and economical evaluation of these value chains is described here.

10.3.1 General description of the value chains

Two main value chains have been identified. The purpose of the value chain 1 (VC1) is to go from castor to a high-value monomer for polyamides with some co-products being used as fuel. Value chain 2 (VC2) starts with oleaginous crops (crambe, safflower) producing high-value monomers and short fatty acids, suitable for fuel application once esterified. VC1 and VC2 have several transformation steps in common, aside from the market for end products. In fact, both value chains have the possibility to start from castor, crambe, and safflower, and some routes have been proposed that combine VC1 and VC2 (Figure 10.9).

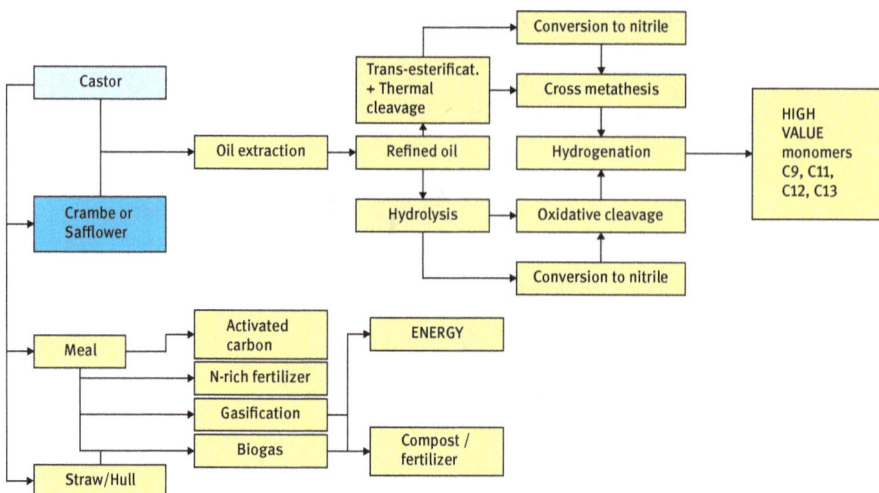

Fig. 10.9: Value chains' general descriptions.

Globally, seven options have been investigated in the EuroBioRef project targeting existing polymers on the market such as polyamides 11 and 12 or new polymers such as polyamides 9 and 13.

Option 1: Castor → PA12 through metathesis base case.

Option 1

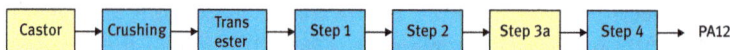

Option 2: Castor → PA12 through metathesis best case (less steps → lower capital cost).

Option 2

Option 3: Castor → PA12 through hydroformylation (high catalyst turnover number).

Option 3

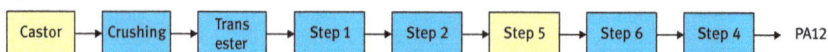

Option 4: Castor → PA11 and PA12 through oxidative cleavage (no need of expensive catalyst).

Option 4

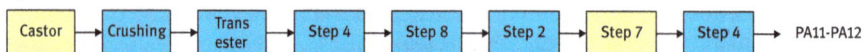

Options 5: Safflower → PA9 through oxidative cleavage.

Option 5

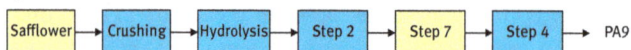

Options 6: Crambe → PA9–PA13 through oxidative cleavage.

Option 6

Crambe → Crushing → Hydrolysis → Step 2 → Step 7 → Step 4 → PA9-PA13

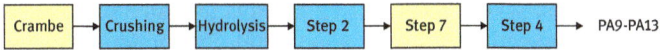

Option 7: Safflower (high oleic) → PA11 through metathesis.

Option 7

Safflower (high oleic) → Crushing → Trans ester → Step 3c → Step 3b → Step 4 → PA11

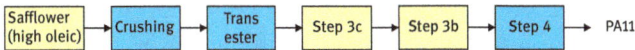

10.3.2 Oil production

A specific scenario has been proposed for the castor production in Madagascar with a complete biomass valorization (Figure 10.10).

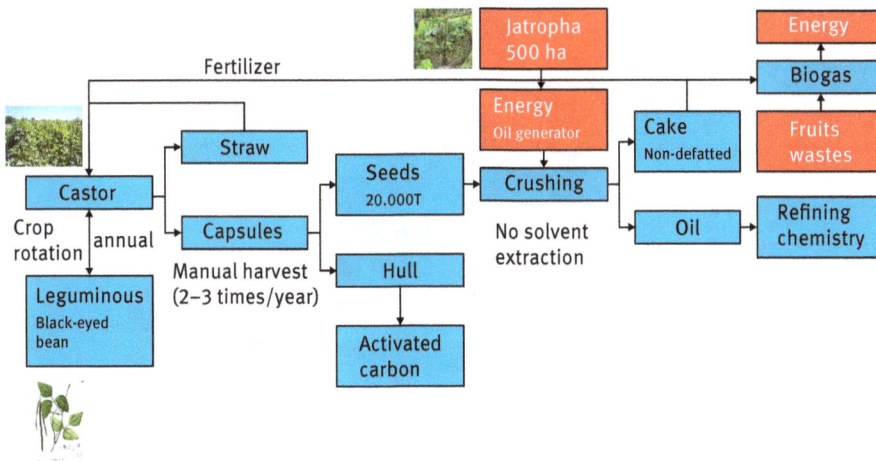

Fig. 10.10: Madagascar castor scenario.

The key issues for the Madagascar scenario are identified. The scenario is based on annual crop rotation with a leguminous plant such as black-eyed bean in order to get savings in terms of fertilizers, pesticides, and herbicides. This is also a way to bring more value to the farmers compared to a single leguminous culture. The harvest could be manual, optimizing the seed yield (two to three harvests a year). The straw would be left on the field as a carbon source. At small to medium scale (10,000–15,000 ha), solvent extraction could be avoided for oil recovery. Crushing could be made in Madagascar with some oil generators that could be powered by jatropha oil. The oily cake would be partly used as fertilizer. The oily cake could be also used for biogas, knowing that other sources for biogas such as fruit wastes or unused biomass from available land are available in Madagascar. These can be blended for biogas production. The biogas digestate rich in nitrogen could also be used as a fertilizer. The hull would be transformed in activated carbon.

The final monomer application would require additional 30,000 ha of castor cultivation. A European scenario has been investigated based on multicriteria analysis (temperature during germination, precipitation during growing period, altitude, irrigated land, monthly average temperature, germination date). Production of castor could be made in Southern or Eastern Europe. A variety improvement program that would lead to homogeneous ripening of castor and shorter cycling time would be profitable. The production of safflower and crambe has also been mapped in Europe. Safflower would be appropriate for Southern and Eastern Europe, while crambe would be more favorable in Northern and Eastern Europe (Figure 10.11).

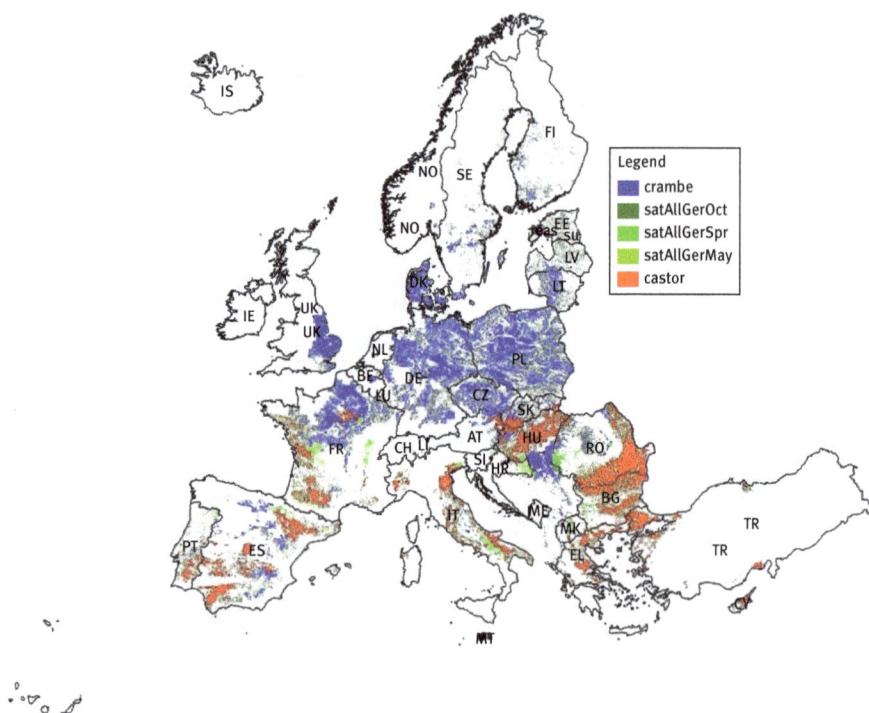

Fig. 10.11: Potential areas for castor, safflower, and crambe cultivation in Europe.

10.3.3 Competition analysis

The worldwide castor oil production is about 600,000 t/year (www.castoroil.in and http://faostat.fao.org/) with more than 80% coming from India. The main applications are lubricants, polymers (polyurethanes, polyamides, polyesters), and cosmetics. The main competitors are India (Wilmar, Jayant, etc.), China, and Brazil (Figure 10.12).

The worldwide safflower oil production is about 130,000 t/year. Main producers are India, USA, Mexico, and Argentina. There is only little production in Europe (Russia and Spain). Price is fluctuating around the world due to trading issues, but it

Detailed world agricultural trade flows

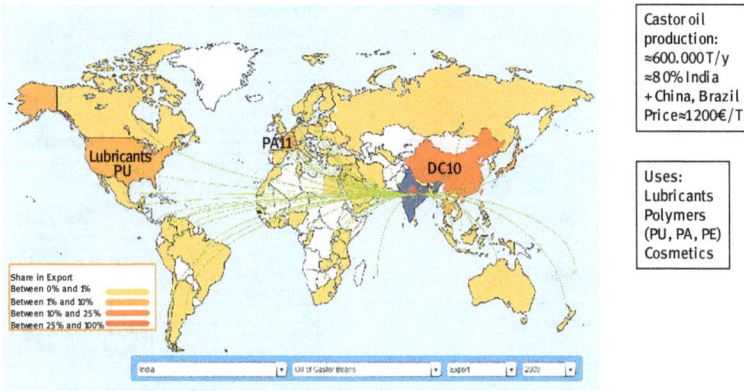

Castor oil
production:
≈600.000T/y
≈80% India
+China, Brazil
Price≈1200€/T

Uses:
Lubricants
Polymers
(PU, PA, PE)
Cosmetics

Fig. 10.12: Castor oil trade flows.

is possible to get safflower oil around $1000–1200/kg (Figure 10.13). Crambe used to be grown in UK, but the production today is rather small. Historically, safflower was grown for the dyes produced from its flowers, but currently, it has little importance due to cheaper synthetic dyes. Its oil (high linoleic) is used in paints and varnishes, as it produces paint that does not yellow with age. Seeds are sold in the birdseed market. EuroBioRef would need to use the high-oleic type of safflower. Markets addressed by crambe oil are much diversified because it is used in pharmaceuticals, lubricants, heat transfer fluids, dielectric fluids, waxes, fish food, and as coating agent.

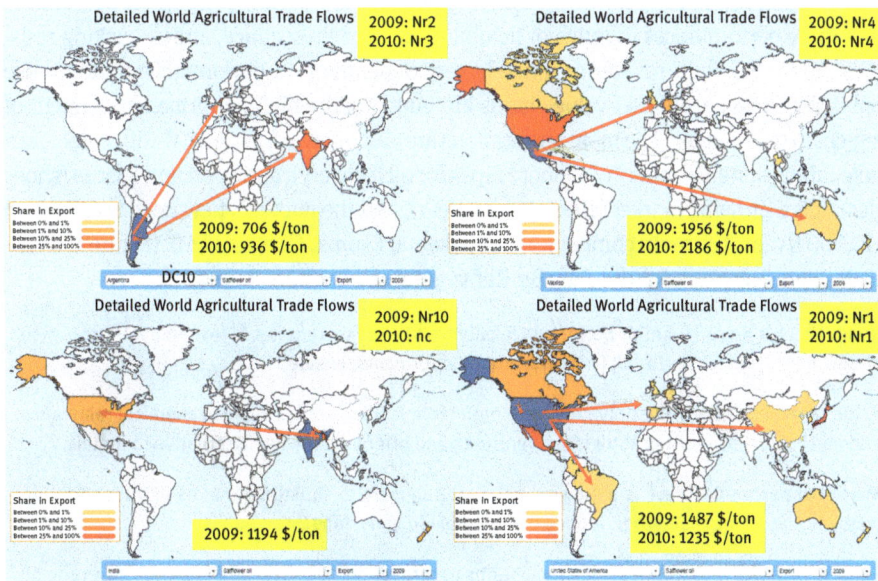

Fig. 10.13: Safflower oil production and trade flows.

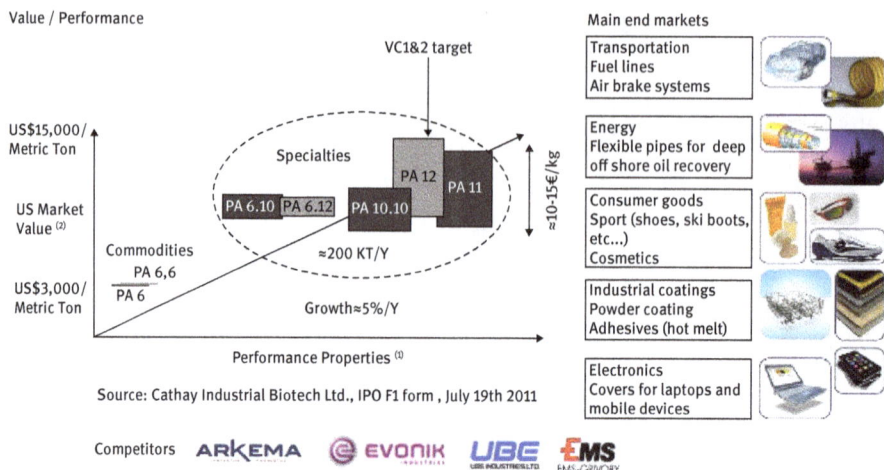

Fig. 10.14: Polyamides market.

Erucic acid is used as a plasticizer, antistatic, and corrosion inhibitor. The main value of crambe is its erucic acid content, and it is therefore in competition with high-erucic-acid rapeseed. Erucamide is used as slip agent in polymers.

The final application targets the polyamides market and will compete with existing polymers: PA12 from butadiene, PA11 from castor oil, PA10,10 from castor oil. PA9 and PA13 are new polymers that would also compete with these existing polymers.

The polyamides market can be segmented in short-chain polyamides (PA6 and PA6,6) and long-chain polyamides (PA6,10, PA6,12, PA10,10, PA11, and PA12). Short-chain polyamides are commodity polymers with huge volumes and low prices. Long-chain polyamides bring more performances in terms of flexibility, moisture resistance, stress cracking resistance, and polar fluid resistance. Price is very dependant on performance and end-use application. The EuroBioRef value chains are targeting the high-performance segment of long-chain polyamides. The main applications are automotive (fuel lines, flexible pipes, air-brake tubing systems), energy (offshore pipes for oil recovery), and sport and leisure (shoes soles). It is a growing market (about 5% per year), and the market's new need is estimated at 10,000 t/year. The main competitors are Arkema, Evonik, Ube, and EMS (Figure 10.14).

Market summary for the EuroBioRef vegetable oil value chains:

Need: There is a need for high-performance polymers based on renewable resources. There is also a growing market for castor oil for polyurethane, lubricants, cosmetics, etc.

Value to the customer: The polymers have high technical properties such as chemical resistance and mechanical properties at high temperatures and offer a renewable alternative to metals.

Market opportunity: There is a growing need for polymers to substitute metals in transportation (cars, airplanes, etc.) and then reduce carbon dioxide emissions.

Impact: Technical polyamides find applications in cars (under the hood), but also sports, crude oil transportation, low-pressure natural gas transportation, electrical industry, etc.

10.3.4 Technology analysis

The EuroBioRef vegetable oil value chains rely both on existing and innovative technologies. Castor oil production is already practiced at industrial scale, and the challenge of the project was to evaluate and adapt this production in Europe and Madagascar/Africa. Transesterification of castor oil and thermal cracking of methylricinoleate is already industrially operated by Arkema, while the purpose of the project was to improve the cracking process, bringing 30% energy saving. Metathesis is a key innovative technology. The final PA12 is an existing polymer but is currently obtained from fossil-based lactam 12 (laurolactam) or amino acid monomers.

Castor, safflower, and crambe plantations are known, but they need to be adapted to European conditions. Oil extraction and hydrolysis rely on commercial technologies. Nitrilation is already carried out at an industrial scale from different fatty acids. The main challenge lies in the oxidative cleavage of the nitrile compound especially with the separation/purification issues of the products.

10.3.4.1 IPscore analysis

The different options investigated in the EuroBioRef project have been ranked with the IPscore tool (www.epo.org) and compared with the current commercial references. This tool was developed by the Danish Patent and Trademark Office and can be used for evaluation of patents and technological development projects. Several options developed in the EuroBioRef project appear more attractive in terms of opportunity/risk than the current commercial technologies (Figure 10.15).

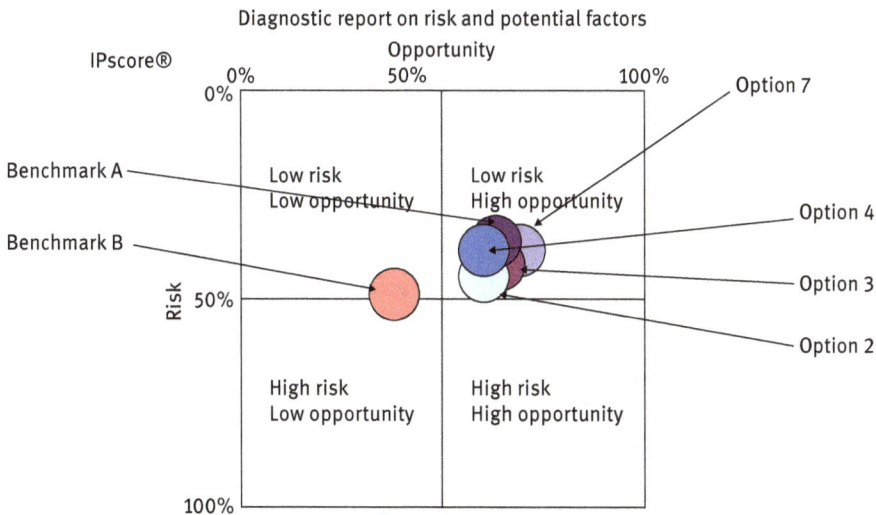

Fig. 10.15: IPscore of the main options developed in the EuroBioRef project.

10.3.4.2 Risk and SWOT analysis

The key technology challenges of the castor value chain have been identified, and the back up solutions defined (Table 10.3). A SWOT (Strengths/Weaknesses/Opportunities/Threats) analysis was also performed including all the options of the castor and safflower/crambe value chain (Table 10.4).

Table 10.3: Risk analysis of the castor value chain.

Risks	Backup
– Castor in Europe: mechanical harvesting, profitability, sustainability	– Castor in Madagascar
– Thermal cleavage: energy and CAPEX demanding	– Ethenolysis and oxidative cleavage options
– Nitrilation of fatty esters: technology not mature yet	– Metathesis with acrylonitrile
– Metathesis: key new technology, catalyst cost	– Hydroformylation or oxidative cleavage options
– Polymerization: new monomer	– Additional step to reach existing monomer

Table 10.4: Global SWOT analysis of EuroBioRef vegetable oil value chains.

Strengths	Weaknesses
– High-performance and high-value products	– Castor adaptation in Europe
– Good seed yield potential	– No mechanical harvesting (castor)
– Partners know-how about castor	– Non-homogeneous ripening (castor)
– Some existing technologies	– Toxicological issues (castor)
– Flexible technologies (PA9, 11, 12, 13 with similar technologies)	– Dedicated crushing units (non-edible crops)
– Diacids also possible	– Metathesis catalyst cost
– Low CAPEX	– Process with several products
– Strong IP for metathesis	– Products separation
– No waste	– New monomer (PA12). Customer homologation
– Energy self-sufficiency	– PA9 and PA13 new polymers

Opportunities	Threats
– Alternative to India castor monopoly	– Costs
– Madagascar complementarities in southern hemisphere	– Competitive markets (oils, existing polymers, potential competing technologies from palm, soybean, canola, etc.)
– Availability of seeds and oils on the market	– Raw material reliability (weather issues, etc.)
– New business for European farmers	
– Polyamides growing market	
– Demand for bio-based polymers (energy saving through metal substitution in transport, i.e., lighter materials)	
– Promotion of biodiversity	

10.3.4.3 Life cycle assessment (LCA)

A multicriteria LCA was made by Quantis. The reference selected is the PA6,6, which was the closest reference in the database. In the castor case, a comparison between the metathesis "base case" and "best case" was achieved (Figure 10.16). The progress along the EuroBioRef project shows significant impact on the resources, climate change, and human health indicators from the base case to the best case (negative impacts means that the biorefinery is performing better than the reference). The ecosystem quality is worse than the reference due to the substitution of a fossil-based polymer by a renewable polymer. This parameter is negative simply because the implementation requires the use of more land for crop cultivation!

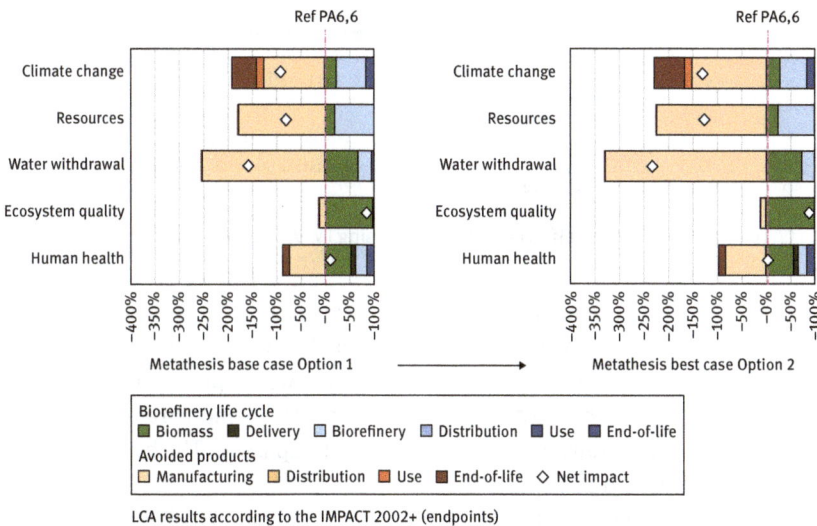

Fig. 10.16: LCA results for VC1 from castor to PA12 using IMPACT 2002+ indicator.

The same analysis has been carried out for safflower and crambe, showing better performances for the resources, climate change, and human health indicators. Globally, the safflower value chain is the most promising from the sustainability point of view (Figure 10.17).

10.3.4.4 Technology Readiness Level (TRL) assessment – actual demonstration status

Good results have been obtained with the castor production in the EuroBioRef project. Field tests are successful in Madagascar up to 20 ha (Figure 10.18). The next step is to check the results over several years including the crop rotation strategy. In Europe, we are still at the R&D level. Castor oil extraction is already an industrial technology

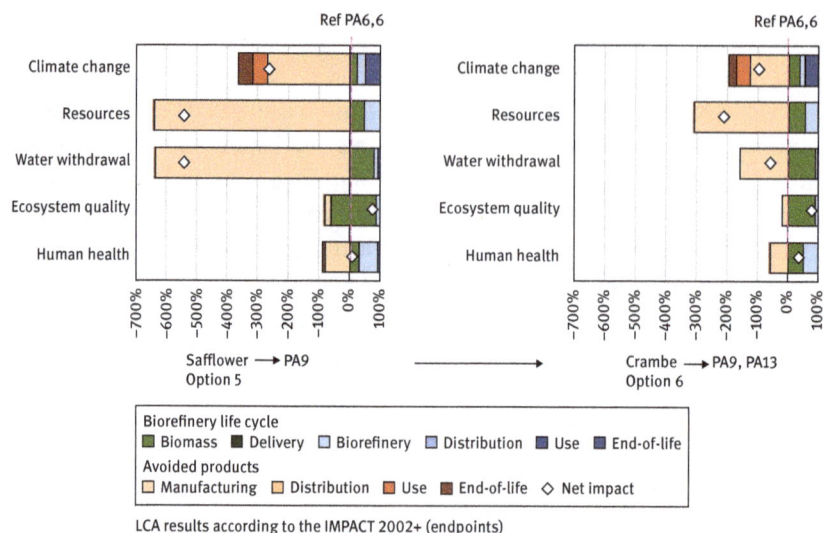

Biorefinery life cycle
■ Biomass ■ Delivery □ Biorefinery □ Distribution ■ Use ■ End-of-life
Avoided products
□ Manufacturing □ Distribution ■ Use ■ End-of-life ◇ Net impact

LCA results according to the IMPACT 2002+ (endpoints)

Interpretation:
LCA results presented according to the biorefinery approach: negative impacts mean that the biorefinery is performing better (from an environmental point of view) than the reference pathway

Fig. 10.17: LCA results for VC2 from safflower and crambe using IMPACT 2002+ indicators.

and we do not foresee any issue for this step as long as the oil composition in the seeds conforms to what already exists on the market. The transesterification step is also an existing industrial technology, and no innovation was expected in the project. Metathesis is the key new technology of the value chain. The demonstration has been performed at 100-L scale, showing that there is no scale-up issue. The hydrogenation has also been demonstrated at pilot scale.

A crambe demonstration field of 10 ha has been established in Poland by UWM while the demonstration field test of a high-oleic variety of safflower has been made in Greece by CRES at 6-ha scale (Figure 10.19). Whereas plantation of safflower and crambe at a larger scale in Europe needs still to be proven, refining oil and hydrolysis to fatty acid are commercial technologies that should apply to the obtained material. The oxidative cleavage technology was demonstrated at a pilot scale.

All the technologies studied in the EuroBioRef project have been ranked on the TRL scale to show the actual demonstration status (Figure 10.20).

(a)   (b)   (c)

Fig. 10.18: Castor production in Madagascar (left) with crop rotation strategy with cowpea (middle) and corn (right).

(a)   (b)

Fig. 10.19: Pictures of crambe production in Poland (left) and safflower production in Greece (right).

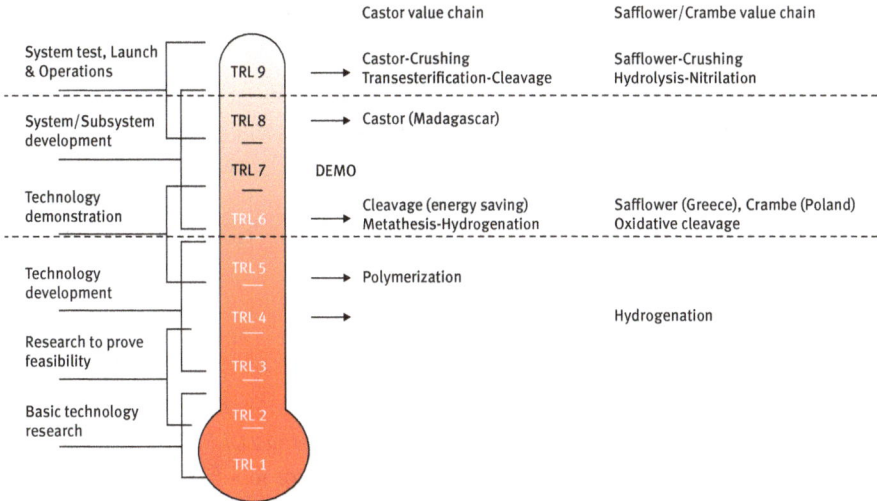

Fig. 10.20: Actual demonstration status of the castor and safflower/crambe value chains in the EuroBioRef project.

Technology summary for the EuroBioRef vegetable oil value chains:

Technology description: From castor seeds production, oil is produced and refined, transesterified, and thermally cracked to produce methylundecenoate. Then, a cross-metathesis step followed by hydrogenation leads to 12-amino-dodecanoic acid, which is a PA12 monomer (commercial petrochemical polymer). Seed meal is used as a fertilizer or converted to biogas/bio-syngas. Castor co-products (hull) are converted to activated carbons or energy.

An alternative option is to transform the castor oil ester into 12-hydroxystearic acid through hydrogenation and hydrolysis, which is then transformed to nitrile. The fatty nitriles undergo oxidative cleavage and hydrogenation reactions to give the PA11 and PA12 monomers.

From safflower and crambe plantations, the corresponding oil is obtained and refined. After hydrolysis, a mixture of fatty acids is obtained. The acids are transformed to nitriles. Reaction of the fatty nitriles with hydrogen peroxide leads to cleavage of the unsaturated nitriles. The obtained acid-nitrile molecule is hydrogenated, leading to the PA9 and PA13 monomer, depending on the type of plant. The produced short fatty acid, after esterification, can be used as fuel or lubricant. Produced biomass is used as fertilizer or converted to energy. Hull and cake are valorized as activated carbon.

10.3.5 Costs analysis

10.3.5.1 Capital expenditure (CAPEX) evaluation

The CAPEX of the castor value chain has been evaluated from literature or existing similar units. The total CAPEX for the best case option is between €120 and 160 million for 10,000 t/year of the final monomer (EU location). The CAPEX for the safflower and crambe biorefineries are in the same range. Additional 150 M€ would be necessary for straw valorization by gasification for energy production.

10.3.5.2 Production costs evaluation

Process Design Center BV estimated the global production costs (variable costs+fixed costs). About 20% cost saving was achieved during the project going from the castor base case (option 1) to the castor best case (option 2) (Figure 10.21). Step 1 is the critical step for cost reduction. The alternative scenarios getting rid of this step could probably make more sense. In the safflower and crambe cases, step 7 appears to be the critical step for cost reduction (Figure 10.22).

10.3.5.3 Job creation

Job creation has been investigated by correlation with literature data related to investment cost of existing biorefineries (Figure 10.23). An average CAPEX of €150 million has been assumed for the castor and safflower/crambe for job creation estimates. A correlation exists between the CAPEX and the jobs announced for different biofuels project. In the EuroBioRef project, the direct jobs evaluated by the partners expected in the plants for the castor or safflower/crambe biorefineries are expected between 200 and 250. This number is well above the correlation in the literature, which is

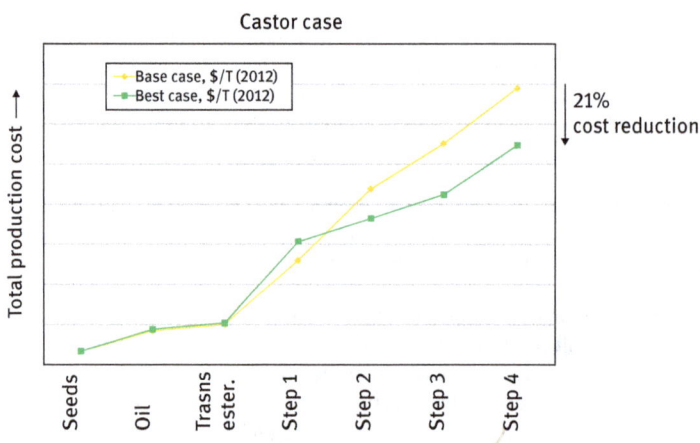

Fig. 10.21: Monomer production costs comparison for the castor cases.

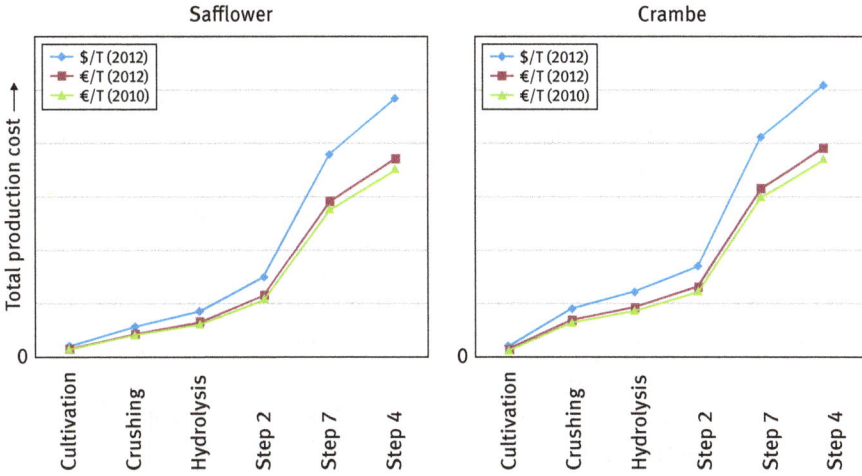

Fig. 10.22: Monomer production costs comparison for the safflower and crambe cases.

probably because high-value products are more job demanding than biofuels. The total number of jobs including indirect jobs and jobs for the biomass production is expected to be 6 to 10 times higher depending on the location.

10.3.5.4 Business model

Potential commercial applications of the technologies developed in the EuroBioRef project include castor seeds, oil and meal (Soabe), polyamides (Arkema), biogas units

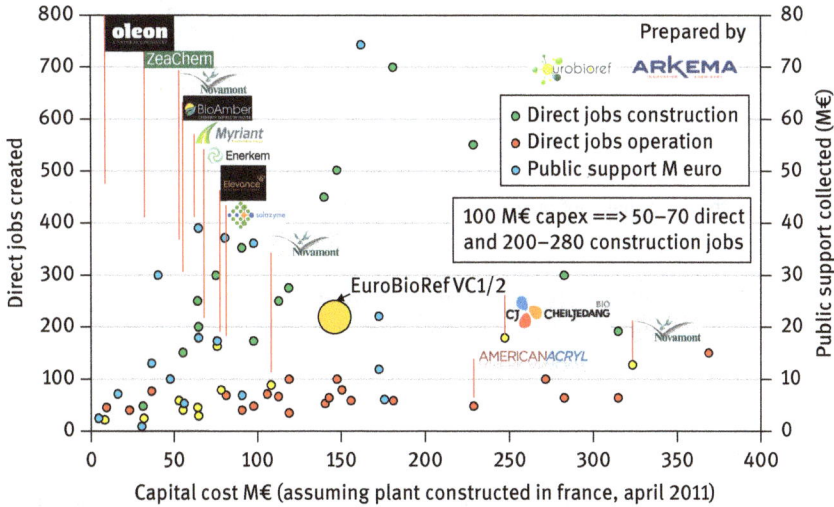

Fig. 10.23: Job creation evaluation for the castor or the safflower/crambe biorefineries.

(BKW), activated carbon (CECA), and metathesis catalyst (Umicore). Concerning the energy application, castor oil can be used for jet fuel once hydrogenated to isoparaffins, and fatty esters (C18, methyl undecenoate, methyl nonenoate, methyl octanoate) can be used in road fuels. Biogas, bio-syngas, and jatropha oil are considered for on-site energy production.

The production of castor seeds could be achieved in Madagascar and Europe, with large crushing unit in Europe. Castor oil would be sold to customers or traders. Polyamide unit could be located next to the raw material or near the customers. The commercial objective is to develop the castor chain in Madagascar and Europe to provide an alternative to India monopoly taking advantage of the growing polyamides market. The project is also an opportunity to develop new access routes to bio-based polyamides, generating at the same time short esters for fuel applications and to enlarge the specialty polyamide product range to PA9 and PA13.

10.3.5.5 Co-location in existing assets

One issue for the startup of a new biorefinery is to manage the ramp-up scenario from scratch to full capacity. Ramp-up of the castor value chain developed in EuroBioRef could take advantage of some existing equipments. Seed crushing could be done in an existing non-edible crop crushing unit (same for safflower and crambe). Transesterification and thermal cleavage of castor oil are already practiced at industrial scale in a Marseille plant (Arkema). We have proposed an option to debottleneck this plant by better valorizing the co-product as fuel. This is a way to minimize the CAPEX and the risks for the first implementation of this value chain (Figure 10.24).

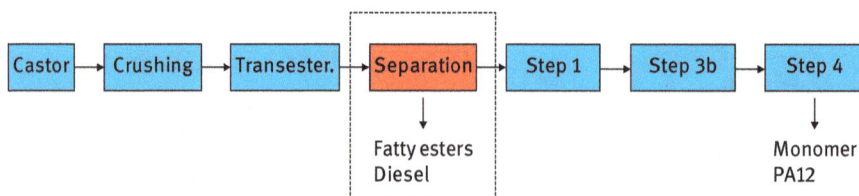

Fig. 10.24: New castor value chain option for co-location scenario.

Concerning safflower and crambe, hydrolysis would be possible in oleochemical plants. Transesterification of safflower oil should be possible in any biodiesel plant.

10.4 Conclusions

Vegetable oil chemistry is a powerful source of valuable chemicals, polymers, and fuels. Industrial applications have been developed for a long time, but recent advances with homogeneous catalytic transformations such as metathesis, oxidative cleavage, and hydroformylation extend the scope of this chemistry process and open new opportunities toward high technical performances from renewable resources.

In the EuroBioRef project, new concepts of biorefineries based on non-edible vegetable oils have been developed. All these biorefineries are chemical driven, while the co-products are used for energy and fuel. Castor, crambe, and safflower have been selected as the most attractive crops for the final targeted applications. Some crop rotation strategies with food crops such as corn or leguminous plants were established in order to promote biodiversity, increase the crops yields, and bring savings in terms of fertilizers, pesticides, and herbicides. The castor cultivation is now demonstrated in Madagascar with field tests at large scale. A broad range of renewable long-chain polyamides is accessible through these biorefineries including existing polymers on the market such as PA11 and PA12 or new polymers such as PA9 and PA13 for main applications in automotive, energy, and sport and leisure. Beyond the use of non-fossil resources, these polyamides are an attractive way to reduce carbon dioxide emissions in transportation through metal substitution by lightweight materials. Typically, a 100-kg weight saving for a vehicle results in a 0.35-L fuel saving per 100 km and 9 g of carbon dioxide emission reduction per kilometer. The new EuroBioRef value chains defined have been completely assessed in terms of risks, competition, life cycle assessment, costs, and job creation. The chemical-driven biorefineries appear more attractive for job creation than pure biofuels projects. Several individual steps of new value chains have been demonstrated during the EuroBioRef project at pilot scale (TRL=6 or more). New crops have a good potential in Europe but need acceptation and variety development programs to secure a profitable cultivation. Even if the combination of vegetable oils and chemistry is not a new story, it will for sure continue to make fascinating advances toward market opportunities.

Bibliography

[1] Malveda, M., Blagoev, M., Funada, C., Major fats and oils industry overview, CEH Industry Overview, IHS, 2012.
[2] Malveda, M., Blagoev, M., Funada, C., Natural fatty acids, CEH Marketing Research Report, IHS, 2012.
[3] Chikkali, S., Mecking, S., Refining of plant oils to chemicals by olefin metathesis, Angew Chem Int Ed. 5 (2012) 5802–5808.
[4] Dubois, J. L., Requirements for the development of a bioeconomy for chemicals, Curr Opin Environ Sustain 2011, 3, 11–14.
[5] Hasenhuettl, G. L., Fats and fatty oils, in Kirk-Othmer Encyclopedia of Chemical Technology, 4th edition, volume 10, Wiley Interscience Publication, 1991, 252–287.
[6] Naughton, F. C., Castor oil, in Kirk-Othmer Encyclopedia of Chemical Technology, 4th edition, volume 5, Wiley Interscience Publication, 1991, 301–320.
[7] Biermann, U., Bornscheuer, U., Meier, M. A. R., Metzger, J. O., Schäfer, H. J., Oils and fats as renewable raw materials in chemistry, Angew Chem Int Ed 50 (2011) 3854–3871.

[8] Köckritz, A., Martin, A., Oxidation of unsaturated fatty acid derivatives and vegetable oils, Eur J Lipid Sci Technol **110** (2008) 812–824.

[9] Behr, A., Vorholt, A. J., Hydroformylation and related reaction of renewable resources, Top Organomet Chem **39** (2012) 103–128.

[10] Tayde, S., Patnaik, M., Bhagt, S. L., Renge, V. C., Epoxidation of vegetable oils: a review, Int J Adv Eng Technol **II** (2011) 491–501.

[11] Sharma, S., Kundu, P. P., Addition polymers from natural oils – a review, Progr Polym Sci **31** (2006) 983–1008.

[12] Mutlu, H., Meier, M. A. R., Castor oil as a renewable resource for the chemical industry, Eur J Lipid Sci Technol **112** (2010) 10–30.

[13] Lin, K. F., Alkyd resins, in Kirk-Othmer Encyclopedia of Chemical Technology, 4th edition, volume 2, Wiley Interscience Publication, 1991, 53–85.

[14] DiBiase, S., Shafer, A., Metathesis: new opportunities from old feedstock, Specialty Chemicals Magazine (May 2012) 16–17.

[15] Elevance Renewable Sciences and Wilmar International Limited begin commercial shipment of specialty chemicals from new world-scale biochemical refinery in Asia, 18 July 2013, http://www.elevance.com.

[16] Bastioli, C., Biorefineries for added value products as opportunity for local competitiveness: the Matrica project, http://ec.europa.eu/research/bioeconomy/pdf/conferences/partnering_regions_20121012/catia_bastioli.pdf, accessed 12 October 2012.

[17] Bieser, A., Borsotti, G., Digioia, F., Ferrari, A., Pirocco, A., Continuous process of oxidative cleavage of vegetable oils, WO2011/080296, 7 July 2011.

[18] Bastioli, C., Borsotti, G., Merlin, A., Milizia, T., Process for the catalytic cleavage of vegetable oils, WO2008/138892, 20 November 2008.

[19] Köckritz, A., Martin, A., Synthesis of azelaic acid from vegetable oil-based feedstocks, Eur J Lipid Sci Technol **113** (2011) 83–91.

[20] Walker, C. T., Cleaning compositions and method of using the same, WO2012/103334, 2 August 2012.

[21] Vasishtha, A. K., Trivedi, R. K., Das, G., Sebacic acid and 2-octanol from castor oil, J Am Oil Chem Soc **67** (1990) 333–337.

[22] Chauvel, A., Lefebvre, G., Castex, L., Procédés de pétrochimie, Editions Technip, Paris 1985.

[23] Van der Steen, M., Stevens, C. V., Undecylenic acid: a valuable and physiologically active renewable building block from castor oil, ChemSusChem **2** (2009) 692–713.

Heiko Lange, Elisavet D. Bartzoka, and Claudia Crestini

11 Lignin biorefinery: structure, pretreatment and use

Abstract: Lignin is the second most abundant polymer in forest biomass. Compared to the ubiquitous use of cellulose, lignin is currently simply wasted. The reason for this lies in the challenging structural features and the still incomplete understanding of the correlation between structural features and polymer characteristics displayed by various lignins. This chapter will introduce the general characteristics and peculiarities of lignin as a biopolymer, present techniques specifically developed for its isolation, and describe promising ways toward its valorization, highlighting possibilities for using lignin in chemistry and material sciences.

11.1 Introduction

The efficient use of all available biomass components is of utmost interest with respect to a sustainable use of renewable resources. The development of biorefinery streams that convert cellulosic biomass into fuels will eventually generate high amounts of residual lignins; therefore, extended efforts are currently underway to valorize the latter. Despite the fact that lignin, the second most abundant component in forest biomass, is nature's dominant aromatic polymer, it has not yet been exploited to its fullest potential. Lignin is a rather complex natural polymer. The straightforward use is severely impeded by several difficulties on the molecular level, and lignin is unfortunately still seen as a rather unpredictable natural polymer with unpredictable properties. Areas of research that could eventually truly benefit from using and incorporating lignin as an already functionalized and polymerized compound seem to be paralyzed by the old paradigm that "lignin is not good for anything but burning it". Prejudices and the admittedly tedious initial effort necessary to get an idea of the nature of the lignin at hand contribute to maintaining lignin in a niche. Methods exist, however, to characterize a given lignin sample. These methods, although quite mature by now, still need further refinement, cross-correlation, and independent validation in order to further improve the processes of acquiring structural data of lignin, being able to analyze structural features in terms of reactivity, and understanding which chemistries are suitable for derivatization, functionalization, and depolymerization of lignin samples from different renewable sources. The isolation of lignin, its structural characterization, and its utilization have been subject to research activities around the globe for decades. Quite some knowledge has been accumulated regarding lignin and its characteristics, and this knowledge is constantly refined with respect to technological advances in adjacent fields such as biorefinery, biotechnology, spectrometry, spectroscopy, etc. The achievements have been summarized in several monographs [1–5]. Nevertheless, fundamental studies on lignin have not yet been concluded.

11.2 Origin and characteristics of lignin

11.2.1 Lignin occurrence and location

Lignin, derived from *lignum*, Latin for wood, was introduced describing a not narrowly classified material early as 1819 by de Candolle, but not until Payen, in 1838, the term *lignin* was connected closely to a substance. Schulze, in 1865, chemically defined lignin as a polymer that was different from the cellulose components. [5]. Lignin accounts for 15–35% of the dry mass of wood, depending on the type of wood; it is thus the second most abundant natural polymer in forest biomass after cellulose [6]. The rather hydrophobic polymer lignin is present in plant cell walls, where it chemically and physically links the other two major matrix components of the cell walls, cellulose and hemicelluloses [7]: it is located in form of a mixture together with hemicellulose between the cellulose fibrils [1–8]. The tight interplay among these three major plant biopolymers results in an increased impermeability, mechanical strength, and rigidity of the plant cell walls and thus serves to give stability to the plants; it also gives the cells greater resistance to microbial attacks. The different functions of lignin in the plant cause its distribution to vary significantly within the different parts of the plant, i.e., among stem, branching points, branches, and leafs, and among the different walls of the plant cells themselves [9]. The concentration of lignin in the middle lamella and the primary wall is higher than the concentration in the secondary wall. Nonetheless, the majority of the total amount of lignin present in the plant, 75–85%, is located in the secondary wall, due to its considerably larger volume. The amount of lignin present in the plant varies from species to species, ranging from ca. 15% in monocots over ca. 20% in hardwoods to ca. 28% in softwoods and herbaceous angiosperms [2, 7].

11.2.2 Lignin structure

The native structure of lignins suggests that it could play a central role as a new chemical feedstock, particularly in the formation of supramolecular materials and aromatic chemicals. It is the only aromatic large-volume renewable feedstock. Lignin, however, shows a heterogeneous composition and, to the best of the current knowledge, lacks a defined primary structure. It is a random three-dimensional

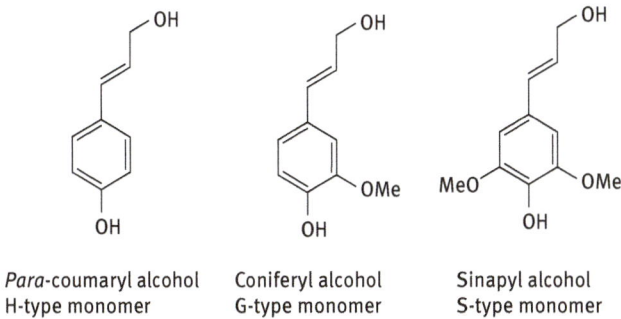

| *Para*-coumaryl alcohol | Coniferyl alcohol | Sinapyl alcohol |
| H-type monomer | G-type monomer | S-type monomer |

Fig. 11.1: Structure of lignin subunits.

phenyl-propanoid (C9) polyphenol mainly linked by arylglycerol ether bonds among the monomeric phenolic *p*-coumaryl alcohol (H), coniferyl alcohol (G), and sinapyl alcohol (S) units; Figure 11.1 shows the structure of lignin subunits. Gymnosperms have a lignin that consists almost entirely of G (G-lignin); dicotyledonous angiosperms lignin is a mixture of G and S (GS-lignin), and monocotyledonous lignin is a mixture of all three units (GSH-lignin). All lignins contain small amounts of incomplete or modified monolignols [10]. Lignin structure is the result of a biosynthetic pathway that occurs via oxidative radicalization of monolignols followed by radical coupling of two monomer radicals that form a dehydrodimer (Figure 11.2). Coupling is favored at monolignol β-positions, resulting in arylglycerol-β-aryl ether (β-O-4′), pinoresinol (β-β′), phenylcoumaran (β-5′), spirodienone (SD), and diphenylethane (β-1′) dimers formation. In principle, dilignol coupling at positions 4 and 5 could occur, yielding diaryl ether (4-O-5′) and diphenyl (5-5′) dimers, as shown in Figure 11.2.

In a subsequent step, the dimer is newly dehydrogenated to a phenoxy radical, and then it can couple with another monomer radical in an end-wise coupling mode [11]. Coupling of two lignin oligomers yields 4-O-5′ and 5-5′ coupling motifs. In turn, 5-5′ subunits undergo α-β-O-4-4′ coupling to dibenzodioxocine units (DBDO) [12]. The phenylpropane (C9) units are thus attached to one another by a series of characteristic linkages (β-O-4′, β-5′, β-β′, β-1′, SD, 5-5′, DBDO, and 4-O-5′). Since lignin is a polydisperse polymer with – according to current knowledge – no extended sequences of regularly repeating units, its composition is generally characterized by the relative abundance of H/G/S units and by the distribution of interunit linkages in the polymer.

Recent advances in the structural characterization of lignin showed that isolated lignins can be low-molecular-weight polymer with a linear structure [13–15]. Figure 11.3 shows a general picture of the lignin structure and the main interunit lignin bonding patterns.

Initial formation of reactive radicals (selection)

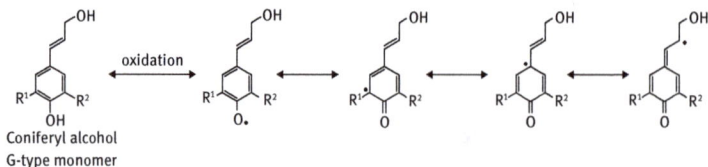

Coniferyl alcohol
G-type monomer

Head-to-tail coupling reactions
β–O–4'-formation (H, G, S)

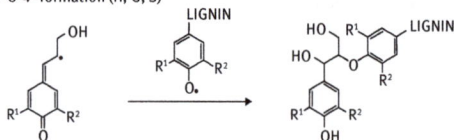

| H-type: R¹ = R² = H |
| G-type: R¹ = H, R² = H = OMe |
| S-type: R¹ =R² = OMe |

β–5'-formation (H, G)

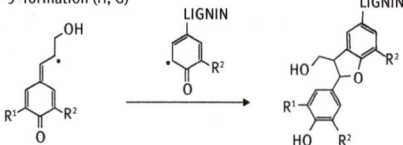

Tail-to-tail coupling reactions
β–β'-formation (H, G, S)

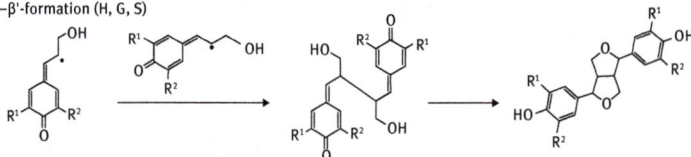

Spirodienone formation and subsequent β–1'-formation (H, G,S)

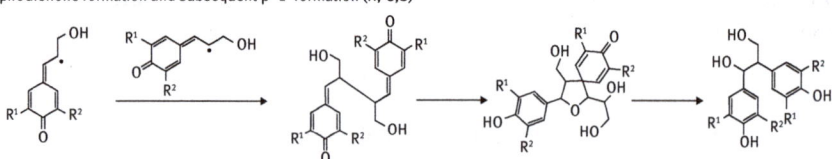

Formation-of 4–O–5', 5–5' and dibenzodioxocine (DBDO) motifs
4–O–5'-formation (H, G)

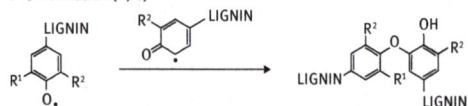

5–5'-formation (H, G) and subsequent dibenzodioxocine (DBDO) formation (H, G,S)

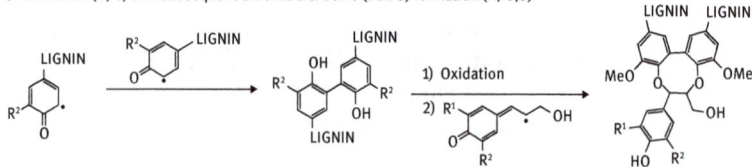

Fig. 11.2: Formation of typical dimeric structures found in lignins.

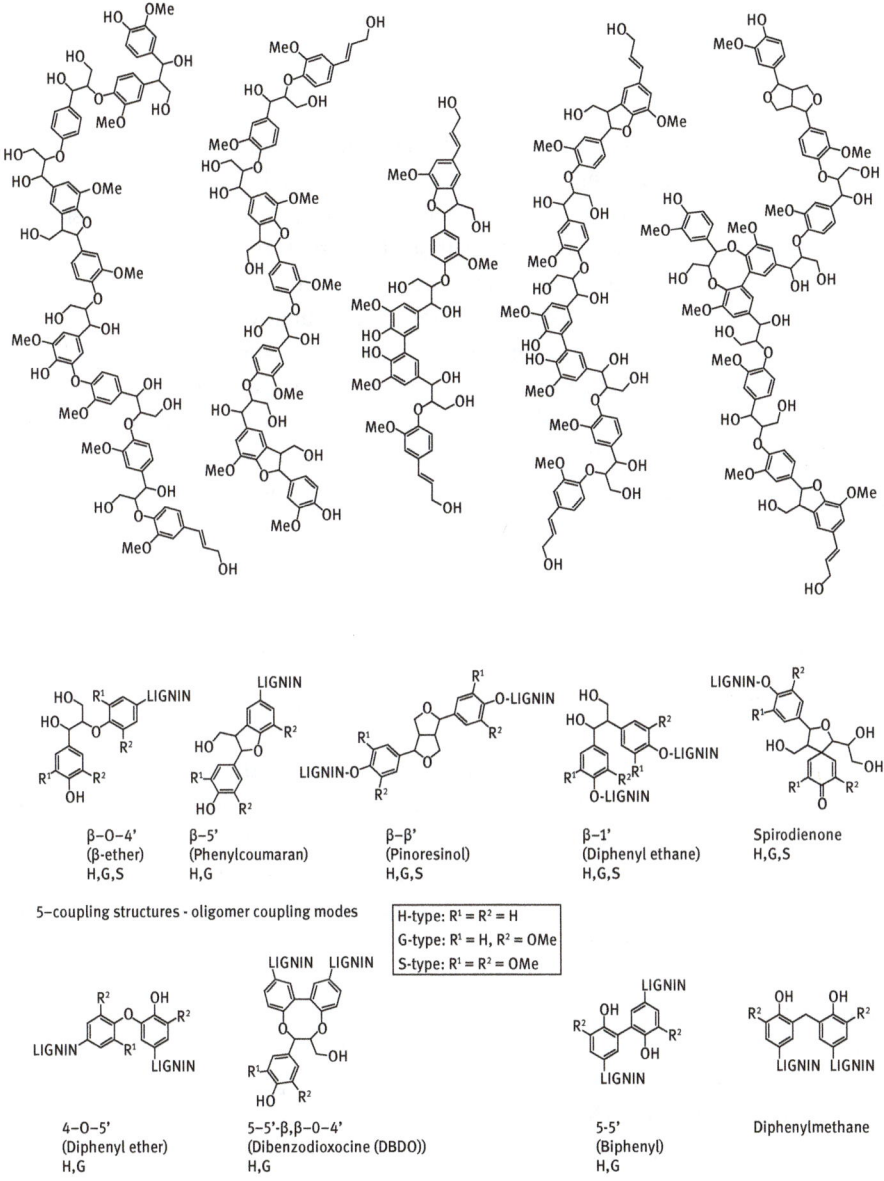

Fig. 11.3: General picture of lignin structure and main interunit lignin bonding patterns.

11.3 Potential sources of biorefinery lignin

11.3.1 Lignin pretreatment

The pretreatment of lignin is an important initial step in biorefinery operation. It separates the principal components of the biomass and degrades the extended polymer to smaller compounds. Pretreatment occasionally causes other chemical transformations, depending on the pretreatment method. The structure of the isolated lignin is highly dependent on the pretreatment nature. Efficient biomass transformation is at the basis of the success of biorefinery streams; consequently isolation/pretreatment methods that result in consistent types of lignin of high quality and purity are highly desirable. Several different lignin sources, derived from a specific form of biomass pretreatment, could be potentially used as feedstocks for lignin valorization in a biorefinery. These sources could originate either from pretreatments in the pulp and paper industries (i.e., kraft or lignosulfonate) or new feedstocks specific to the biorefinery scheme (i.e., organosolv) [16].

11.3.2 Kraft lignin process

Kraft pulping of lignocellulosic biomass is the main chemical pulping process. The process comprises the treatment of the biomass at 150–180°C with sulfide, sulfhydryl, and polysulfide at high pH. Solubilized lignin is localized in the spent pulping liquor ("black liquor") along with most of the hemicellulose in the wood. Lignin contained in black liquor is currently used as a fuel for the kraft mill after concentration. Lignin is an important fuel for paper and pulp manufacturers because it contributes heavily to a pulp mill's energy self-sufficiency. Kraft lignin, however, may be recovered and isolated from the black liquor by precipitation upon lowering the pH.

Approximately 70–75% of kraft-isolated lignin is chemically sulfonated. During kraft pulping, most of the β-O-4′ lignin subunits are cleaved, while β-1′ and β-5′ subunits undergo the formation of stilbene and aryl enol ether subunits in the residual lignin. Kraft lignin is soluble in alkali and in strongly polar organic solvents. Its average molecular weight (M_n) is generally between 1000 and 3000 Da, but it exhibits a polydispersity typically between 2 and 4. Polydispersity and functional group analysis suggest that the average monomer molecular weight is around 180 Da. A "molecular formula" of $C_9H_{8.5}O_{2.1}S_{0.1}(OCH_3)_{0.8}(CO_2H)_{0.2}$ has been reported for softwood kraft lignin, and a model structure for kraft lignin has been reported (Figure 11.4). Nearly 4% by weight are typically free phenolic hydroxy group motifs. Kraft pulping of wood constitutes potentially the largest source of the of lignin for biorefinery processes [17]. Table 11.1 shows current commercial lignin productions and productions that are currently being established based on the implementation of newer biorefinery technologies [18].

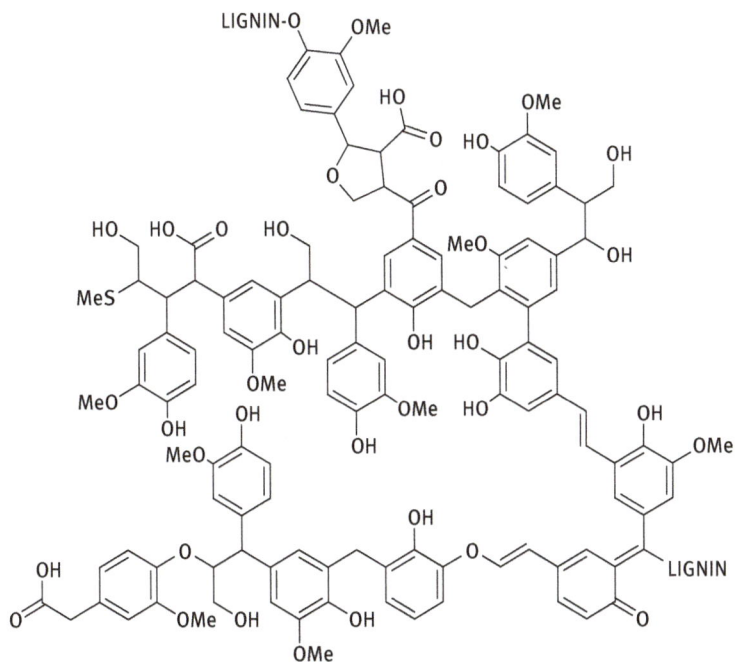

Fig. 11.4: Model structure for kraft lignin.

11.3.3 The sulfite pulping process

The sulfite cooking process is based on the use of aqueous sulfur dioxide and a base. It is carried out between pH 2 and pH 12, depending on the cationic composition of the pulping liquor. Most sulfite processes are acidic and use calcium and/or magnesium as the counter-ion. Higher pH sulfite pulping is generally done with sodium or ammonium counter-ions. Because of the nature of the sulfite process, the isolated lignin contains considerable amount of sulfur in the form of sulfonate groups present in the aliphatic side chains. Precipitation of calcium lignosulfonate with excess lime (the Howard process) [19] is the simplest recovery method, and up to 95% of the lignin of the liquor may be recovered. Sulfite lignin has a higher average molecular weight than kraft lignin. M_n values of 1000 Da and even up to 140,000 Da have been claimed, although values of 5000 to 20,000 Da are more common. The polydispersity of lignosulfonates (4–9) is higher than that of kraft lignin, and these lignins have a higher sulfur content (3–8%). Lignosulfonates are generally soluble in water throughout almost the entire pH range; they are also soluble in some highly polar organic solvents. Approximate "molecular formulas" of $C_9H_{8.5}O_{2.5}(OCH_3)_{0.85}(SO_3H)_{0.4}$ for softwood sulfite lignin and $C_9H_{7.5}O_{2.5}(OCH_3)_{1.39}(SO_3H)_{0.6}$ for hardwood sulfite lignin have been claimed; a model lignosulfonate structure has been reported (Figure 11.5) [20].

Table 11.1: Current commercial productions of lignin, and planned productions using new technologies.

Entry	Process	Established	Scale (ktpa)	Sulfur containing	Supplier (location; distribution)
1	Sulfite (lignosulfonates)	n.n.	~1000	yes	Borregaard LignoTech (NO; worldwide)
2	Kraft	1940s	40	yes	Meadwestvaco (US; worldwide)
3	Kraft	n.n.	~50	yes	Stora Enso (USA; n.n.)
4	Kraft	n.n.	~10	yes	West Fraser (USA; n.n.)
5	Kraft	2015	n.n.	yes	Stora Enso (FI; n.n.)
6	Kraft	2015	n.n.	yes	West Fraser (USA; n.n.)
7	Kraft (LignoBoost)	2013	20	yes	Domtar (USA; worldwide)
8	Soda (sulfur-free)	n.n.	5–10	no	Greenvalue (CH, IND; worldwide)
9	Organosolv (sulfur-free)	n.n.	~3	no	Lignol innovations (CAN; worldwide)
10	Organosolv (sulfur-free)	2009	n.n.	no	DECHEMA (DE; n.n.)
11	Organosolv (sulfur-free)	2008	n.n.	no	Dedini (BR; worldwide)
12	Organosolv (CIMV process)	2006	n.n.	no	CIMV (FR: worldwide)
13	Organosolv (Plantrose)	After 2014	n.n.	no	Renmatix (USA; n.n.)
14	Organosolv (evolUTIA)	After 2014	n.n.	no	Tennera (USA; n.n.)
15	Organosolv (Alcell)	Pilot plant since 1990s	1–2	no	Lignol Innovations (CA; n.n.)
16	Steam explosion	After 2014	n.n.	yes	Chemtex (IT; n.n.)
17	Steam explosion	After 2014	n.n.	yes	ENEA (IT; n.n.)
18	Hydrolysis lignin	After 2014	n.n.	yes	Chemtex (IT; n.n.)
19	Hydrolysis lignin	After 2014	n.n.	yes	Inbicon (DE; n.n.)

11.3.4 Organosolv lignin

Organosolv pulping is a general term for the separation of wood components through treatment with organic solvents. Such operations normally give separate process streams of cellulose, hemicellulose, and lignin. A wide variety of solvents and combinations thereof have been proposed for organosolv pulping. Many include acids or alkaline substances to enhance pulping rates. The best known process is the Allcel process [21], which uses ethanol or ethanol-water as a solvent. Organosolv processes offer several possible advantages as sources for lignin for biorefinery processes. In general, the processes result in separate and easily isolated streams of cellulose, hemicellulose and lignin. Specifically, organosolv lignin can be easily separated from the pulping solvents

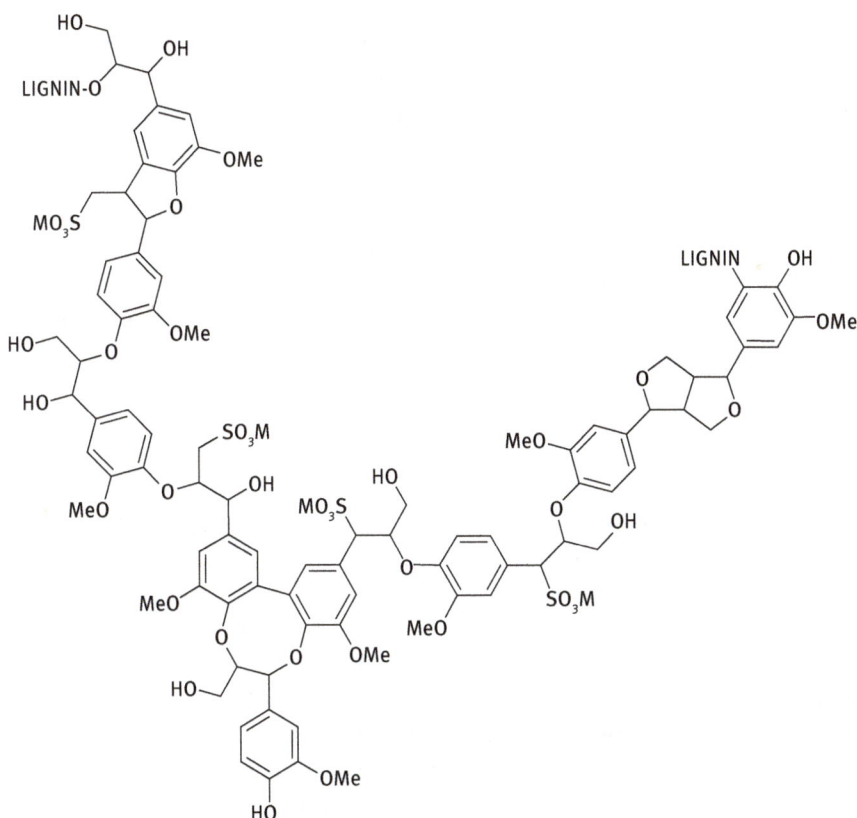

Fig. 11.5: Model structure for lignosulfonate.

either by solvent removal and recovery or a combination of precipitation with water accompanied by distillation to recover the solvent. Most organosolv lignins are insoluble in water between pH 2 and pH 7, but will dissolve in alkali and many polar organic solvents. M_n values are typically less than 1000 Da, and polydispersities may range from 2.4 to 6.4. An approximate molecular formula of $C_9H_{8.53}O_{2.45}(OCH_3)_{1.04}$ and a calculated monomer molecular weight of 188 Da have been reported [22].

11.3.5 Pyrolysis process

Pyrolytic processes (thermal decompositions occurring in the absence of oxygen) can be used to produce a lignin stream for potential use in biorefinery processes. They require relatively high temperatures (723 K) and short vapor residence times, up to 2 s. The by-products from biomass pyrolysis are char and gas, which are used within the process to provide the process heat requirements. There are no waste streams other than fuel gas and ash. The main disadvantage lies in the high carbohydrate consumption required to fuel the process.

The largest difference between pyrolytic lignins and the lignin in biomass are the very low molecular weights of pyrolytic lignins, indicating the high degree of depolymerization caused by the thermal treatment during pyrolysis and suggesting that pyrolysis may be useful as a technology for the controlled molecular weight reduction of lignin. Scholze et al. [23] reported M_w values of 600–1300 Da and M_n values of 300–600 Da for pyrolytic lignins, indicating the presence of dimeric to nonameric phenolic units. A C8 repeat unit was proposed for pyrolytic lignin rather than the C9 unit normally used, with the emperical formula being $C_8H_{6.3-7.3}O_{0.6-1.4}(OH)_{1-1.2}(OCH_3)_{0.3-0.8}$. These features imply unique opportunities to generate specific aromatic hydrocarbons from pyrolysis lignin that are not available from lignins from other processes [24].

11.3.6 Steam explosion lignin

Steam explosion consists of biomass impregnation with steam (180–230°C) under high pressures (200–500 psi) at short contact times (1–20 min) followed by rapid pressure release. The steam explosion process allows the release of individual biomass components, and the process has generally been used as a method for preparing cellulose pulp. Alkali washing or extraction with organic solvents allows recovery of up to 90% for hardwood lignins. Steam explosion lignin shows a lower molecular weight and higher solubility in organic solvents than kraft lignin. Thus, steam explosion lignin may be an interesting candidate for selective conversion of lignin to a relatively narrow fraction of mixed phenols. Since the steam explosion process provides separated cellulose and lignin streams, it is a potentially attractive biorefinery process focused on fuel ethanol, chemicals, and lignin derivatives [25].

11.3.7 Other processes

Several other methods have been developed to isolate lignins from biomass. Among them, diluted acid treatments suffer from low yields and corrosion disadvantages. Pulping can also be done via alkaline oxidation using, e.g., oxygen or hydrogen peroxide. The delignification rates of these processes are, however, slow. Both acid and alkaline treatments provide lignins similar to organosolv lignins with low molecular weight and high solubility in organic solvents [26].

11.4 The use of lignin in current and future biorefinery schemes

Industrial processes for the exploitation of wood for the production of wood pulp for paper making and, more recently, modern saccharification process for the production of bioethanol from lignocellulosic materials are able to use the cellulose and hemicellulose fraction of wood. In 2007, the US Energy and Independence Act was introduced, requiring an annual production of 79 billion liters of second-generation biofuels by 2022. This would imply the side production of 62 million tons of lignin per year [27], which would exceed the current world market for lignin in specialty products [28]. Although the lignin resulting from the business cycle of ethanol has an energy content sufficient to support the energy needs of the system, it is a remnant of work that is currently not valued appropriately. Although low-market volume chemical additives can be derived from lignin, the amount of lignin from an industrial cellulosic ethanol plant will range from ~100,000 to 200,000 t/year; this scale disparity, however, is expected to shape lignin valorization research and development [29]. Currently, with the exception of the vanillin process and the cement dispersion applications, it is burned as low-grade boiler fuel. Considering its complexity and richness of functional groups, there are many possible alternatives for converting it instead into aromatics, agrochemicals, polymers, and high-performance materials such as carbon fibers. In fact, the value of lignin is the key stage in the development of processes for the production of renewable energy [30]. As mentioned before, the complexity of lignin is the main obstacle to its use because there are no industrial processes available that can produce materials from chemically regular or simple molecules, using commercially available products [31].

The two major industrial sources of lignin are composed of lignosulfonate and kraft lignin. Alcell lignin, produced from hardwoods, is available in quantities from the seed industry, representing about 1500 t/year [32]. All applications and industrial production of lignin that have been used to date aimed at low value-add. Basically, its polyphenolic structure has not been fully exploited for the production of raw materials for the polymer industry, nor for the development of new materials. This is mainly in the case of composite materials and the low compatibility of lignin with hydrophobic plastics. Lignin-based composite materials show low performance due to such low compatibility. In addition, numerous studies have clearly shown that lignins exhibit interesting antioxidant, antibacterial, and antiviral activities. The biological properties, however, have not yet been properly exploited, neither [33].

The use of lignin is nowadays limited to thermovalorization processes as filler in composites, a component in binders and coatings, or, to a lower extent, as surfactant/dispersant additives, whereas its potential as a source of valuable phenols in the production of high value-added biopolymers in alternative to petrol chemistry is largely unexploited. Thus, novel processing methods and product concepts are required to extend the role of lignin for future biomass and biofuel application in emerging platforms such as the biorefinery [34]. Lignin valorization to chemicals is an important tool for economic profitability of the biorefinery. Lignin exploitation can be divided into three categories (Figure 11.6): (1) power, green fuels, and syngas; (2) macromolecules; and (3) aromatics and other chemicals [3]. The first option focuses on the use of lignin as a carbon source using aggressive means to break down its polymeric structure [35].

The second option aims at valorizing the macromolecular lignin structure via incorporation in high-molecular-weight applications [36]. In the third option, the macromolecular lignin backbone is broken down to monomeric compounds. Figure 11.6 shows possible new product opportunities from lignin transformations in these three fields.

As stated before, the main drawback for lignin valorization consists in its wide structural diversity and heterogeneity. The impossibility to get reproducible lignin batches from sequential biomass treatments results in a scarcely defined material that can be

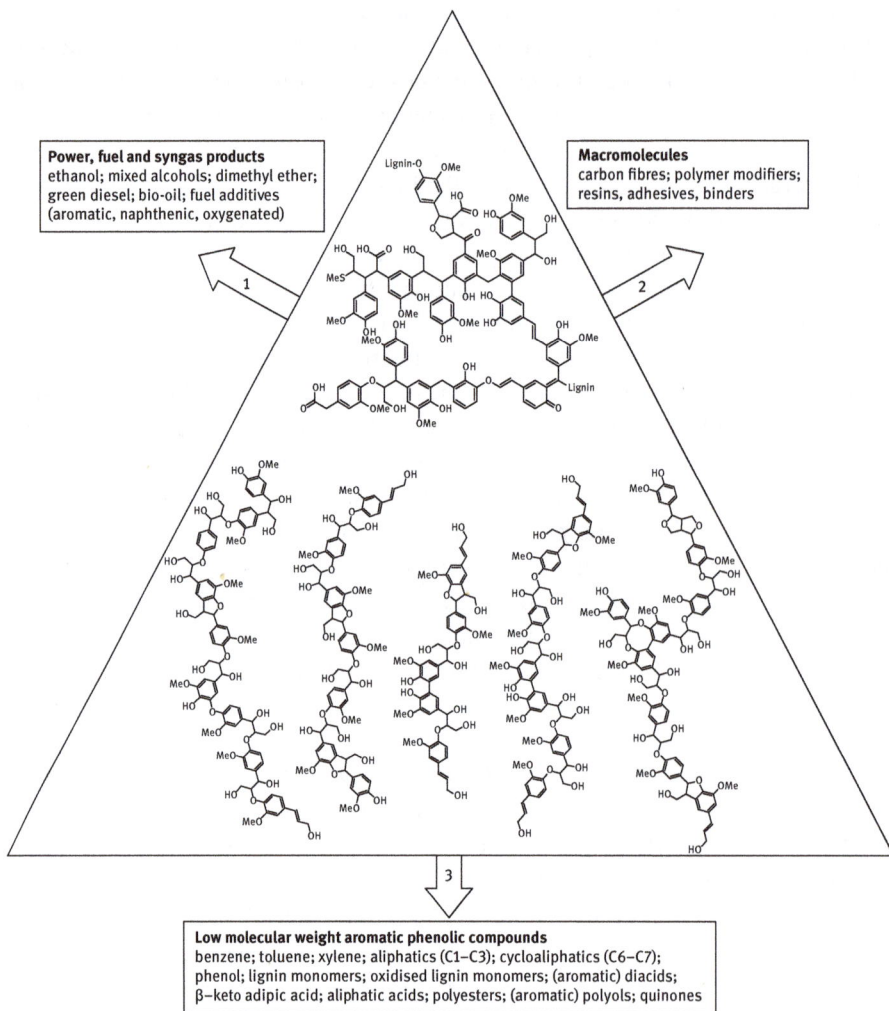

Power, fuel and syngas products
ethanol; mixed alcohols; dimethyl ether; green diesel; bio-oil; fuel additives (aromatic, naphthenic, oxygenated)

Macromolecules
carbon fibres; polymer modifiers; resins, adhesives, binders

Low molecular weight aromatic phenolic compounds
benzene; toluene; xylene; aliphatics (C1–C3); cycloaliphatics (C6–C7); phenol; lignin monomers; oxidised lignin monomers; (aromatic) diacids; β–keto adipic acid; aliphatic acids; polyesters; (aromatic) polyols; quinones

Fig. 11.6: Possible valorization products from lignin according to the three main approaches to generate these valorization products.

hardly added any value to. Several studies are in course in order to develop lignin frac-tionation and purification processes with the aim to get reproducible lignin streams possessing defined molecular weight ranges and solubility characteristics. More specifi-cally, fractional precipitation, fractional solubilization, and membrane filtration approa-ches have been recently reported for the purification of kraft lignin streams [37, 38].

11.4.1 Power – green fuels – syngas

Lignin is currently used primarily for process heat, power, and steam [39]. Lignin gasification produces syngas (carbon monoxide/hydrogen), the addition of a second phase that uses the technology of water-gas shift (WGS) allows the production of a hydrogen flow "pure" to conform with carbon dioxide. Hydrogen can be used to produce electricity (fuel-cell applications) or by hydrogenation/hydrolysis. Lignin-derived syngas can be subjected to Fischer-Tropsch (FT) conditions to produce green diesel. The technical requirements for FT include economical purification of syngas streams and catalyst and process improvements to reduce unwanted (by-)products.

Some lignin fractions originating from a biorefinery process are not expected to be suitable for material applications; these can still be valuable for conversion into fuels and chemicals. Additionally, hydrodeoxygenation catalysis for the production of fuels from lignin-derived intermediates suffers from the inconvenience of catalyst deactivation that is due to high coke formation, hydrothermal instability, and catalyst sintering [40].

The dry biomass can be converted to a liquid known as pyrolysis oil or bio-oil by fast pyrolysis. Bio-oils obtained as a product are generally not stable to changes in viscosity and oxidation, and this makes them awkward to be used as fuel and chemi-cal products. Pyrolysis oils may be included in some of the oil refinery processes after pretreatment and stabilization. The underlying process includes preconditioning prior to stabilization of the pyrolysis oil [41].

11.4.2 Macromolecules

The polymer and polyelectrolyte properties of lignin are important for all current com-mercial uses of lignin. Targeted applications include the use as emulsifiers, binders and antidispersants. In fact, nearly 75% of lignin products are within these appli-cations. Without modification of lignin, these applications remain low-added-value ones. Medium-term conversion technologies will be focused on improving of their performance by targeted lignin functionalizations.

Lignin has been used as low-cost hydroxyl component for the formation of poly-urethanes (PUs) and polyesters (PE) [42–44]. The PU characteristics can be varied according to different possible applications [45–48]. Generally, lignins represent the less elastic part of a PU.

Current research results show a general stability of lignin as polymeric component for PEs. In fact, lignin carries alcoholic and acidic functionalities at the same time, and as such, it has been used for the formation of PE [49] using application-dependent stability-introducing acyl chlorides [42].

Tailoring of lignin molecular weights by oxypropylation allows to tune the characteristics of a copolymer incorporating this modified lignins, prior to the use in PU and PE formation [50, 51].

Other polymer blends containing lignin have been reported [52], such as phenol-formaldehyde adhesives [53]. In this case, lignin is used to partially replace the phenol needed for the polymerization and to convey new properties such as conductivity. The incorporation of lignin helps to reduce the amount of phenol of up to 60%. Also, lignin-epoxy adhesives were grafted and studied extensively [54, 55].

Lignin-polyolefin blends show increased mechanical strength, stability to UV light, and biodegradability [52, 56, 57]. The characteristics of the lignin starting material strongly affect the stability of the blends, whose characteristics also change with varying amounts of lignin, i.e., UV-absorbing films have been reported [58] or low-cost films with up to 50% lignosulfonate content [58].

Lignin poly(vinyl chloride) (PVC) blends have been produced with increased thermal strength and light stability [59–61]. Although the functional groups present in the two starting materials can easily undergo proton-donor/proton-acceptor interactions, complete miscibility is not automatically granted. Successfully grafted PVC-lignin blends showed an improvement of polymer characteristics in terms of yield and breaking strength.

Lignin blends have also been realized with poly(vinyl acetate) (PVAc), poly(vinyl alcohol) (PVA), and polysaccharides (e.g., starch, gums). In all applications, the limiting factor is the solubility and thermal stability of the lignin used [52].

Thermoplastics can also be produced by the use of lignin [62, 63].

Another important sector in the exploitation of lignin is the production of carbon fibers [64, 65], which have been a research object since the first example was reported in 1969 [66]. In fact, lignin could, in principle, be an ideal precursor for carbon fibers. Unfortunately, up to now, the development of lignin-based carbon fibers is limited by the limited knowledge of the fundamental lignin chemistry involved in the process. The production of lignin-based carbon fibers is based on the extrusion of lignin filaments from melts or swollen gels. A second step requires oxidation of lignin fibers using an air atmosphere. This implies the thermal stabilization of the fibers. Pyrolysis under inert atmosphere serves to remove of volatile hydrocarbons, CO, CO_2, and H_2O. The carbon precursor and processing methodology strongly affect the final morphology of the carbon fibers. Currently, lignin-based carbon fibers display poor mechanical properties with respect to commercial ones due to their porosity and lack of a graphitic structure. This behavior is related to the original lignin structure that has a partial globular morphology that undergoes, under thermal pyrolysis, the formation of rigid oxidized segments and cross-linking. The cross-linking process is able to increase the overall char content in a pyrolyzed polymer, but on the other side, decreases the carbon order. New chemical modifications of lignin are needed in order to develop linear fibers forming lignins.

11.4.3 Aromatics and chemicals

Lignin is the only renewable source of an important and high-volume class of compounds – the aromatics. It is easy to conclude that direct and efficient conversion of lignin to discrete molecules or classes of high-volume, low-molecular-weight aromatic molecules is an attractive goal. As petroleum resources diminish and prices increase, this goal is very desirable, and is perhaps the most challenging and complex of the lignin technology barriers. Bringing high-volume aromatics efficiently from a material as structurally complex and diverse as lignin thus represents a challenging but viable long-term opportunity [67, 68].

Bulk commodity low molecular weight aromatic molecules could be easily and directly used by conventional petrochemical processes. Development of the required aggressive and nonselective chemistries is part of the long-term opportunity but is likely to be achievable sooner than highly selective depolymerizations. Recent studies on the hydroliquefaction of lignin suggest that this concept represents a valid alternative [69, 70]. However, the depolymerization of lignin represents a challenge due to the high variety of C-C and C-O bonds present in lignin that yield to highly heterogeneous products arrays. Such streams are difficult to separate and upgrade, given the heterogeneity of low-molecular-weight species generated in a selective depolymerization process, which often possess diverse functional groups. Another problem bound to depolymerization of lignin lies in the tendency of low-molecular-species to undergo repolymerization to more recalcitrant products.

To date, thermochemical treatments and homogenous and heterogeneous catalysis and biocatalysis have been used to perform lignin depolymerization and upgrade. Lignin hydrogenolysis and hydrodeoxygenation at high hydrogen pressure in single- or two-step processes in the presence of transition metal catalysts has been long used [71].

The most studied reductive depolymerization and upgrading strategies require the use of high temperature and pressure in the presence of catalysts originally developed for the processing of petroleum for the removal of sulfur and nitrogen [72, 73]. A major problem in lignin hydrogenation is connected with catalysts poisoning by biomass-derived streams. Additionally, the heterogeneity of lignin-derived low-molecular-weight products limits the yields of single compounds. Future development of more robust catalysts that have been already investigated on model compounds will eventually overcome these problems [74].

Oxidative lignin upgrade processes have also been widely studied. These processes involve lignin depolymerization, side-chain oxidations and fragmentation reactions [75–80]. The products obtained range from monomeric aromatic compounds such as aromatic acids and aldehydes to aromatic ring opening products such as muconates and bicarboxylic acids. In this case, the development of efficient product separation methods is the key step for the development of valuable products streams. Figure 11.7 shows possible pathways in lignin oxidation. [81, 82].

In nature, lignin is selectively oxidized by white-rot fungi. These species excrete a pool of ligninolytic enzymes that activate both dioxygen and hydrogen peroxide for the degradation of lignins. Among them, laccases and peroxidases,

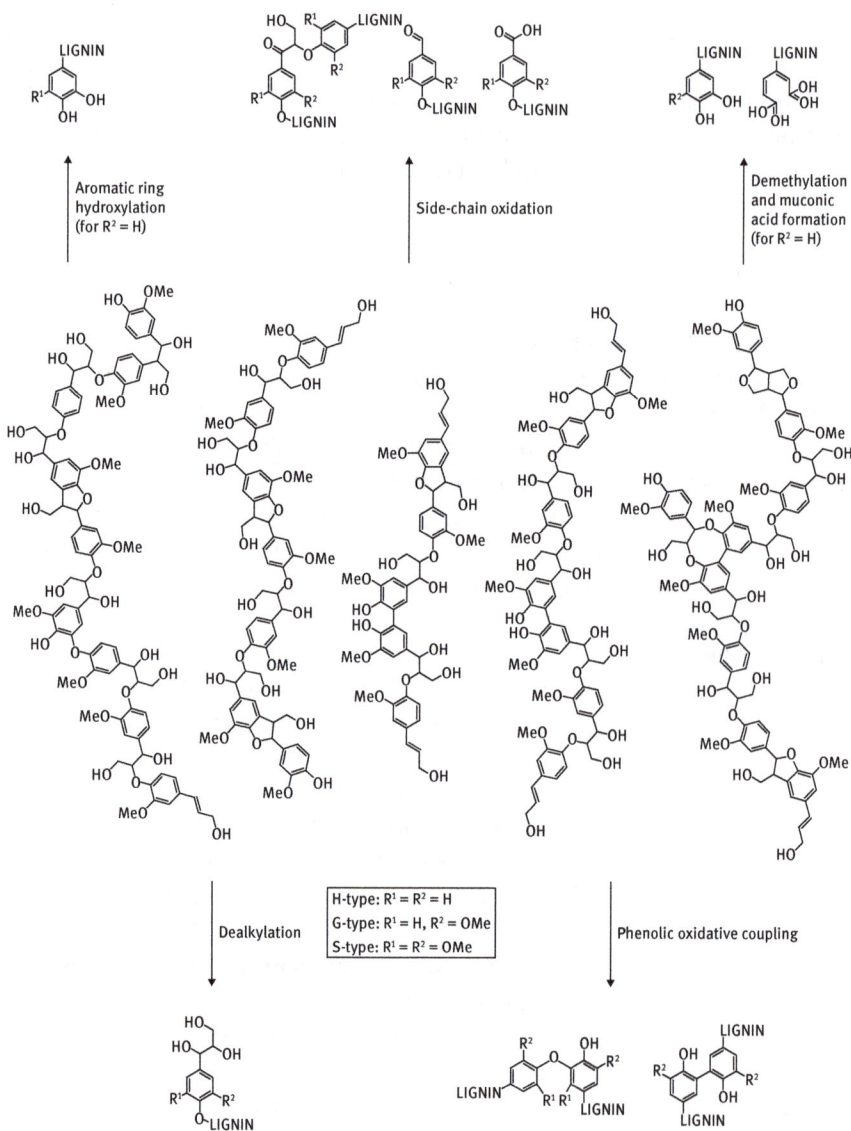

Fig. 11.7: Oxidative pathways in lignin valorization.

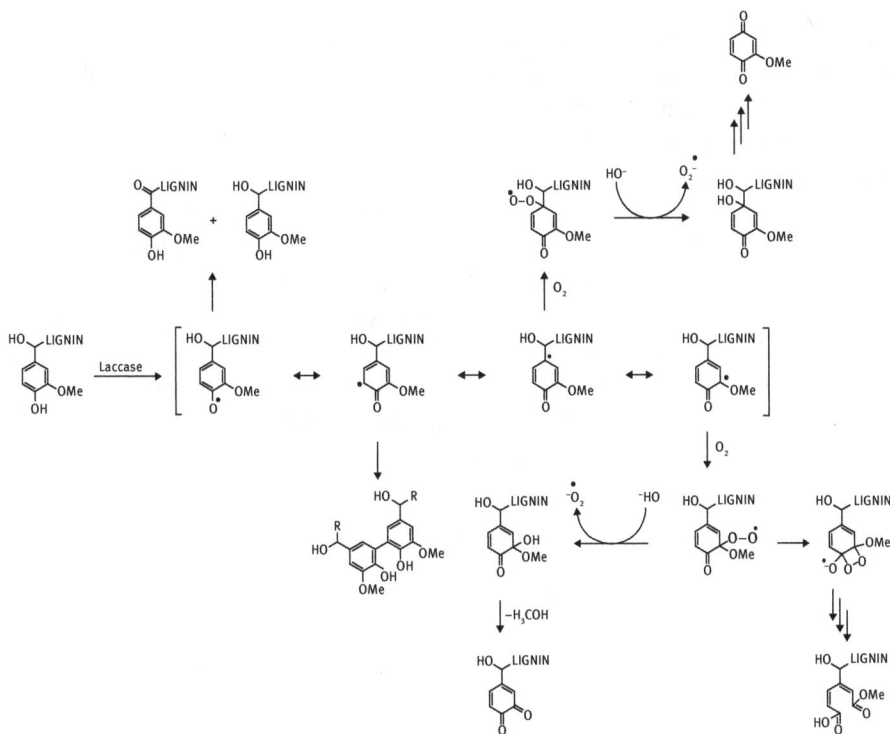

Fig. 11.8: Oxidation of lignin by laccases.

more specifically Mn-peroxidase and lignin peroxidases, are the most active ones. Several studies have been carried out in order to establish their reaction mechanisms and to evaluate their potential with respect to the development of biotechnological routes to lignin oxidation either in the presence or in the absence of oxidation mediators. Strategies for enzyme immobilization and use in mixtures to reproduce natural cascades have also been reported [15, 78, 83, 84]. Figure 11.8 shows the lignin oxidation pathway by laccase.

Lignin pyrolysis yields low-molecular-weight compounds that in the vapor phase condense into higher-molecular-weight oligomers [85].

The present production of vanillin from lignin constitutes an enlightening precedent for the development of innovative advances [86].

11.4.4 Applications in nanomaterials

11.4.4.1 Capsules

Nanoparticles, micelles, or capsules possess a great potential for the development of new materials. Recently, lignins were used to produce capsules capable of encapsulating either hydrophobic [87] or hydrophilic compounds [88]. Lignin capsules for hosting hydrophobic substances were produced by exploiting the amphiphilic character of kraft lignin: when dissolved in oil-in-water emulsions, the lignin polymers preferentially locate at the oil-water interface, where they can be further intermolecularly cross-linked to form capsules [87]. This cross-linking can be achieved using ultrasonification for radical activation or by reacting diglycidyl ether-terminated poly(ethylene glycol) with the lignin polymers. The capsules can be filled with hydrophobic molecules, which in turn can be released upon addition of surfactants.

When treating lignin-containing aqueous solutions that were emulsified by the aid of an surfactant in an organic phase comprised of cyclohexane containing 2,4-toluene diisocyanate, a lignin-based PU is formed at the interface between the aqueous and the organic phases [88]. Hydrophilic substances could be encapsulated upon formation and subsequently upon enzymatic degradation of the lignin shell.

11.4.4.2 Nanoparticles

Emulsion-based techniques can also be exploited for generating lignin-based beads, given that the lignin-containing solution constitutes the dispersed phase: Suspensions of black liquor in a mixture of oil and chlorobenzene in the presence of an emulsifier allowed the polymerization of the lignin using epichlorhydrin as cross-linker; spherical beads with diameters between 300 and 450 µm are obtained [89]. Less defined bead sizes were obtained when an alkaline lignin solution was dispersed in 1,2-dichloroethane and the lignin intermolecularly cross-linked in the aqueous droplets – either by mediation of epichlorhydrin [90] or by radical polymerization processes involving acrylic acid moieties that were attached to the lignin before [91]. Attaching further functionalization to these beads, e.g., sulfonhydrazine groups [90, 91] makes them attractive for applications in the wine industry, to remove carbonyl-bearing compounds responsible for binding of sulfur dioxide in wines [92].

Several recent publications reported the preparation of lignin-based nanoparticles. Starting from lignin solutions in ethylene glycol, the formation of lignin nanoparticles is achieved by gradually acidifying the solutions using mineral acids such as hydrochloric acid [93]. The nanoparticles could be cross-linked using glutaraldehyde and were found to be stable over a wide pH range when redispersed in water. The so-produced nanoparticles do not display cytotoxic activity when yeast or microalgae were exposed to them.

The gradual addition of a non-solvent, in this case water, to a solution of acetylated lignin in tetrahydrofuran (THF) resulted in the formation of nano-sized colloidal spheres that were water-dispersible and stable up to pH 12 [94]. Further increase in pH led to the hydrolysis of the acetyl groups, thus releasing free phenolic groups, which prevented the colloidization because of electrostatic repulsion.

When alkaline lignin solutions are added to cationic polyelectrolyte solutions of poly(diallyldimethylammonium chloride) (PDDAC), the cationic polymer is adsorbed on the lignin aggregates due to the display of overall negative surfaces stemming from the alkaline conditions chosen for solubilizing the lignin. Similar to process of creating Layer-by-Layer (LbL) protections for enzymes [78], the now positively charged surface of the lignin-PDDAC hybrids interacted with negatively charged natural latex rubber particles to associate further in solution, resulting in natural rubber materials reinforced with finely dispersed nano-sized lignin particles. These new lignin-based hybrid particles exhibit enhanced mechanical and thermal properties [95].

A simple physical method to modify lignin, here commercially available wheat straw and Sarkanda grass lignin, by ultrasonic irradiation in order to obtain nanoparticles has just been published, in which it could be shown that the compositional and structural changes of the lignin in the produced nanoparticles are not significantly modified by the intensity of the ultrasound applied but depend mainly on the nature of the lignin used [96].

11.4.4.3 Microporous and nanoporous materials

Porous substances called aerogels are seen as promising materials for a range of applications, e.g., insulating material, adsorbents, catalysts, etc. [97, 98]. They are commonly prepared by sol-gel polymerization of a resin, followed by solvent exchange and drying. Most often, the resins used are of polyphenolic type, and as indicated during the discussion of the lower-value applications, lignin been considered a potential source of phenols for this kind of resins.

Replacing part of resorcinol by lignin is enough for producing resorcinol-lignin-formaldehyde resins in aqueous solutions of NaOH [99]. Similarly formulated phenol-lignin-formaldehyde and tannin-lignin-formaldehyde resins have been produced [100, 101]. The maximum amount of lignin as replacement for the phenolic part cannot exceed 50%. In any case, the replacement of simple phenols by lignin generates an increase in the pore size, with the formation of macropores (>50 nm) to the detriment of mesopores (2–50 nm). Even if this causes an increase in the overall porosity and consequently a decrease in the bulk density, the total surface area is reduced, thus limiting the final material properties, such as the adsorption capacity, or the insulating performances [99–101]. Recently, lignin-based aerogels were produced without the use of formaldehyde and in the absence of any other phenol [102]. Lignin was cross-linked using oligo(ethylene glycol) or (propylene glycol) terminated with glycidyl moieties. Long-chain cross-linkers appear to favor the gel formation, probably because they can more easily react with two different lignin molecules, thus favoring the network formation.

Blocks of porous polymeric lignin hybrids can be obtained using black liquor directly, using the high internal phase emulsion (HIPE) technique [103]. Again, epichlorhydrin is used as cross-linker in the presence of a surfactant. After the addition of oil, the mixture is heated to generate lignin cross-linking in the continuous phase via reaction with epichlorhydrin. The obtained monoliths are then washed and display materials with an interconnected porous morphology. Depending on the nature and amount of surfactant, different void sizes can be obtained, ranging from about 5 to 20 μm in diameter [103].

11.5 Conclusions and perspectives

Lignin is the second most abundant plant biopolymer and the only renewable aromatic material source in nature. Its potential is to date largely unexploited because of large heterogeneity and diversity depending on specific botanical origin and isolation process. The emergence of new biorefinery streams that will produce high amounts of lignins have provided a new momentum to the studies for its upgrade.

Three main approaches to lignin valorization are focused on its transformation into biofuels, its use as a macromonomer, and its depolymerization to yield monomeric aromatic or carboxylic compounds. Development of new catalysts and chemical functionalization processes is needed to achieve these targets. The recent development of new analytical techniques like quantitative 2D-NMR allows to acquire structural information that were inaccessible a few years ago. For these reasons, today, it is possible to develop a specific lignin chemistry that will allow a rational approach to specific chemical modifications and functionalizations necessary for the development of new and advanced products and materials.

Bibliography

[1] Adler, E., Lignin chemistry – past, present and future, Wood Sci Technol **11** (1977) 169–218.
[2] Sarkanen, K. V., Ludwig, C. H., Lignins: Occurrence, Formation, Structure and Reactions, Wiley-Interscience, New York, NY, 1971.
[3] Hu, T. Q., Chemical Modification, Properties, and Usage of Lignin, Kluwer Academic/Plenum Publishers, New York, NY, 2002.
[4] Heitner, C., Dimmel, D., Schmidt, J., Lignin and Lignans: Advances in Chemistry, 1st edition, CRC Press, Boca Raton, FL, 2010.
[5] Sjöström, E., Wood Chemistry: Fundamentals and Applications, Academic Press, San Diego, CA, 1993.
[6] Lewis, N. G., Yamamoto, E., Lignin: occurrence, biogenesis and biodegradation, Annu Rev Plant Physiol Plant Mol Biol **41** (1990) 455–496.
[7] Norman, G., Lewis, Laurence B, Davin, Simo Sarkanen Lignin and Lignan Biosynthesis: Distinctions and Reconciliations, ACS Symposium Series, volume 697, American Chemical Society, Washington, DC, 1998, 1–27.
[8] Glasser, W. G., Sarkanen, S., Lignin: Properties and Materials, ACS Symposium Series, volume 397, American Chemical Society, Washington, DC, 1989.

[9] European Liaison Committee for Cellulose and Paper, The 8th International Symposium on Wood and Pulping Chemistry, June 6–9, 1995, Helsinki, Finland, Proceedings, Congrex, Blue & White Conferences, 1995.

[10] Boerjan, W., Ralph, J., Baucher, M., Lignin biosynthesis, Annu Rev Plant Biol **54** (2003) 519–546.

[11] Sjostrom, E., Haglund, P., Janson, J., Changes in cooking liquor composition during sulphite pulping, Sven Papperstidn **65** (1962) 855–869.

[12] Brogdon, B. N., Dimmel, D. R. J., Fundamental study of relative delignification, Wood Chem Technol **16** (1996) 261–297.

[13] Sette, M., Wechselberger, R., Crestini, C., Elucidation of lignin structure by quantitative 2D NMR, Chem Eur J **17** (2011) 9529–9535.

[14] Crestini, C., Melone, F., Saladino, R., Novel multienzyme oxidative biocatalyst for lignin bioprocessing, Bioorg Med Chem **19** (2011) 5071–5078.

[15] Crestini, C., Melone, F., Sette, M., Saladino, R., Milled wood lignin: a linear oligomer, Biomacromolecules **12** (2011) 3928–3935.

[16] Smith, B. R., Rice, R. W., Ince, P. J., Pulp Capacity in the United States, General Technical Report FPL-GTR-139, USDA Forest Service, 2000.

[17] Fredheim, G. E., Braaten, S. M., Christensen, B. E., Molecular weight determination of lignosulfonates by size-exclusion chromatography and multi-angle laser light scattering, J Chromatogr A **942** (2002) 191–199.

[18] Lake, M., What are we going to do with all this lignin? 2014, http://www.frontiersinbiorefining.org/Documents/Session_4B/What%20To%20Do%20With%20All%20This%20Lignin_MichaelLake.pdf.

[19] Howard, G. C., Process of treating waste sulphite liquor, US1699845 A, 1929.

[20] Buchholz, R. F., Neal, J. A., McCarthy, J. L., Some properties of paucidisperse gymnosperm lignin sulfonates of different molecular weights, J Wood Chem Technol **12** (1992) 447–469.

[21] Pye, E. K., Lora, J. H., The Alcell process, a proven alternative to kraft pulping. Tappi J **74** (1991) 113–118.

[22] Braaten, S. M., Christensen, B. E., Fredheim, G. E., Comparison of molecular weight and molecular weight distributions of softwood and hardwood lignosulfonates, J Wood Chem Technol **23** (2003) 197–215.

[23] Scholze, B., Hanser, C., Meier, D., Characterization of the water-insoluble fraction from fast pyrolysis liquids (pyrolytic lignin): part II, GPC, carbonyl groups, and 13C-NMR. J Anal Appl Pyrolysis **58–59** (2001) 387–400.

[24] McDonough, T. J., The chemistry of organosolv delignification, Tappi J **76** (1993) 186–193.

[25] Tanahashi, M., Tamabuchi, K., Goto, T., Aoki, T., Karina, M., Higuchi, T., Characterization of steam-exploded wood, 2: chemical changes of wood components by steam explosion, Wood Res Jpn **75** (1988) 1–12.

[26] Aziz, S., Sarkanen, K., Organosolv pulping – a review, Tappi J **72** (1989) 169–175.

[27] Langholtz, M., Downing, M., Graham, R., Baker, F., Compere, A., Griffith, W., Boeman, R., Keller, M., Lignin-Derived Carbon Fiber as a Co-product of Refining Cellulosic Biomass, SAE Int J Mater Manuf **7** (2014) 115–121.

[28] Stewart, D., Lignin as a base material for materials applications: chemistry, application and economics, Ind Crops Prod **27** (2008) 202–207.

[29] Wyman, C. E., Potential synergies and challenges in refining cellulosic biomass to fuels, chemicals, and power, Biotechnol Prog **19** (2003) 254–262.

[30] Avellar, B. K., Glasser, W. G., Steam-assisted biomass fractionation, I. Process considerations and economic evaluation, Biomass Bioenergy **14** (1998) 205–218.

[31] Shevchenko, S. M., Beatson, R. P., Saddler, J. N., The nature of lignin from steam explosion/enzymatic hydrolysis of softwood, Appl Biochem Biotechnol **79** (1999) 867–876.

[32] Heitz, M., Capek-Ménard, E., Koeberle, P. G., Gagné. J., Chornet, E., Overend, R. P., Taylor, J. D., Yu, E., Fractionation of Populus tremuloides at the pilot plant scale: optimization of steam pretreatment conditions using the STAKE II technology, Bioresour Technol **35** (1991) 23–32.

[33] Sun, Y., Cheng, J., Hydrolysis of lignocellulosic materials for ethanol production: a review, Bioresour Technol **83** (2002) 1–11.

[34] Torget, R. W., Kim, J. S., Lee, Y. Y., fundamental aspects of dilute acid hydrolysis/fractionation kinetics of hardwood carbohydrates, 1. Cellulose hydrolysis. Ind Eng Chem Res **39** (2000) 2817–2825.

[35] Bozell, J. J., Hoberg, J. O., Dimmel, D. R., Heteropolyacid catalyzed oxidation of lignin and lignin models to benzoquinones, J Wood Chem Technol **20** (2000) 19–41.

[36] Bozell, J. J., Hoberg, J. O., Dimmel, D. R., Catalytic oxidation of para-substituted phenols with nitrogen dioxide and oxygen, Tetrahedron Lett **39** (1998) 2261–2264.

[37] Cui, C., Sun, R., Argyropoulos, D. S., Fractional precipitation of softwood kraft lignin: isolation of narrow fractions common to a variety of lignins, ACS Sustain Chem Eng **2** (2014) 959–968.

[38] Sevastyanova, O., Helander, M., Chowdhury, S., Lange, H., Wedin, H., Zhang, L., Ek, M., Kadla, J. F., Crestini, C., Lindström, M. E., Tailoring the molecular and thermo-mechanical properties of kraft lignin by ultrafiltration, J Appl Polym Sci **131** (2014).

[39] DeBons, F. E., Whittington, L. E., Improved oil recovery surfactants based on lignin, J Petrol Sci Eng **7** (1992) 131–138.

[40] Wang, H., Male, J., Wang, Y., Recent advances in hydrotreating of pyrolysis bio-oil and its oxygen-containing model compounds, ACS Catal **3** (2013) 1047–1070.

[41] Zhang, Q., Chang, J., Wang, T., Xu, Y., Review of biomass pyrolysis oil properties and upgrading research, Energy Convers Manag **48** (2007) 87–92.

[42] Bonini, C., D'Auria, M., Emanuele, L., Ferri, R., Pucciariello, R., Sabia, A. R., Polyurethanes and polyesters from lignin, J Appl Polym Sci **98** (2005) 1451–1456.

[43] Gandini, A., Belgacem, M. N., Guo, Z.-X., Montanari, S., Lignins as macromonomers for polyesters and polyurethanes, in Hu, T. Q., editor, Chemical Modification, Properties, and Usage of Lignin, Kluwer Academic/Plenum Publishers, New York, NY, 2002, 57–80.

[44] Thielemans, W., Can, E., Morye, S. S., Wool, R. P., Novel applications of lignin in composite materials, J Appl Polym Sci **83** (2002) 323–331.

[45] Saito, T., Perkins, J. H., Jackson, D. C., Trammel, N. E., Hunt, M. A., Naskar, A. K., Development of lignin-based polyurethane thermoplastics, RSC Adv **3** (2013) 21832–21840.

[46] Cateto, C. A., Barreiro, M. F., Ottati, C., Lopretti, M., Rodrigues, A. E., Belgacem, M. N., Lignin-based rigid polyurethane foams with improved biodegradation, J Cell Plast **50** (2013) 81.

[47] Cateto, C. A., Barreiro, M. F., Rodrigues, A. E., Belgacem, M. N., Kinetic study of the formation of lignin-based polyurethanes in bulk, React Funct Polym **71** (2011) 863–869.

[48] Cheradame, H., Detoisien, M., Gandini, A., Pla, F., Roux, G., Polyurethane from kraft lignin, Br Polym J **21** (1989) 269–275.

[49] Vilela, C., Sousa, A. F., Fonseca, A. C., Serra, A. C., Coelho, J. F. J., Freire, C. S. R., Silvestre, A. J. D. The quest for sustainable polyesters – insights into the future, Polym Chem **5** (2014) 3119–3141.

[50] Cateto, C. A., Barreiro, M. F., Rodrigues, A. E., Belgacem, M. N., Optimization study of lignin oxypropylation in view of the preparation of polyurethane rigid foams, Ind Eng Chem Res **48** (2009) 2583–2589.

[51] Nadji, H., Bruzzèse, C., Belgacem, M. N., Benaboura, A., Gandini, A., Oxypropylation of lignins and preparation of rigid polyurethane foams from the ensuing polyols, Macromol Mater Eng **290** (2005) 1009–1016.

[52] Feldman, D., Lignin and its polyblends – a review, in Hu, T. Q., editor, Chemical Modification, Properties, and Usage of Lignin, Kluwer Academic/Plenum Publishers, New York, NY, 2002, 2002, 81–99.

[53] Abdelwahab, N., Nassar, M., Preparation, optimisation and characterisation of lignin phenol formaldehyde resin as wood adhesive, Pigment Resin Technol **40** (2011) 169–174.

[54] Feldman, D., Banu, D., Khoury, M., Epoxy-lignin polyblends, III. Thermal properties and infrared analysis, J Appl Polym Sci **37** (1989) 877–887.

[55] Feldman, D., Banu, D., Luchian, C., Wang, J., Epoxy-lignin polyblends: correlation between polymer interaction and curing temperature, J Appl Polym Sci **42** (1991) 1307–1318.

[56] Cazacu, G., Pascu, M. C., Profire, L., Kowarski, A. I., Mihaes, M., Vasile, C., Lignin role in a complex polyolefin blend, Ind Crops Prod **20** (2004) 261–2673.

[57] Kharade, A. Y., Kale, D. D., Lignin-filled polyolefins, J Appl Polym Sci **72** (1999) 1321–1326.

[58] Casenave, S., Aït-Kadi, A., Riedl, B., Mechanical behaviour of highly filled lignin/polyethylene composites made by catalytic grafting, Can J Chem Eng **74** (1996) 308–315.

[59] Hofmann, G. H., Polymer blend modification of PVC, in Walsh, D. J., Higgins, J. S., Maconnachie, A., editors. Polym Blends Mix, Springer, Netherlands, 1985, 117–148.

[60] Feldman, D., Banu, D., El-Raghi, S., Poly(vinyl chloride)-lignin blends for outdoor application in building, J Macromol Sci Part A **31** (1994) 555–571.

[61] Feldman, D., Banu, D., Contribution to the study of rigid PVC polyblends with different lignins, J Appl Polym Sci **66** (1997) 1731–1744.

[62] Hilburg, S. L., Elder, A. N., Chung, H., Ferebee, R. L., Bockstaller, M. R., Washburn, N. R., A universal route towards thermoplastic lignin composites with improved mechanical properties, Polymer **55** (2014) 995–1003.

[63] Nägele, H., Pfitzer, J., Ziegler, L., Inone-Kauffmann, E. R., Eckl, W., Eisenreich, N., Lignin matrix composites from natural resources – ARBOFORM®, in Kabasci, S., editor, Bio-Based Plastics: Materials and Applications, Wiley & Sons, New York, NY, 2013, 89–115.

[64] Chatterjee, S., Jones, E. B., Clingenpeel, A. C., McKenna, A. M., Rios, O., McNutt, N. W., et al., Conversion of lignin precursors to carbon fibers with nanoscale graphitic domains, ACS Sustain Chem Eng **2** (2014) 2002–2010.

[65] Kadla, J. F., Kubo, S., Gilbert, R. D., Venditti, R. A., Lignin-based carbon fibers, in Hu, T. Q., editor, Chemical Modification, Properties, and Usage of Lignin, Kluwer Academic/Plenum Publishers, New York, NY, 2002, 121–137.

[66] Otani, S., Fukuoka, Y., Igarashi, B., Sasaki, K., Method for producing carbonized lignin fiber, US Patent 3461082, 1969.

[67] Tuck, C. O., Pérez, E., Horváth, I. T., Sheldon, R. A., Poliakoff, M., Valorization of biomass: deriving more value from waste, Science **337** (2012) 695–699.

[68] Clark, H. J., Deswarte, F. E. I., Farmer, T. J., The integration of green chemistry into future biorefineries, Biofuels Bioprod Biorefining **3** (2009) 72–90.

[69] Brand, S., Susanti, R. F., Kim, S. K., Lee, H., Kim, J., Sang, B.-I., Supercritical ethanol as an enhanced medium for lignocellulosic biomass liquefaction: influence of physical process parameters, Energy **59** (2013) 173–182.

[70] Akhtar, J., Amin, N. A. S., A review on process conditions for optimum bio-oil yield in hydrothermal liquefaction of biomass, Renew Sustain Energy Rev **15** (2011) 1615–1624.

[71] Pandey, M. P., Kim, C. S., Lignin depolymerization and conversion: a review of thermochemical methods, Chem Eng Technol **34** (2011) 29–41.

[72] Zakzeski, J., Bruijnincx, P. C. A., Jongerius, A. L., Weckhuysen, B. M., The catalytic valorization of lignin for the production of renewable chemicals, Chem Rev **110** (2010) 3552–3599.

[73] Di Giuseppe, A., Crucianelli, M., De Angelis, F., Crestini, C., Saladino, R., Efficient oxidation of thiophene derivatives with homogeneous and heterogeneous MTO/H_2O_2 systems: a novel approach for, oxidative desulfurization (ODS) of diesel fuel, Appl Catal B Environ **89** (2009) 239–245.

[74] Alonso, D. M., Wettstein, S. G., Dumesic, J. A., Bimetallic catalysts for upgrading of biomass to fuels and chemicals, Chem Soc Rev **41** (2012) 8075–8098.

[75] Lange, H., Decina, S., Crestini, C., Oxidative upgrade of lignin – recent routes reviewed, Eur Polym J **49** (2013) 1151–1173.

[76] Das, L., Kolar, P., Sharma-Shivappa, R., Heterogeneous catalytic oxidation of lignin into value-added chemicals, Biofuels **3** (2012) 155–166.

[77] Crestini, C., Crucianelli, M., Orlandi, M., Saladino, R., Oxidative strategies in lignin chemistry: a new environmental friendly approach for the functionalisation of lignin and lignocellulosic fibers, Catal Today **156** (2010) 8–22.

[78] Crestini, C., Perazzini, R., Saladino, R., Oxidative functionalisation of lignin by layer-by-layer immobilised laccases and laccase microcapsules, Appl Catal Gen **372** (2010) 115–123.

[79] Crestini, C., Pastorini, A., Tagliatesta, P., Metalloporphyrins immobilized on motmorillonite as biomimetic catalysts in the oxidation of lignin model compounds, J Mol Catal Chem **208** (2004) 195–202.

[80] Crestini, C., Pastorini, A., Tagliatesta, P., The immobilized porphyrin-mediator system Mn(TMePyP)/clay/HBT (clay-PMS): a lignin peroxidase biomimetic catalyst in the oxidation of lignin and lignin model compounds, Eur J Inorg Chem **2004** (2004) 4477–4483.

[81] Crestini, C., Caponi, M. C., Argyropoulos, D. S., Saladino, R., Immobilized methyltrioxo rhenium (MTO)/H_2O_2 systems for the oxidation of lignin and lignin model compounds, Bioorg Med Chem **14** (2006) 5292–5302.

[82] Crestini, C., Pro, P., Neri, V., Saladino, R., Methyltrioxorhenium: a new catalyst for the activation of hydrogen peroxide to the oxidation of lignin and lignin model compounds, Bioorg Med Chem **13** (2005) 2569–2578.

[83] Salvachúa, D., Prieto, A., López-Abelairas, M., Lu-Chau, T., Martínez Á. T., Martínez, M. J., Fungal pretreatment: an alternative in second-generation ethanol from wheat straw, Bioresour Technol **102** (2011) 7500–7506.

[84] Perazzini, R., Saladino, R., Guazzaroni, M., Crestini, C., A novel and efficient oxidative functionalization of lignin by layer-by-layer immobilised horseradish peroxidase, Bioorg Med Chem **19** (2011) 440–447.

[85] Patwardhan, P. R., Brown, R. C., Shanks, B. H., Understanding the fast pyrolysis of lignin, ChemSusChem **4** (2011) 1629–1636.

[86] Araújo, J. D. P., Grande, C. A., Rodrigues, A. E., Vanillin production from lignin oxidation in a batch reactor, Chem Eng Res Des **88** (2010) 1024–1032.

[87] Tortora, M., Cavalieri, F., Mosesso, P., Ciaffardini, F., Melone, F., Crestini, C., Ultrasound driven assembly of lignin into microcapsules for storage and delivery of hydrophobic molecules, Biomacromolecules **15** (2014) 1634–1643.

[88] Yiamsawas, D., Baier, G., Thines, E., Landfester, K., Wurm, F. R., Biodegradable lignin nanocontainers, RSC Adv **4** (2014) 11661–11663.

[89] Chen, G.-F., Liu, M.-H., Adsorption of L-lysine from aqueous solution by spherical lignin beads: kinetics and equilibrium studies, BioResources **7** (2012) 298–314.

[90] Saidane, D., Barbe, J.-C., Birot, M., Deleuze, H., Preparation of functionalized kraft lignin beads, J Appl Polym Sci **116** (2010) 1184–1189.

[91] Saidane, D., Barbe, J-C., Birot, M., Deleuze, H., Preparation of functionalized lignin beads from oak wood alkaline lignin, J Appl Polym Sci **128** (2013) 424–429.

[92] Blasi, M., Barbe, J.-C., Maillard, B., Dubourdieu, D., Deleuze, H., New methodology for removing carbonyl compounds from sweet wines, J Agric Food Chem **55** (2007) 10382–10387.

[93] Frangville, C., Rutkevičius, M., Richter, A. P., Velev, O. D., Stoyanov, S. D., Paunov, V. N., Fabrication of environmentally biodegradable lignin nanoparticles, ChemPhysChem **13** (2012) 4235–4243.

[94] Qian, Y., Deng, Y., Qiu, X., Li, H., Yang, D., Formation of uniform colloidal spheres from lignin, a renewable resource recovered from pulping spent liquor, Green Chem **16** (2014) 2156–2163.

[95] Jiang, C., He, H., Jiang, H., Ma, L., Jia, D. M., Nano-lignin filled natural rubber composites: preparation and characterization, EXPRESS Polym Lett **7** (2013) 480–493.

[96] Gilca, I. A., Popa, V. I., Crestini, C., Obtaining lignin nanoparticles by sonication, Ultrason Sonochem **23** (2015) 369–375.

[97] Hüsing, N., Schubert, U., Aerogels – airy materials: chemistry, structure, and properties, Angew Chem Int Ed **37** (1998) 22–45.

[98] Gesser, H. D., Goswami, P. C., Aerogels and related porous materials, Chem Rev **89** (1989) 765–788.

[99] Chen, F., Xu, M., Wang, L., Li, J., Preparation and characterization of organic aerogels by the lignin-resorcinol-formaldehyde copolymer, BioResources **6** (2011) 1262–1272.

[100] Grishechko, L. I., Amaral-Labat, G., Szczurek, A., Fierro, V., Kuznetsov, B. N., Celzard, A., Lignin-phenol-formaldehyde aerogels and cryogels, Microporous Mesoporous Mater **168** (2013) 19–29.

[101] Grishechko, L. I., Amaral-Labat, G., Szczurek, A., Fierro, V., Kuznetsov, B. N., Pizzi, A.,Celzard, A., New tannin-lignin aerogels, Ind Crops Prod **41** (2013) 347–355.

[102] Perez-Cantu, L., Liebner, F., Smirnova, I., Preparation of aerogels from wheat straw lignin by cross-linking with oligo(alkylene glycol)-α,ω-diglycidyl ethers, Microporous Mesoporous Mater **195** (2014) 303–310.

[103] Forgacz, C., Birot, M., Deleuze, H., Synthesis of porous emulsion-templated monoliths from a pulp mill by-product, J Appl Polym Sci **129** (2013) 2606–2613.

Raf Roelant, Fabrizio Cavani, Carla S.M. Pereira, and
Alírio E. Rodrigues

12 Utilization of existing assets

Abstract: This chapter proposes the integration of bio-based technology in existing plants. Low capital costs and relatively low risks are shown to promise a fast industrial deployment of novel bio-based technologies. Two case studies are performed. The first, the production of maleic anhydride from bio-*n*-butanol in a phthalic anhydride plant, is an example of integration downstream. A techno-economic evaluation reveals a need for further catalyst development and reaction engineering. The second case study, the production of 1,1-dibutoxyethane co-located with an *n*-butanol plant, is an example of integration upstream. Simulated moving-bed membrane reactor technology, which is still in a low development stage, is evaluated qualitatively. The case is promising, but remains to be demonstrated.

12.1 Introduction

12.1.1 Why utilizing existing plants to implement bio-based technology?

Just like a petroleum refinery, a biorefinery consists of a collection of processes exchanging product and energy streams. The construction of a stand-alone biorefinery generally involves the simultaneous implementation of each of the processes it consists of, because one process generally depends on the other. The construction of a stand-alone biorefinery therefore often requires prohibitively high capital costs. Moreover, as biorefineries usually rely on a multitude of relatively new technologies, there is a relatively high risk of an overall operational failure. An elementary probability calculation illustrates this. If a biorefinery relies on five new technologies, each failing at operation with a probability of $p = 10\%$, the chance of an overall failure (meaning a failure of at least one of the new technologies) is already at an unacceptable level of $1 - (1 - p)^5 \cong 41\%$.

The preceding paragraph shows two ways of increasing the chance of implementing bio-based processes:
1. Reduce capital costs.
2. Reduce the number of novel technologies to be implemented simultaneously.

This chapter is devoted to a strategy to accomplish both of these reductions at the same time. This is done by implementing bio-based technology by an addition to an existing plant, which itself need not be bio-based. Compared to the case where a stand-alone biorefinery needs to be constructed, first, using the existing infrastructure, utility systems, and logistics, an important saving on capital costs is obtained. Second, by using already implemented process steps, only one or two innovative technologies need to be supplemented, which maintains the operational risk at an acceptable level. A (partial) retrofit of the existing plant may be needed, but this is not necessarily so.

The two advantages mentioned above are supplemented by others:

3. Skilled personnel are already available on site.
4. The market risks of the existing plant are mitigated by adding variety in the raw material feedstock or product portfolio. This adds a degree of freedom allowing the adaptation of the production to the market situation.
5. To supplement novel bio-based technology, those plants that are facing a poor economy, for example, due to a fading product market or a rise of the raw materials price, can be targeted. By offering new prospects to such plants, jobs and production can be preserved.

The integration of bio-based technology into existing plants also presents some drawbacks:

1. Sometimes a product that has an incomplete and possibly variable degree of renewability is obtained. A partially renewable product is much more difficult to market than a fully renewable one.
2. The envisaged integration may face the reluctance of the management of the existing plant.

Despite these drawbacks, the authors believe that the utilization of existing assets will prove crucial as a first step in the deployment of novel bio-based technology.

12.1.2 Kinds of integration scenarios

There are several ways in which bio-based technology can be integrated into existing plants. A first distinction to be made is the one between downstream and upstream integration (see Figure 12.1).

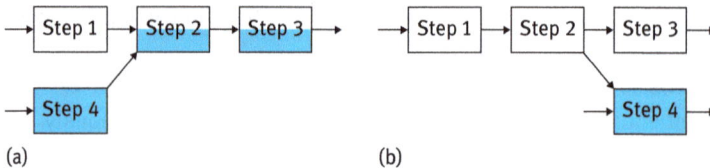

(a)   (b)

Fig. 12.1: Different ways in which bio-based technology (in blue) can be integrated into an existing process: (a) integration downstream and (b) integration upstream. Partially blue boxes represent process steps possibly needing a retrofit.

In the case of downstream integration (see Figure 12.1a), a bio-based raw material is converted into a stream that can be processed by the existing plant. If the bio-based stream is not exactly the same as an already present intermediate or raw material stream, the process steps downstream need retrofitting and the product(s) can be different. This will be illustrated by case study 1, described below. Ultimately, the integration scenario involves a mere retrofit. In Figure 12.1a, this would mean that Step 4 is absent and a bio-based stream is directly fed to a plant retrofitted for the purpose.

In the case of upstream integration, a product or intermediate stream of an existing plant is processed by co-located (partially) bio-based technology (see Figure 12.1b). This integration model normally does not involve any retrofitting. Case study 2 will provide an example of this kind of integration.

12.2 Case study 1: maleic anhydride production in a phthalic anhydride plant

12.2.1 General

Maleic anhydride (MA) is a raw material for coatings and polymers, food acidulants (malic acid, fumaric acid), and agricultural chemicals. The European market of MA is structurally undersupplied. Most MA is currently produced by catalytic partial oxidation of fossil n-butane. As yet, there is no production of renewable MA. This case study investigates the production of renewable MA from bio-n-butanol. This idea is particularly promising because n-butanol is much more reactive than n-butane and because the partial oxidation of n-butanol is less exothermic than the one of n-butane.

Since there is a great interest for using bio-butanol as a fuel component, it can be foreseen that there will be considerable trade volumes of this bio-alcohol in the future and that it will be produced by various suppliers. Figure 12.2 gives a present-day worldwide overview of producers and technology providers. Both n-butanol and isobutanol are produced. Note that, as yet, there is no commercial production of either of the isomers in Europe.

As a fast track to the production of MA from bio-n-butanol, a downstream integration scenario was evaluated. This scenario is represented schematically in Figure 12.3. n-Butanol obtained by fermentation of bio-based saccharides is subjected to partial oxidation in a retrofitted phthalic anhydride (PA) plant. Such a plant produces PA by partial oxidation of fossil o-xylene. As a very interesting feature, the PA plant is already co-producing MA as a by-product. Therefore, infrastructure to recover and purify the MA is already available.

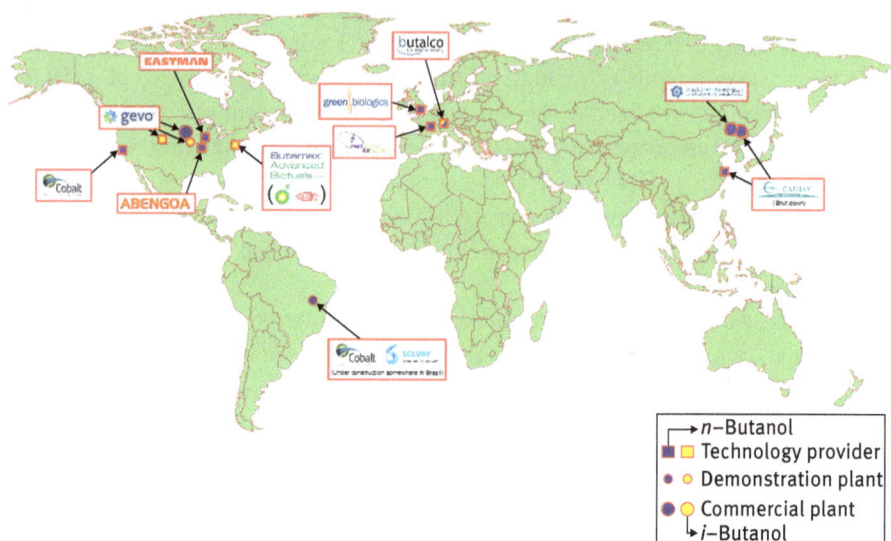

Fig. 12.2: Bio-butanol technology providers and producers in the world.

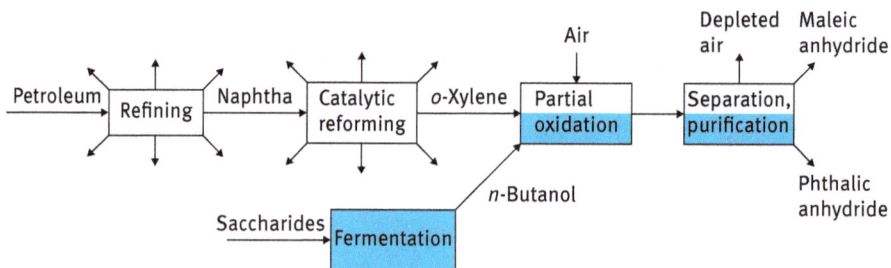

Fig. 12.3: Integration of bio-butanol partial oxidation to MA in a PA plant.

PA, a raw material for polymer plasticizers, chemicals, and dyes, is being banned because exposure to it or its derivatives is suspected to have adverse health effects in humans. Currently, there is already an oversupply of 100,000 metric tons per year in Europe. Therefore, the evaluated integration scenario has the potential to save jobs. Figure 12.4 shows the many PA plants operational in Europe.

12.2.2 Technological aspects of the retrofit

A 50-kta PA plant was studied as a typical small-capacity plant targeted for retrofitting. Figure 12.5a shows a typical process flow diagram of such a plant. o-Xylene is partially oxidized to PA under a large excess of air in a multitubular fixed-bed catalytic reactor. Overoxidation leads to MA and CO_2 as by-products. Also, some PA is

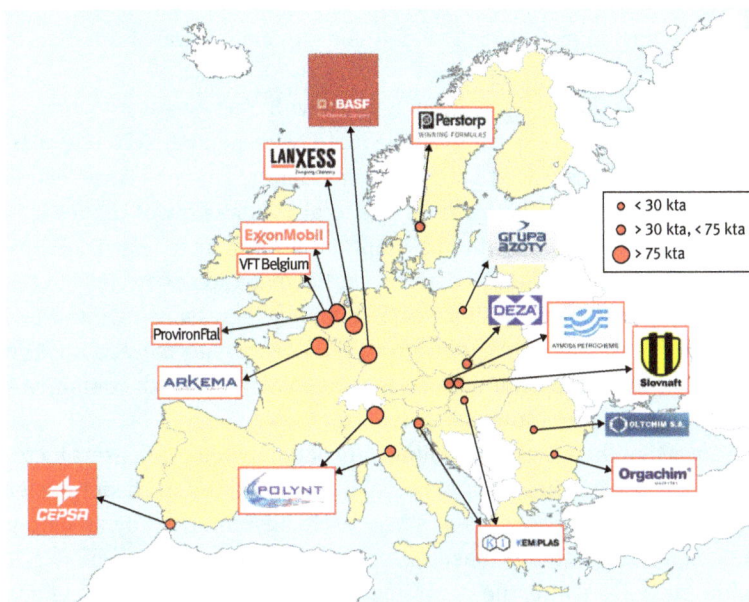

Fig. 12.4: PA plants in Europe. Member states of the European Union are indicated in pale yellow.

hydrated undesirably to phthalic acid. Crude PA is recovered from the product gas mixture through switch condensers. The phthalic acid, which is recovered along with it, is reconverted into PA in a decomposer vessel. The PA is finally purified in two distillation columns. The non-condensed fraction from the switch condensers is washed by a solvent to recover the MA. After evaporation of the solvent, the crude MA is purified by distillation.

As a retrofit scenario, the plant described above is made to co-produce MA from *n*-butanol. Half of the reactor tubes are fed by an *n*-butanol/air mixture instead of an *o*-xylene/air mixture. The PA catalyst in these tubes is replaced by a new catalyst combining dehydration and partial oxidation functionalities, allowing the conversion of *n*-butanol into MA. As the current process already coproduces MA, its already present separation and purification infrastructure can be put to use. Nevertheless, changes are necessary.

In regard to the catalyst type, a bifunctional system is needed in order to facilitate both steps involved in the transformation of *n*-butanol into MA: (i) the dehydration of the alcohol into butenes, a reaction that is catalyzed by Brønsted acidity, and (ii) the oxidation of butenes to MA catalyzed by several classes of materials, among which those showing the best performance are based either on metal molybdates or on vanadyl pyrophosphate [1]. The latter is also used industrially for the one-pot oxidation of *n*-butane to MA and is known to show acidic features, besides redox properties suited for the selective transformation of C4 hydrocarbons into MA [2]. Therefore, vanadyl pyrophosphate is an excellent candidate for the one-pot transformation of bio-*n*-butanol into MA.

Figure 12.5b shows the process flow diagrams with the necessary changes indicated in red. The reactor is compartmentalized into two equal parts. In one compartment, the catalyst bed is replaced with the new catalyst. The *n*-butanol feed is diluted in air up to the same volume fraction. The total volume flow of the feeds is kept equal. Note that this is a conservative assumption. Indeed, as the partial oxidation of *n*-butanol to MA is less exothermic than the partial oxidation of *o*-xylene to PA, it may be possible to feed more MA by diluting it less and/or increasing its flow. Now, as the load is assumed equal, the total cooling duty of the reactor decreases a bit. The two effluent streams of the reactor are mixed immediately because both contain MA and PA, albeit in different concentrations.

Simulation shows that the necessary cooling duty of the switch condensers increases a bit compared to the original scenario. Therefore, some additional heat transfer area is needed. The decomposer consists of a heat exchanger followed by an empty vessel separating vapor and liquid. It has to process a feed that contains more MA than in the original case. Therefore, the vapor load increases and an extra vessel has to be added in order to ensure a smooth operation. As the vapor stream released by the decomposers contains a higher concentration of MA, it is routed to the washing column instead of the top section of the lights removal column for PA. The liquid effluent of the decomposer vessels is sent to the PA purification section as before. As the production of PA is lower than in the original case, the load of the PA purification section decreases. Therefore, the existing columns can still be used, although possibly with adapted internals.

The mixed vapor streams from switch condensers and decomposers are subjected to absorption to recover the MA. As the production of MA is clearly increased compared to the original case, a higher absorption capacity is needed, which represents the main investment cost of the retrofit scenario. Correspondingly, the recovery of the solvent and the purification of MA require extra equipment. The purification of MA is performed in two distillation columns instead of the original one.

12.2.3 Economic evaluation

An economic evaluation was performed for several cases corresponding to different catalyst performances. Case 1 represents a performance currently proven in the laboratory and through pilot tests (see Table 12.1). Cases 2 and 3 represent future projections, where the selectivity to MA has been improved through catalyst development.

Simulation, shortcut equipment design, and cost estimation have allowed estimating the capital investment costs associated with the retrofit scenario for each of the three cases (see Table 12.1). As a next step, the production cost of MA has been estimated, which was possible through an inventory of differential energy and material

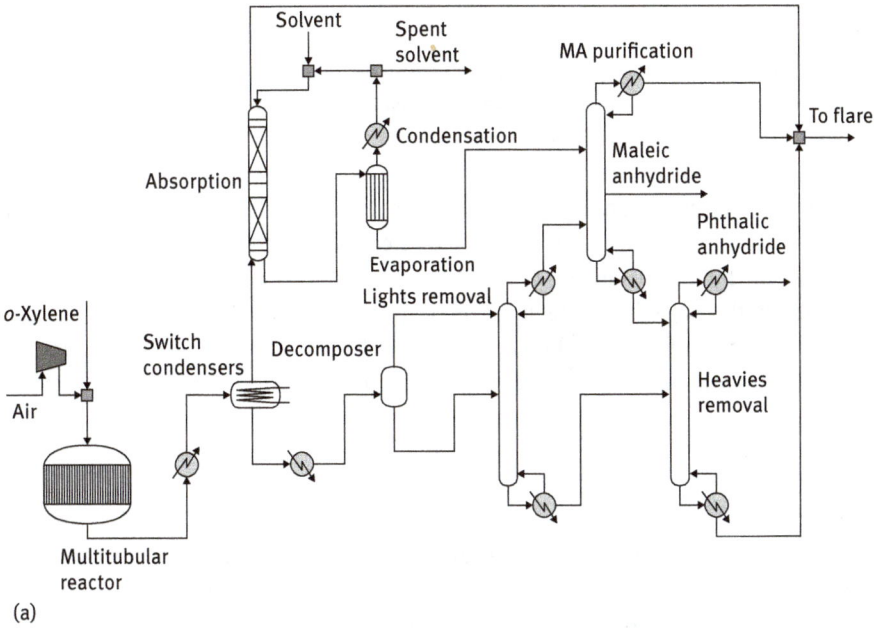

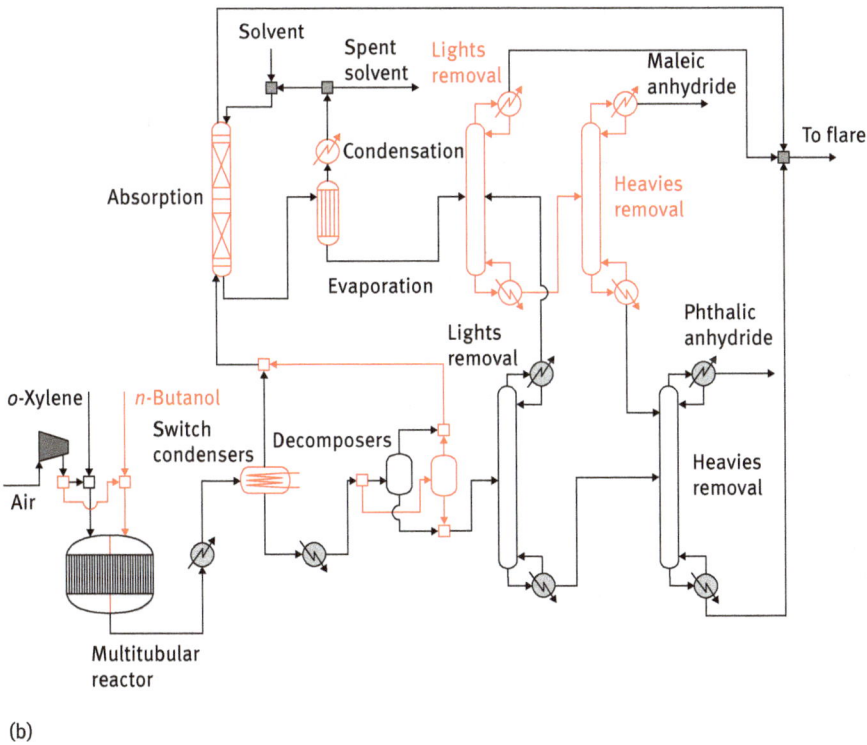

Fig. 12.5: A typical small-scale PA production plant: (a) original process and (b) process retrofitted for the coproduction of MA from *n*-butanol.

streams (including a decreased production of PA and a decreased consumption of o-xylene) for each of the cases relative to the base case (original PA plant). Each of these streams represents a cost or a revenue. Note that the production cost estimate, shown in Table 12.1, includes a contribution for the depreciation of the capital cost.

The capital investment figures in Table 12.1 are much smaller than what would be needed to build a stand-alone MA plant with the same capacity. The capital investment cost does not vary a lot as a function of the selectivity of the reaction from n-butanol to MA. Meanwhile, the estimates of the MA production cost do show a great variety as a function of this selectivity. The numbers have to be compared to the market value of MA, which is about €1.5/kg at the time of writing. Clearly, the catalytic performance of the catalyst should be significantly improved to make the retrofit scenario economical.

12.2.4 Demonstration status

Table 12.2 shows an overview of technologies relevant to the integration scenario with their technology readiness level (TRL) and risk of industrial implementation. There are several technologies to produce bio-n-butanol, from which only the acetone-butanol-ethanol (ABE) fermentation is applied commercially.

Table 12.1: Summary of the retrofit scenarios evaluated on their industrial economics.

	Case 1	Case 2	Case 3
Carbon selectivity from n-butanol			
MA	0.39	0.60	0.80
PA	0.12	0.08	0.04
Butene	0.01	<0.01	<0.01
Acrylic acid	0.08	0.05	0.03
Acetic acid	0.08	0.05	0.03
Formaldehyde	0.01	<0.01	<0.01
CO	0.15	0.10	0.05
CO_2	0.15	0.10	0.05
Heavy ends	0.01	<0.01	<0.01
Production (10^3 metric tons per year)			
MA	9.4	14.1	18.4
PA	27.4	26.7	26.0
Capital investment cost (million €)	18.8	17.3	17.8
MA production cost (€/kg)	2.59	1.81	1.46

Note that the total capital investment cost represents the costs needed to perform the retrofit of the plant. The capital investment cost of the original plant is therefore not included.

Table 12.2: Summary of relevant technologies with their TRL.

Step	TRL	Risk	Status
ABE fermentation from starch	9	Low	Operational at industrial scale: technology applied by Cathay Industrial Biotech (China), promoted by Green Biologics (UK), etc.
Other fermentation technologies to n-butanol (e.g., Green Biologics, Tetravitae, Cobalt)	7	Moderate	Pilot experiments successful, but no industrial implementation
n-Butanol produced catalytically from bio-ethanol	6	Moderate	Pilot experiments performed by Abengoa
Purified n-butanol partial oxidation to MA	7	Low	Successful tests in pilot unit

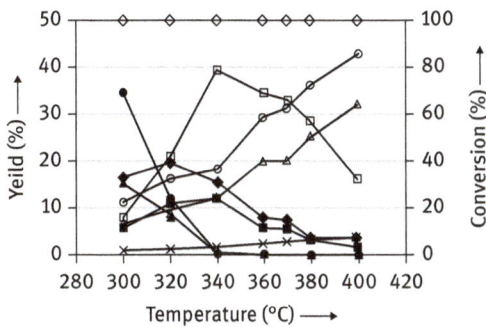

Fig. 12.6: n-Butanol conversion and yields to products as a function of temperature. Catalyst: DuPont vanadyl pyrophosphate. Conditions: feed 1 mol% 1-butanol, 20% O_2, remainder N_2; W/F 1.33 g s mL^{-1}. Symbols: n-butanol conversion ($\Diamond$). Selectivity to MA ($\square$), 1-butene ($\blacktriangle$), 2-butenes ($\bullet$), acetic acid+acrylic acid ($\blacklozenge$), CO ($\triangle$), CO_2 (O), PA ($\blacksquare$), and "lights" ($\times$).

The catalytic partial oxidation of bio-n-butanol to MA is already at a high TRL because it has been proven experimentally in the laboratory and in a pilot unit. As an example, Figure 12.6 shows the effect of the reaction temperature on n-butanol conversion and on selectivity to the products. In these laboratory-scale experiments, an inlet feed stream containing 1 mol% n-butanol in air was used, while the catalyst was made of a commercial vanadyl pyrophosphate sample, formerly used by DuPont for the oxidation of n-butane to MA.

Under the conditions examined, the conversion of n-butanol was complete over the entire range of temperatures investigated; products obtained were MA, light acids (acrylic and acetic acids), carbon oxides, and PA; butenes formed at temperatures lower than 340°C only. PA may form by means of a Diels-Alder reaction between the intermediately formed butadiene and MA. Maximum selectivity to MA was observed at 340°C (39%); at a higher temperature, the selectivity declined, with a concomitant increase of CO and CO_2 formation. Experiments carried out in a pilot unit available at the Orgachim factory in Ruse (BG) gave quite similar results, with a maximum yield to MA close to 40%.

12.2.5 Strengths, weaknesses, opportunities, and threats analysis

An analysis of the strengths, weaknesses, opportunities, and threats (SWOT) associated with the integration scenario has been performed. The result is shown in Table 12.3.

12.2.6 Value creation

MA is used in the production of coatings and polymers, food acidulants, and agricultural chemicals. The overall capacity is greater than 3 Mt/year. The food industry especially may be in favor of using bio- instead of fossil-derived chemicals.

As the PA market is shrinking, retrofitting European PA plants for an increased co-production of MA, for which there is still a growing market, is expected to save jobs. For the evaluated retrofit scenario (50 kt/year PA plant), the number of full-time jobs is estimated at about 28 (some 19 of them operators). The capital investment costs required for this are estimated at about €18 million.

Table 12.3: SWOT of the MA production from bio-*n*-butanol by a retrofit of a PA plant.

Strength	Weaknesses
– Minimal investments required, since MA is already a by-product in PA production. – The reactor does not need to be changed.	– The optimal reaction conditions for the production of PA from *o*-xylene are not the same as the optimal ones for the production of MA from *n*-butanol. However, they cannot be made very different if they share the same reactor.
Opportunities	**Threats**
– *n*-Butanol is a potential biofuel and can therefore become widely available. – Quite a lot of PA producers are facing a fading market.	– Possible future increase in the price of bio-*n*-butanol. – If a better catalytic selectivity cannot be obtained, the retrofit will not be economical. – Impurities in bio-*n*-butanol may cause some currently unexpected complication of the purification of both MA and PA.

12.2.7 Concluding remarks

In order to obtain an economical case, a further catalyst screening and optimization is necessary for the conversion of *n*-butanol to MA. Furthermore, in order to obtain an economical case, it would help to increase the load of the multitubular reactor with *n*-butanol. This would mean that the degree of dilution with air is decreased and/or that the volume flow of the feed mixture is simply increased. It remains to be investigated whether this is possible.

12.3 Case study 2: 1,1-dibutoxyethane production added to an *n*-butanol plant in Europe

12.3.1 General

Acetals are applied as starting materials for agricultural chemicals [3] and intermediates for pharmaceutical products such as vitamins, analgesics, and antifungal medication [4, 5]. They are also used in the design of synthetic perfumes in order to increase their oxidation resistance and the perfume's lifetime [6], in the flavoring of food [6], and to contribute to the aroma of alcoholic beverages, such as spirit drinks [7]. Nowadays, 1,1-dibutoxyethane (DBE) and other similar acetals such as dibutoxymethane derived from *n*-butanol and formaldehyde are produced at a small scale, mainly for applications in the flavor and fragrance industries and as solvents (cleaning, aerosols, adhesives, etc.). Producers and production technology owners include Lambiotte & Cie, Kreussler [8], FutureFuel, Kairav Chemofarbe Industries, Sanofi-Aventis [9], Inoue Perfumery, and Godavari Biorefineries.

Acetals have also been identified as oxygenates for diesel blending due to their capability to reduce emissions, mainly of particulate matter, and to facilitate the combustion of the final products without decreasing the ignition quality, maintaining, or even increasing the cetane number [10, 11]. Moreover, DBE can be produced from renewable resources such as *n*-butanol and acetaldehyde (derived from ethanol). Therefore, DBE can provide part of the answer to the European Renewable Energy Directive (2009/28/EC), which states that by 2020, at least 10% of all transportation fuels must be derived from renewable sources.

The goal is to use bio-based raw materials for the production of DBE. However, as already mentioned in caste study 1, there is no production of bio-butanol yet at an industrial scale in Europe. An alternative scenario would be a retrofit of an existing bio-ethanol plant, but that would require the introduction of several new technologies: (1) conversion of the fermentation to *n*-butanol or a Guerbet-like ethanol-to-butanol conversion, (2) ethanol selective oxidation to acetaldehyde, (3) *n*-butanol plus acetaldehyde conversion to DBE. The simultaneous introduction of these new technologies represents very high risks. Consequently, in this case study, a DBE production plant is added to a fossil-based *n*-butanol plant (operating through hydroformylation of petrochemical propylene) instead, as shown in Figure 12.7.

Existing *n*-butanol plants in Europe and their production capacity are highlighted in Figure 12.8. Bio-*n*-butanol can be supplied as an alternative raw material, either by importing or by co-locating a bio-butanol plant in the future. Acetaldehyde has to be imported except if it is available on site. If possible, bio-based acetaldehyde should be used. For example, SEKAB, a Sweden-based company, could provide such bio-acetaldehyde (see Figure 12.8).

12.3.2 Technological aspects

DBE can be synthesized through a liquid-phase reversible acetalisation in acidic medium:

$$2\,n\text{-Butanol} + \text{Acetaldehyde} \xleftarrow{\quad H^+ \quad} 1,1\text{-DBE} + \text{Water}$$

Due to the chemical equilibrium limitation of this reaction (55% equilibrium conversion is achieved at 20°C and stoichiometric ratio of reactants [12]), the production of DBE is enhanced by applying multifunctional reactors allowing simultaneous

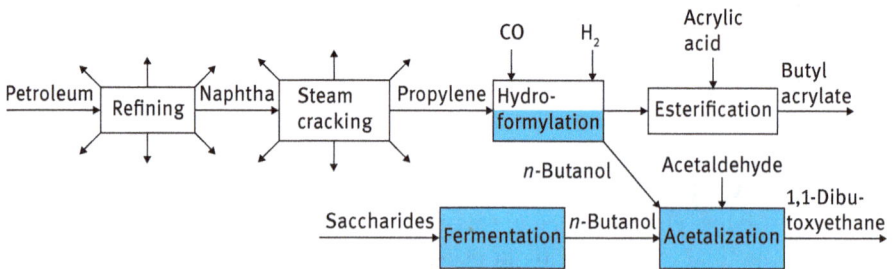

Fig. 12.7: Integration of 1.1-DBE production at an *n*-butanol plant. The fossil *n*-butanol can be potentially replaced by bio-*n*-butanol.

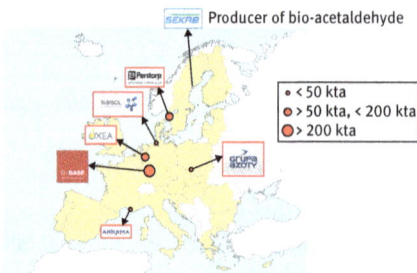

Fig. 12.8: *n*-Butanol producers in Europe. Also indicated is SEKAB, a producer of bio-acetaldehyde.

reaction and water removal. The simulated moving-bed membrane reactor (PermSMBR) is a newly developed multifunctional reactor that combines reaction with two separation techniques: adsorption and membranes [13]. This technology was successfully implemented for the production of DBE leading to an almost complete conversion, high productivity, and high DBE purity [14]. This technology has been selected for this case study.

The PermSMBR is a hybrid technology that combines the SMBR with membranes. The SMBR is implemented in well-known, but nontrivial, SMB equipment [15], where the columns are packed with a solid catalyst with adsorptive properties or with a mixture of solid catalyst and adsorbent particles. The standard SMBR configuration comprises two inlet streams (feed and desorbent) and two outlet streams (extract and raffinate), and the countercurrent solid movement is simulated by a synchronous shift of these streams by one column in the direction of the fluid at regular time intervals called the switching time. If the feed comprises two reactants (A and B), in which, for instance, A is used as desorbent and A and B react to give two products, C and D, the latter being more adsorbed than the former, then a mixture of D and A is obtained in the extract and a mixture of C and A in the raffinate.

The production of DBE by PermSMBR technology was evaluated with the commercial ion exchange resin, Amberlyst-15 (A15) as catalyst and selective adsorbent to water. *n*-Butanol was used as desorbent to water. Hydrophilic commercial tubular membranes (from Pervatech BV, The Netherlands) were applied in order to enhance the water removal [14, 16].

In this case, the membranes considered are water selective, and so it was possible to eliminate the extract stream (where water is removed). The integrated PermSMBR (see Figure 12.4) has been proven to be the more efficient way of producing DBE, leading to a significantly lower desorbent consumption than the other configurations [16], and it is therefore the selected configuration to be co-located with an existing *n*-butanol plant.

As can be observed in Figure 12.9, besides the permeate stream, only one outlet stream is obtained, the raffinate that comprises mainly DBE and *n*-butanol. The product DBE can be readily removed from the raffinate by atmospheric distillation. As the remaining stream is essentially *n*-butanol, it can be recycled without further treatment (see Figure 12.10). The small amount of water recycled does not significantly affect the performance of the PermSMBR reactor and is removed through the membrane along with the reaction water. The small amount of acetaldehyde recycled is reacted away by acetalization.

12.3.3 Demonstration status

The production of DBE by PermSMBR is in the development stage. Although the PermSMBR itself has not yet been proven experimentally, its two constituent elements

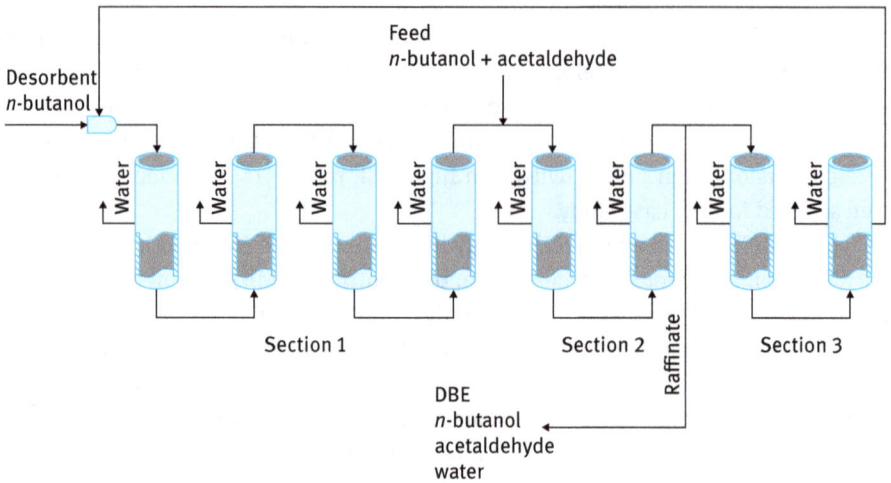

Fig. 12.9: Evaluated 1,1-DBE production scheme by PermSMBR.

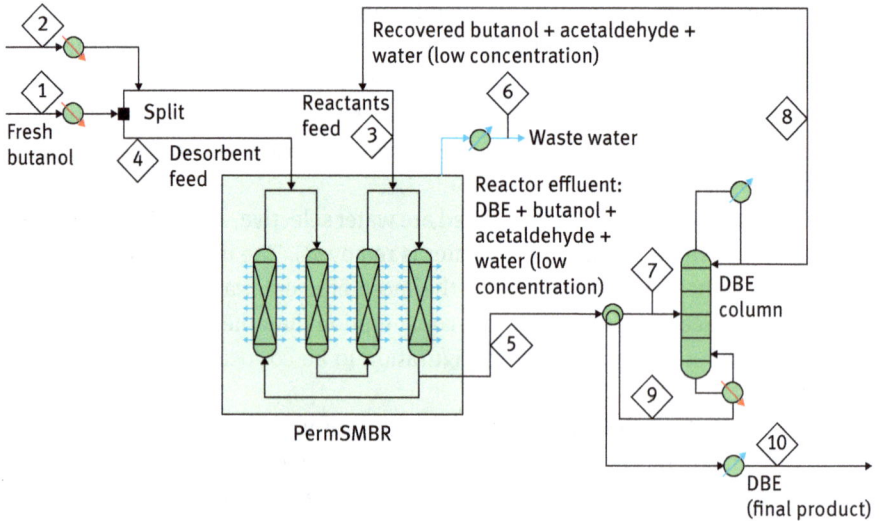

Fig. 12.10: Simplified process scheme for the production of 1,1-DBE.

have: the bare SMBR technology and the pervaporation membranes. The characteristics of the latter were measured from a pervaporation membrane laboratory-scale prototype. PermSMBR as a combination of SMBR and membranes was evaluated using a mathematical model relying on the experimental results. This simulation has shown its benefit. The removal of DBE by distillation of the raffinate stream has been simulated. Given that the components involved are well known and that the column is operated at atmospheric pressure, the simulation results can be assumed reliable.

Recall that the production of DBE studied is a co-location of a PermSMBR plant with a fossil *n*-butanol plant. However, at a later stage, a bio-butanol plant can partially or fully replace this source of *n*-butanol. Such a plant can be based on well-known ABE technology, but there are alternatives with a lower TRL.

A summary of the different steps required to implement the production of DBE with their corresponding TRL is presented in Table 12.4.

12.3.4 Strengths, weaknesses, opportunities, and threats analysis

The SWOT involved in the considered business model are enumerated in Table 12.5.

12.3.5 Value creation

The production of DBE in an existing *n*-butanol plant in Europe, especially in Sweden, where a producer of bio-based acetaldehyde is also available, will lead to (i) savings on CAPEX since infrastructure, utilities, and logistics are already available on site as well as skilled personnel; (ii) mitigation of operational and market risks; (iii) creation/preservation of jobs and production. The success of the business case depends mainly on the price of *n*-butanol.

As far as the production technology is concerned, PermSMBR has been shown to decrease the cost of utilities by a factor of 5 compared to a more conventional production in a fixed-bed reactor.

Table 12.4: Summary of relevant technologies with their TRL.

Step	TRL	Risk	Status
ABE fermentation from starch	9	Low	Operational at industrial scale: technology applied by Cathay Industrial Biotech (China), promoted by Green Biologics (UK), etc.
Other fermentation technologies to *n*-butanol (e.g., Green Biologics, Tetravitae, Cobalt)	7	Moderate	Pilot experiments successful, but no industrial implementation
n-Butanol produced catalytically from bio-ethanol	6	Moderate	Pilot experiments performed by Abengoa
1-Butanol and acetaldehyde to 1,1-DBE by PermSMBR	3	Moderate	SMBR and pervaporation separately proven on bench scale Simulation has shown the benefit of combining both techniques

Table 12.5: SWOT of the 1,1-DBE production by PermSMBR co-located with a petrochemical *n*-butanol plant.

Strength	Weaknesses
– The product has a potentially high value that is non-carcinogenic, biodegradable, (partially) renewable, and has low toxicity. – The production by PermSMBR technology is highly energy efficient. – The technology is highly flexible. PermSMBR can be applied to produce other acetals and other products such as ethyl lactate, which decreases the economic risk.	– DBE is a novel component: the market needs to be developed. – There is no production of bio-based *n*-butanol yet in Europe. – The PermSMBR technology represents high capital costs. – There is an operational risk, since no SMBRs are currently operated at commercial scale.
Opportunities	**Threats**
– Possible future ban on chlorinated solvents can create a market for DBE. – The technology offers a solution for more demanding products than DBE, including products with low thermal stability or products with high boiling point.	– Possible future increase in the price of bio-*n*-butanol. – Running a PermSMBR plant requires specific expertise, which may not be found easily.

The application of (partially) renewable DBE has several environmental benefits. Not only can it decrease our carbon footprint, it can also replace toxic and carcinogenic products such as chlorinated dry-cleaning agents. If applied in diesel blends, DBE will reduce the engine emissions of pollutants. Moreover, it will contribute to the settled goal of a 10% share of transportation fuels in Europe to be derived from renewable sources by 2020.

12.3.6 Concluding remarks

1,1-DBE is promising as a renewable product. It has several (potential) applications in the perfume, agricultural, pharmaceutical, food industries, and as a solvent. It also has a tremendous potential impact as a diesel additive.

The production of DBE is to be co-located with the production of *n*-butanol. DBE can be produced from bio-based *n*-butanol, but as this component is not yet commercially available in Europe, a transitional phase in which fossil *n*-butanol is being applied is being foreseen. In a later stage, a bio-butanol plant can be co-located.

The PermSMBR multifunctional reactor has been evaluated as an alternative to conventional fixed-bed reactor technology. The hybrid technology results from the integration of the SMBR with membranes. The use of PermSMBR technology decreases the cost of utilities by a factor of 5 compared to a more conventional production in a fixed-bed reactor. The SMBR was already used to produce DBE, and the membranes were applied to perform the water separation from the remaining components involved in the synthesis of DEB. These are important steps to validate the PermSMBR concept. However, the next phase must involve the experimental validation of a full PermSMBR setup.

12.4 Perspective view

The integration in existing plants can be a first step in the deployment of many novel biomass-based process technologies. Indeed, such an integration represents low capital costs and low risks compared to the implementation of a stand-alone integrated biorefinery.

Integration of a bio-based process with an existing process downstream often causes a need to retrofit the latter. The production of MA from bio-n-butanol in a retrofitted PA plant was evaluated as a first case study. With pilot tests having been performed, this integration scenario has already obtained a high TRL. However, further catalyst development and reaction engineering is needed to render the concept economical.

Integration upstream usually does not need retrofitting. The production of DBE was evaluated as a second case study. As there are no commercial bio-n-butanol production units in Europe, a co-location with a petrochemical n-butanol was considered. Later on, the production of bio-n-butanol can be co-located to replace the fossil source of n-butanol. PermSMBR technology has been shown to have a high energy efficiency but needs further development.

Acknowledgment: The authors wish to express their gratitude to Linda Simeonova and Plamen Alexandrov from Ruse Chemicals, Bulgaria, who contributed to the first case study: MA production in a PA plant.

Bibliography

[1] Cavani, F., Centi, G., Manenti, I., Riva, A., Trifirò, F., Oxidation of 1-butene and butadiene to maleic anhydride, 1. Role of oxygen partial pressure, Ind Eng Chem Prod Res Dev **22** (1983) 565–570.

[2] Ballarini, N., Cavani, F., Cortelli, C., Ligi, S., Pierelli, F., Trifirò, F., Fumagalli, C., Mazzoni, G., Monti T. VPO catalyst for n-butane oxidation to maleic anhydride: a goal achieved or a still open challenge?, Top Catal **38** (2006) 147–156.

[3] Iwasaki, H., Kitayama, M., Onishi, T., Process for producing acetals, EP Patent 0,771,779 A1, 1996.

[4] Hoffmann-LaRoche, Manufacture of unsaturated aldehydes, GB Patent 797,200, 1958.

[5] Iwai, H., Fujigaski, J., Antifungal fragrance composition, EP Patent 1,214,879 A2, 2002.

[6] Kohlpaintner, C., Schulte, M., Falbe, J., Lappe, P., Weber, J., Aldehydes, aliphatic and araliphatic, in Ullmann's Encyclopedia of Industrial Chemistry, Wiley-VCH, Weinheim, 1999.

[7] Kelly, J., Chapman, S., Brereton, P., Bertrand, A., Guillou, C., Wittkowski, R., Lenartowicz, P., Kiddie, R., Durante, P., Garcia, A., Maignial, L., Williams, M., Low, A. D., Vidal, J. P., Richards, A. T., Bourrier, M., Cuatero, M., Grimm, M., Lees, M., Lamoureux, T., Smith, P., Swanson, W., Smith, A., Davies, R. J., Wardle, K., Terwel, L., Lopes, J. M. S., Clutton, D., Hampton, I. J., Maynard, P., Hiero, J. R. G., Frank, W., Bauer-Christoph, C., Klingemann, K., Senf, D. R., Liadouze, I., Bolkas, M. S., Martin, J. D., Munoz, M. J. V., Conchie, E. C., Malandain, A., Leclerc, A., Pineau, M., Barboteau, P., Lafage, M., Laurichesse, D., Airchinnigh, M. N. A., McGowan, S., Cresto, B., Bossard, A., Gas chromatographic determination of volatile congeners in spirit drinks: Interlaboratory study, J AOAC Int **82** (1999) 1375–1388.

[8] Meyer, C., Eigen, H., Seiter, M., Use of diether compounds for chemically cleaning textile, leather, or fur goods, US Patent 2012084928 (A1), 2012.

[9] Arnold, D., Hierholzer, B., Wloch, H., Muck, K.-F., Natural circulation reactor and use for producing linear and cyclic acetals, US Patent 5,955,041, 1999.

[10] Nord, K. E., Haupt, D., Reducing the emission of particles from a diesel engine by adding an oxygenate to the fuel, Environ Sci Technol **39** (2005) 6260–6265.

[11] Boennhoff, K., Obenaus, F., 1,1-Diethoxyethane as diesel fuel, DE Patent 2 911, 1980.

[12] Graça, N. S., Pais, L. S., Silva, V. M. T. M., Rodrigues, A. E., Oxygenated biofuels from butanol for diesel blends: synthesis of the acetal 1,1-dibutoxyethane catalyzed by amberlyst-15 ion-exchange resin, Ind Eng Chem Res **49** (2010) 6763–6771.

[13] Silva V. M. T. M., Pereira, C. S. M., Rodrigues, A. E., Simulated moving bed membrane reactor, new hybrid separation process and used thereof, PT Patent 104496, WO Patent 2010/116335, 2009.

[14] Pereira, C. S. M., Silva, V. M. T. M., Rodrigues, A. E., Green fuel production using the PermSMBR technology, Ind Eng Chem Res **51** (2012) 8928–8938.

[15] Broughton, D. B., Gerhold, C. G., Continuous sorption process employing fixed bed of sorbent and moving inlets and outlets, US Patent 2 985 589, 1961.

[16] Pereira, C. S. M., Silva, V. M. T. M., Rodrigues, A. E., Coupled PermSMBR – process design and development for 1,1-dibutoxyethane production, Chem Eng Res Des **92** (2014) 2017–2026.

Michele Aresta

13 Biogas from wet biomass: basic science and applications

Abstract: This chapter discusses the utilization of wet residual biomass for the production of energy in the form of bio-gas ($CH_4 + CO_2$). The steps of the conversion are presented and the role of several enzymes in the complex biotechnological process is highlighted. The potential of such technology for closing the cycle and reducing the production of waste wet biomass is demonstrated with some practical applications.

13.1 Introduction

Non-cellulosic fresh vegetal materials [the so-called fruit-vegetal-garden (FVG) residues], animal manure, water treatment sludge, some residues of the food industry, and other similar materials are characterized by a high water content that makes their thermal treatment economically and energetically disadvantageous. Conversely, such wet waste biomass (WB) is suitable for the generation of energy products like methane via an anaerobic digestion process, a technology that during last years has found a renewed interest for the benefits it generates. In fact, besides producing energy with a quasi-zero impact on greenhouse gas emission, it avoids disposal/land filling of large volumes of organic waste, thus reducing their potential polluting impact on the environment.

An important aspect of anaerobic digestion is that the energy input factor (energy in/energy out) is less than 1, making the overall energetic balance positive (at least for medium- to high-concentrated waste and well-designed and managed plants), giving to such technology a premium over other biological or non-biological treatments.

Anaerobic digestion is an interesting route to the valorization of waste and contributes to (i) reduction of land filling, which is under strict limitation in many countries, (ii) reduction water and soil pollution, (iii) water recovery and reutilization, while producing usable energy that would be lost in case of land filling. Several key principles are known since quite long time.

The conversion of WB can also take place in presence of oxygen (aerobic treatment or composting). The two approaches, "aerobic" and "anaerobic" treatment, are shortly discussed and compared in next paragraph.

13.1.1 The "aerobic" and "anaerobic" processes for wet WB

Wet biomass can be treated either under "aerobic" or "anaerobic" conditions. The aerobic treatment of WB (said composting) is based on a partial oxidation of volatile compounds (an exergonic reaction) carried out in air, without energy recovery. The microorganisms respiration process, occurring in aerobic environments, ultimately leads to carbon dioxide (CO_2), water (H_2O), and biomass production (bacterial cells). An effective total aerobic degradation may occur only with soluble materials and in non-concentrated systems: O_2 availability is a key factor, often more important than the substrate composition. Composting is not aimed at the total degradation of WB but to a partial degradation of solid organic substances that requires proper aeration in order to improve the kinetics [1]. The energy produced in the respiration/oxidation phase causes a temperature increase in the mass of up to 80–90°C, but the composting mass is usually maintained at a temperature around 60°C by aeration and dissipation of heat to the atmosphere. The cost of the process is essentially an operational cost (OPEX) and depends on the aeration frequency [2]. The investment (CAPEX) is moderate and due to the cost of the area and construction of a leachate collection/treatment plant and machinery for aeration. The degree of maturation of the compost is checked by monitoring either the internal (to the mass) temperature or the concentration of organic radicals present in it: a fading of the increase of both witnesses the consumption of oxidable material. The compost is mature usually in 20–30 days: it is used as soil additive.

The anaerobic digestion instead converts organic carbon into "biogas", i.e., a mixture of methane and CO_2, from which the energy-rich methane can be separated. The process takes place in a closed and controlled bioreactor: its cost is a capital investment (CAPEX) in addition to OPEX. The overall efficiency is quite variable and not very high (30–55%) and depends on the amount of low biodegradable solid fraction (see below) present in the raw biomass (cellulose, lignin, and hemicellulose are not easily biodegraded). The process has long retention times (20–30 days) [3]. Biogas technology is being continuously improved by optimizing the process parameters and reactor geometry [4, 5] and with process integration. The methane separation technology is mature and continuously improved for its efficiency [5]. The produced gas can be locally used for thermal or electric energy production or else immitted into methanoducts.

WB treatment must take place close to the area where it is produced for two main reasons:

- The low energy/volume density of the raw materials makes the transportation quite expensive.
- HRB (Humid Residual Biomass) is a good substrate for bacterial growth and is therefore easily attacked by microorganisms with partial conversion that would occur in the containers during long-term transport.

The main theoretical limits to the application of an anaerobic process are the following:

- Incomplete conversion of the substrate: often, more than 50% of the organic material (the polymeric fraction) is not degraded.
- Medium or long retention time.
- Formation and persistence of some acids that may be polluting agents.
- Bacteria may need some nutrients that are not available in the original substrate. Their growth may be slow because of the scarce energy available.
- Permanence of ammonia (NH_3) and other N compounds.

A preliminary question to answer is whether or not the WB is suited for biogas production. The answer to such question can be given either by a laboratory-scale test or by simple calculations.

The transformation of an organic molecule into methane and carbon dioxide during the anaerobic digestion process can be simply described by Eq. (13.1) in which a, b, c, and d represent either the stoichiometric indexes of the elements in a well-defined compound or the average elemental composition of a mixture of compounds and $\alpha 1$, $\alpha 2$, $\alpha 3$, and $\alpha 4$ are the stoichiometric coefficients of species in Eq. (13.1) that represents the conversion of the compound/mixture into CH_4, CO_2, NH_3.

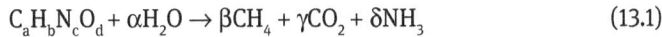

$$C_a H_b N_c O_d + \alpha H_2O \rightarrow \beta CH_4 + \gamma CO_2 + \delta NH_3 \tag{13.1}$$

By applying a mass balance to Eq. (13.1), one can obtain Eqs. (13.2i)–(13.2iv).

$$\begin{aligned}
&\text{i)} \quad a = \beta + \gamma \\
&\text{ii)} \quad b + 2\alpha = 4\beta + 3\delta \\
&\text{iii)} \quad d + \alpha = 2\gamma \\
&\text{iv)} \quad c = \delta
\end{aligned} \tag{13.2}$$

Solving the system of Eqs. (13.1) and (13.2), one obtains Eq. (13.3), which represents a molar ratio of the species CH_4, H_2O, and NH_3 obtained from the raw material:

$$C_a H_b O_c N_d + \left(a - \frac{b}{4} - \frac{c}{2} + \frac{3d}{4}\right) H_2O \rightarrow \left(\frac{4a + b - 2c - 3d}{8}\right) CH_4 + \left(\frac{4a - b + 2c + 3d}{8}\right) CO_2 + dNH_3 \tag{13.3}$$

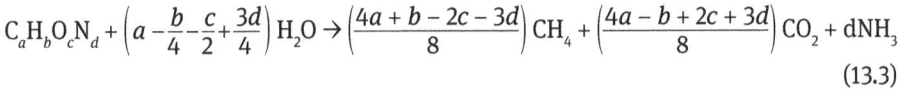

Keeping in mind that the molar volume of gases in standard conditions is 22.415 L and considering the atomic mass of the reagents and products, the volume of the biogas is obtained using Eq. (13.4) and the

$$B_{0\text{-biogas}}\left[\frac{m_n^3}{kg_{vs}}\right] = \frac{\left[\left(\frac{4a + b - 2c - 3d}{8}\right) + \left(\frac{4a - b + 2c + 3d}{8}\right)\right] \cdot 22.415}{12a + b + 16c + 14d} = \frac{22.415a}{12a + b + 16c + 14d} \tag{13.4}$$

the volume of methane will be given by Eq. (13.5), while the methane abundance in the gas phase,

$$B_{0\text{-methane}}\left[\frac{m_n^3}{kg_{vs}}\right] = \frac{\left(\frac{4a + b - 2c - 3d}{8}\right) \cdot 22.415}{12a + b + 16c + 14d} \tag{13.5}$$

(P_{CH_4}) will be given by Eq. (13.6).

$$P_{CH_4}\left[\frac{m_{n\text{-}CH_4}^3}{m_{n\text{-biogas}}^3}\right] = \frac{B_{0\text{-methane}}}{B_{0\text{-biogas}}} = \left(\frac{4a + b - 2c - 3d}{8}\right) \tag{13.6}$$

From Eqs. (13.5) and (13.6), it is easy to deduce that the methane content in the biogas will increase with the increase in H/C ratio in the raw material.

The equations above allow to check quickly whether the biomass is suited for biogas production or not. Biogas that contain low percentage of methane (< 40%) are not interesting.

13.2 Structure of raw materials used for biogas production

In this section, the constituents of WB that maybe involved in biogas production are briefly described: more detailed information on the properties and behavior of materials like cellulose and lignin can be found in other chapters of this book. The texture and structural features of the raw materials have a great importance in biogas production and determine the extent at which such compounds can be digested by bacteria and converted into biogas, as well as the composition of the biogas. Noteworthy, compounds such as cellulose, hemicellulose, and lignin may not be converted by bacteria and fungi, and will constitute the solid residue of the digestion, while proteins will produce ammonia that needs to be eliminated. A pretreatment of polymeric materials may be useful for making available larger amounts of monomers, thus increasing the methane production.

13.2.1 Cellulose

Cellulose is a polymer of D-glucose with monomeric units linked in a β-1,4 mode (Figure 13.1). The three free hydroxyl groups present in each monomeric unit form hydrogen bonds that confer the particular resistance to the fiber. It may be worth to remind that the energy of a classic hydrogen bond, such as those in cellulose, is equal to ca. 7 kcal/mol. An extended network of H-bonds may represent a very high energy barrier to overcome for any transformation one wishes to carry out on cellulose, and such operations will require harsh conditions.

Cellulose hydrolysis is carried out by both extracellular, and intracellular enzymes. The formers perform the hydrolysis, whose products are transferred inside

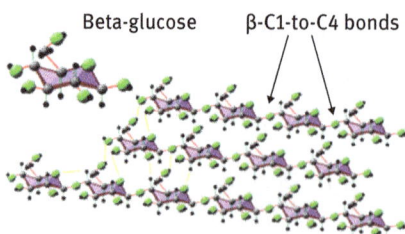

Fig. 13.1: Cellulose structure.

the cell where they follow the catabolic path. Three different cellulolytic enzyme activities can be distinguished [6]:

- Endoglucanases attack the internal cellulose chain. It is active on "amorphous", but not on crystalline, cellulose (it is worth to recall that natural cellulose is 70% crystalline)/
- Exoglucanases attack the terminal part of the chain.
- β-Glucosidases hydrolyze cellobiose and cellodextrins produced by previous reactions. The final monomeric product is glucose.

Endoglucanases are able to degrade crystalline cellulose only in combination with exoglucanase [7]. It must be noted that the products of hydrolysis are not used only by methanogenic bacteria, but by a much larger class of bacteria present in the digester [8]. Therefore, cellulose hydrolysis can turn to be either the process limiting the whole anaerobic digestion or one of the reactions playing a fundamental role in it. Noteworthy, the accumulation or scarcity of a product (e.g., glucose, cellobiose) will reflect on the rate of the processes in which it is produced-consumed through a kind of feedback mechanism.

13.2.2 Hemicellulose

Hemicellulose, one of the main constituents of plants, is a polymer with short-branched chains (Figure 13.2) hydrolyzed essentially by extracellular enzymes [7].

13.2.3 Lignin

Lignin, as a vascular tissue, plays an important role in building the plants cellular walls. Its complex structure contains aromatic units like guaiacilic (I), pepperilic (II), and syringilic (III) moieties that co-polymerize (Figure 13.3).

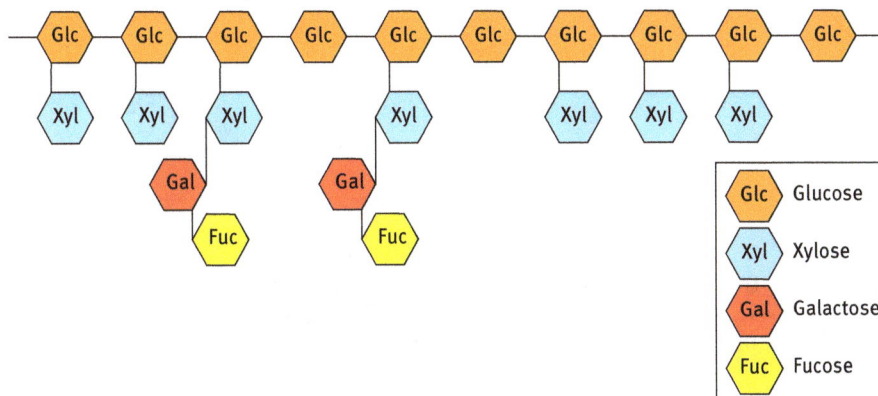

Fig. 13.2: Hemicellulose structure.

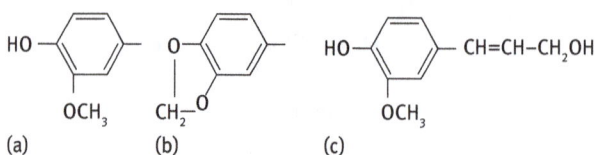

HO—⟨ ⟩— (a) OCH₃

O—⟨ ⟩— (b) CH₂—O

HO—⟨ ⟩—CH=CH–CH₂OH (c) OCH₃

Fig. 13.3: Phenyl derivatives belonging to lignin structure.

Chemico-physical treatments are the most indicated for lignin depolymerization.

13.2.4 Pectin

Pectin is present either as constituent of cellular walls in plant or in intercellular layers. It has a linear structure with the galacturonic acid units bound by a α-1,4 bond; the carboxylic groups are methylated (Figure 13.4). The enzymes that cause pectin degradation are pectinesterase, polyglycanohydrolase, and polylyase [7].

13.2.5 Starch

Starch, the main supply constituent for plants, exists in two structural forms (Figures 13.5 and 13.6), namely, amylose, a linear homopolysaccharide with d-glucose units linked through an α-1,4 bond, and amylopectin, bearing every 25 monomeric units [9] lateral chains with α-1,6 bonds.

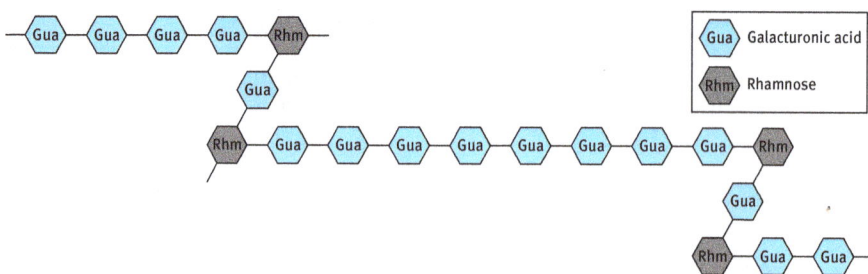

Fig. 13.4: Pectin structure.

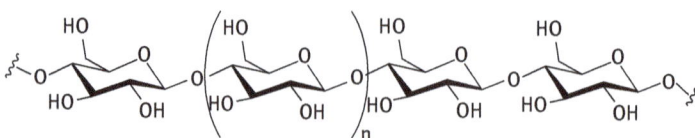

Fig. 13.5: Amylose structure.

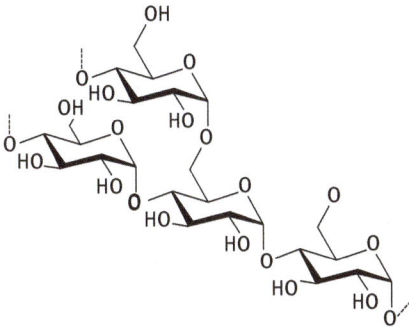

Fig. 13.6: Amylopectin structure.

Four enzymatic systems are involved in starch degradation: (i) α-amylase (α-1,4 glucan glucanohydrolase), (ii) β-amylase (β-1,4-glucan maltohydrolase), (iii) amyloglucosidase, and (iv) debranching enzymes [10].

13.2.6 Lipids

Lipids are derivatives of glycerol esterified in two positions with long-chain monocarboxylic acid. The third position can be used to bind another fatty acid (triglycerides), a phosphate group (phosphatides or phospholipids), or else a sugar unit (glycolipids).

Lipid hydrolysis in rumens is different from that in anaerobic digesters. While in the former, lipids degradation leads to fatty acids that are not further degraded but directly absorbed by the intestine, in the latter, the fatty acids are further degraded by means of "obligate hydrogen-producing acetogen" (OHPA) bacteria [11].

It is useful to recall that unsaturated fatty acids undergo a rapid hydrogenation reaction [12]. Saturated fatty acids undergo a β-oxidation that removes two C-atoms from the carboxylic end. The products are a fatty acid with a C_{n-2} chain, acetic acid (as AcetilCoA), and 4H. In case the chain has an odd number of carbon atoms, propionic acid is also produced as end product. Glycerol itself enters the glycolysis path through the glycerol-1-P-dehydrogenase enzyme.

13.2.7 Proteins

Proteins are linear sequences of amino acids characterized by peptide bonds, namely -NH-CO-. The polymer may also bear sulfide -SH and disulfide -S-S- moieties. They can be simple or conjugated (containing inorganic groups). Their three-dimensional structure shows different forms stabilized by hydrogen bonds and disulfide bridges. Proteins degradation occurs via de-amination, trans-amination, and de-carboxylation reactions that bear to the formation of free NH_3 or CO_2 and are competitive and governed by the pH value of the medium.

13.3 The phases of biogas production

The FVG biomass conversion into biogas encompasses a number of phases, as categorized below.
  i)   depolymerization
  ii) acidogenesis
  iii) acetate formation
  iv) methanogenesis
  v)   methanation of CO_2

Each of them requires specialized bacterial communities and a complex metabolic food chain [13–15]. In the whole process, H_2 and organic carboxylic acids, such as acetic acid, are formed: it is important to maintain a low H_2 partial pressure as key biological reactions for biogas production may not occur, for thermodynamic reasons, under high H_2 pressure [13]. The anaerobic digestion of fatty acids, alcohols, and organic compounds is accomplished through a syntrophy between H_2-producing and H_2-consuming methanogenic archea [15] that favor the best use of the energy content of primary substrates [16].

The enzymes involved in the biogas production process are known, and the role of iron, nickel, and cobalt during the anaerobic digestion of a sludge has been elucidated. The above metals play a key role in anaerobic metabolism during the methanogenic digestion. In fact, they constitute the active center in several enzymes, each playing a specific role in the complex methane production process (Table 13.1).

Table 13.1: Metal enzymes involved in the conversion of CO_2 or H_2.

Enzyme/coenzyme	Metal in the active site	Reaction catalyzed
Conversion of CO_2		
Formate dehydrogenase	W	$CO_2 \rightarrow HCOO^-$
THF	Ni (in F-430 factor of	$CO_2 \rightarrow {}^-CH_3/CH_4$
Methanofurane	CH_3-S-CoM)	
Tetrahydromethanopterin CH_3-S-CoM		
methyl reductase		Methyl transfer
Methyl transferase (cobalamin)	Co	
CODH	Ni, Fe	$CO_2 \rightarrow CO$ or CH_3COOH
Dihydrogen formation/consumption		
Hydrogenases	Fe	$H^+ \rightarrow H_2$ (and $H_2 \rightarrow H^+$)
Hydrogenases	Ni, Fe	$H_2 \rightarrow H^+$ mainly Ni
Hydrogenases	Ni, Fe, Se	

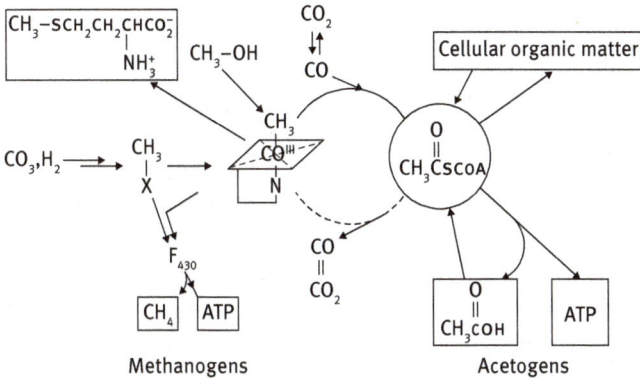

Fig. 13.7: The methanogenesis scheme.

In fact, nickel is the active center of the methyl-coenzyme M reductase (known as F_{430}) and of several H_2-consuming hydrogenases [17, 18] and acetate-forming enzymes [19–22]. Iron is present in several hydrogenases (H_2 uptake or evolution) and, as Fe_4S_4 protein, in carbon monoxide dehydrogenase (CODH), which is responsible of the formation of acetic acid [20, 22, 23]. Cobalt is part of cobalamin, a methyl-transfer catalyst [24]. Several enzymes synergistically work for the production of methane and carbon dioxide during the anaerobic digestion of WB or sludge [25] (Figure 13.7) It is noteworthy that anaerobic digestion is largely applied in the treatment of process or municipal water, leading to the recovery of carbon (as CH_4 and CO_2) and energy (CH_4), with water streams cleaning and potential reusability.

13.3.1 Anaerobic digestion: a nature-based biotechnology

The anaerobic digestion of WB is a typical example of transfer at the industrial scale of a natural process. In fact, methanogenesis is a common process occurring in quite different environments in nature such as oceanic and lagoon sediments and animal intestine (particularly in rumens). The anaerobic digestion may occur also under microaerobic conditions: it is carried out thus by strict anaerobic and facultative bacteria, the latter growing both in anaerobic and aerobic conditions.

The whole process occurs through different phases as listed above. Hydrolytic bacteria are responsible of the hydrolysis of polymeric organic compounds like carbohydrates, lipids, proteins. Such hydrolytic phase is followed by the acidogenesis during which organic acids, alcohols, neutral compounds, and hydrogen are produced. The products above are converted into acetate, hydrogen, and carbon dioxide by the OHPA bacteria. Acetate is converted into methane during the methanogenic phase, while H_2 converts CO_2 into methane during the methanation phase.

Acetate, the substrate that is mostly used by methanogens, is also produced by a fourth bacterial class called homoacetogenic bacteria, which can ferment a wide spectrum of substrates. OPHA and homoacetogens are generally called "transitional" bacteria.

As we have already discussed, the methane yield of the anaerobic digestion mainly depends on the yield of the hydrolysis of the organic fraction. Lignin, for instance, under anaerobic conditions, is hardly biodegraded. The difficulty is essentially due to the lack of specific hydrolytic enzymes in anaerobic bacteria that would be able to hydrolytically cleave the ethereal bonds present in lignin and the oxygen demand typical of hydrolytic enzymes. For this reason, it may be useful to treat the organic fraction with specific hydrolytic agents (fungi, other microorganisms) before the anaerobic digestion is started. This procedure may increase the methane yield and reduce the residual solid fraction.

Methane production also depends on the biodegradable organic fraction composition: "reduced" substrates (like proteins and lipids) give better methane yield than "oxidized" ones (sugars). Different processes regulate the methanation speed. If the substrate is rich of polymeric materials like cellulose, the rate-determining step is the hydrolysis. If the substrate is soluble, it is the methanation to determine the overall rate [26]. In general, one observes the following trend for the rate of processes: $k_{hydrolysis} < k_{acidogenesis} < k_{methanogenesis}$. Several parameters are used to describe the stability and efficiency of an anaerobic process, for example:
- methane production
- methane volumetric rate (MVR)
- organic substance degradation rate (ODR)
- culture stability
- thermal efficiency

The chemico-physical changes in the biodegradation of a substrate are typical of an exergonic process. While the biodegradability of waste, i.e., the fraction that can be converted into biogas, depends on the degradation thermodynamics, the biogas daily yield depends on the kinetics of the process. A compound that is not biodegradable or may require a long induction time for biodegradation is termed "refractory".

13.3.2 Hydrolytic bacteria and acidogenesis

The first reaction occurring in a digester is the depolymerization of substrates such as homopolysaccharides (cellulose, starch), heteropolysaccharides (hemicellulose), pectins, proteins, lipids. The anaerobic degradation of these polymers requires the action of different enzymes able to attack their terminal- or internal-functional groups. Table 13.2 lists some of the hydrolytic bacteria [1, 7], either specific or polyvalent.

Table 13.2: Bacteria involved in the hydrolysis-acidogenesis phase [7].

	C	H	Lg	Pc	S	Lp	Pr
Anaerovibrio						×	
Bacteroides amylophilus				×			×
Bacteroides fibrisolvens	×	×		×		×	
Bacteroides succinogens		×		×	×		×
Butyrivibrio fibrisolvens	×	×					
Clostridium multifermentans				×			×
Clostridium thermocellum	×						×
Ruminococcus albus	×	×					
Ruminococcus flavefaciens	×	×					
Succinomas amylotica					×		

C, cellulose; H, heminocellulose; Lg, lignin; Pc, pectin; S, starch; Lp, lipids; Pr, proteins.

Table 13.2 shows that none of the listed bacteria is able to hydrolyze lignin that is thus considered as "non-degradable" through an anaerobic process. As cellulose, hemicellulose, and lignin are bound together to form the lignocellulose matrix, the relative percentage of lignin will make such matrix more or less degradable [27]. Hydrolytic reactions are followed by acidogenesis leading to the formation of soluble extracellular intermediates, i.e., acetic, propionic, or butyric acid, usually produced at low concentration but with a high turnover [28]. However, methane production is not generally influenced by the eventual loss of acids through the effluent. In fact, the high production rate leads to a quick re-establishment of the equilibrium conditions. The presence of hydrolytic bacteria has been ascertained in biosystems such as human colon [29], rat cecum [30], horse intestine [31], guinea pig cecum [32], estuary sediments [33], and in soil [34].

13.3.2.1 Transitional bacteria

It has long been clear that growing bacteria on a single substrate or on multiple substrates (the latter is typical of a digester) may produce different products from both quantitative and qualitative points of view [35, 36] due to growth rate variation, pH, and concentration of the substrate used as energy source [37]. All such parameters affect the bacterial flora composition and the extracellular enzymes concentration [38–44]. Fermentation (ethanol methane production) is a disproportionation reaction in which carbon in a formal oxidation state zero (as in the HCOH unit), increases its oxidation state to +4 in CO_2 and reduces to −4 in CH_4. During the anaerobic fermentation, energy is derived essentially from oxidation reactions in which molecules other than oxygen are used as electron acceptors. A narrow class of bacteria use either nitrates or sulfates as electron acceptors, but the greatest part reduces the compounds produced in the hydrolysis-acidogenesis phase or form gaseous-H_2 in combination with hydrogenase enzymes [45].

13.3.2.2 Acetogenesis

As reported above, acetate is the substrate used by methanogenic bacteria, while hydrogen and carbon dioxide are used by methanation bacteria. Acetogenesis reactions are driven by transitional bacteria (OHPA and homoacetogens). Acetate derives from both the hydrolysis of the original substrate (24%, [46]) and from reactions involving other substrates produced during the hydrolysis-acidogenesis phase (propionate, butyrate, lactate, ethanol, methanol) (76%, [47]).

There is an important functional difference between the two classes of bacteria that produce acetate. Whereas hydrolytic bacteria can also produce acetate in the presence of an excess of hydrogen produced by themselves, the OHPA bacteria are able to produce acetate only if H_2 is removed. A high hydrogen concentration in the gaseous phase originates a feedback mechanism that inhibits the OHPA bacteria with a consequent accumulation of organic acids (propionic, butyric) at the expense of the acetic acid. Such interaction between the H_2-producing and H_2-utilizing species is known as "H_2-transfer interspecies" [48]. The reactions involving OHPA bacteria are endergonic, if the substrates and products are in their standard state [48]. It has been shown that within a well-defined range of H_2 (1.6×10^{-6} atm $< P_{H_2} < 5.8 \times 10^{-5}$ atm), the reactions are exergonic. The H_2 partial pressure thus plays an important role in the overall process.

13.3.2.3 Bacterial flora composition

Several OHPA bacteria have been identified, among which are *Syntrophomanas wolfei* (oxidizes butyrate to acetate and hydrogen) [49, 50], *Syntrophomanas wolinii* (oxidizes propionate to acetate, carbon dioxide, and hydrogen) [51], and *Methanobacterium thermoautrophicum* (oxidizes butyrate to acetate and hydrogen) [52]. Bacteria that produce acids different from acetic acid perform the opposite reaction carried out by OHPA bacteria. This rises an issue of "energy": in fact, the two reactions (direct and reverse) cannot be both "exergonic". However, it is possible to assume that if the

Table 13.3: Homoacetogenic bacteria [48].

Organism	Products
Acetobacterium kivui	Acetate
Acetobacterium wierinage	Acetate
Acetobacterium woodii	Acetate
Clostridium aceticum	Acetate
Clostridium formicoaceticum	Acetate
Clostridium thermoaceticum	Acetate
Desulfobulbus propionicus	Propionate
Eubacterium limosum	Acetate, butyrate
Peptostreptococcus products	Acetate, succinate

hydrogen concentration varies with the microsystem considered, the two reactions may be both exergonic in different microenvironments [48]. Table 13.3 provides a list of some "homoacetogenic" bacteria that synthesize acetate (and other volatile organic acids) from H_2 and CO_2, even if at a very low level (1% of the total [46]).

13.4 Role of hydrogenases

As already discussed, the production of biogas from WB or other residual organics, such as monomeric compounds or proteins or polysaccharides, occurs in five inter-connected phases, three of which, namely, depolymerization, acidogenesis, and methanogenesis, are directly consociated [13–15]. In the first phase, oligomers or monomers are formed, such as sugar, amino acids, peptides, and acids from polymeric materials. Such monomers are converted in the acidogenesis phase by fermentative bacteria into the so-called volatile fatty acids (VFAs), that is, acetic, propionic, and butyric acid, and H_2, plus ammonia and CO_2. VFAs can be converted into CO_2 and H_2. In the third step, or methanogenesis, acetic acid is converted by methanogens into CH_4 and CO_2, while in the methanation phase, CO_2 and H_2 give CH_4. In such a complex process, metal enzymes play a key role, as they drive the key reactions such as H_2 formation and conversion (such enzymes are called hydrogenases and indicated as H_2-ases), CO_2 reduction to CO (CODH), and the formation of acetic acid from CO, among others. Table 13.1 lists the enzymes and the relevant active centers involved in H_2 formation-consumption. H_2-ases enzymes are classified by indicating the transition metal present in their active site: three main H_2-ases (FeFe, FeNi, and FeS) are classified plus a Mo-ase involved in nitrogen fixation to afford NH_3.

13.4.1 [FeFe]H_2-ase

The active site in *Clostridium pasteurianum* (Figure 13.8) contains a large unit characterized by an unusual arrangement of two moieties, an Fe_4S_4 iron protein linked, through a cysteine-S, to an Fe_2 cluster in which two octahedral irons bear five -CX groups (CO or CN), one water molecule, and three bridging sulfur groups. Such "large domain" is accompanied by four "small domains" containing either the Fe_4S_4 protein or the Fe_2-non-proteic cluster.

The two iron centers are designated as "proximal" and "distal" depending on their spatial relation to the nearby $[Fe_4S_4]$ cluster and protein backbone. Infrared (IR) spectroscopy [53–56] and electron paramagnetic resonance (EPR) spectroscopy [57–60] have demonstrated the existence of at least four different forms of the [FeFe]-$[Fe_4S_4]$ active site. Two S = 1/2, EPR-active states have been identified, designated as H_{ox} (g = 2.06) and H_{ox} (g = 2.10). Two EPR-silent states, namely H_{ox} and H_{red}, have been identified using IR, the H_{ox} form corresponds to an over-oxidized species, which is

$$[Fe_2(S_2X)CY](CY)_4(H_2O)[Fe_4S_4](S^YCys)_4$$

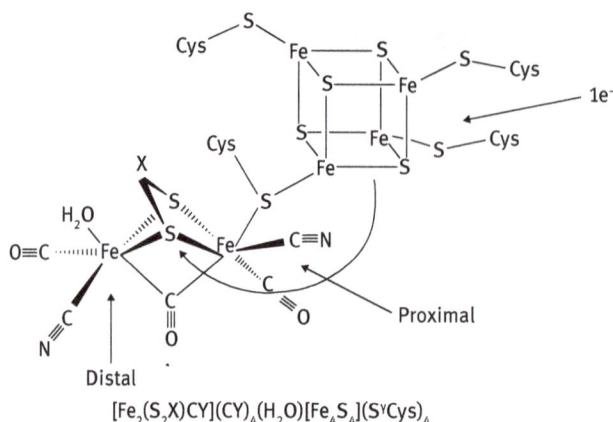

Fig. 13.8: Active center of *Clostridium pasteurianum* hydrogenase.

not active catalyst for H^+ reduction or H_2 oxidation. The H_{ox} form may be reactivated by either electrochemical reduction or using a reducing agent. The active site can undergo a "one-electron" reduction: the electron is initially localized on the $[Fe_4S_4]$ moiety (Figure 13.8) generating a species designated as H_{ox} (g = 2.06). The transfer of the electron from such site to the [FeFe] cluster is performed through a conformational change of the protein superstructure. The second one-electron reduction follows yielding a species designated as H_{red}. Models of the active site of [FeFe]H_2-ase have been built using small molecules [61].

Iron-only hydrogenases have been isolated from several microorganisms [62, 63] and shown to be able to both produce and consume dihydrogen [Eq. (13.7)].

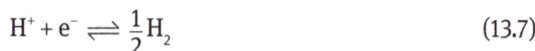

$$H^+ + e^- \rightleftharpoons \frac{1}{2}H_2 \tag{13.7}$$

13.4.2 [FeS]H₂-ase

The iron-sulfur cluster-free H_2-ase are H_2-utilizing enzymes, which activate dihydrogen for use in catabolic processes within the cell, but do not catalyze H^+ reduction or H_2 oxidation.

13.4.3 [NiFe]H₂-ase and [Fe-Ni-Se]-ase

Nickel-iron hydrogenases [NiFe] (Figure 13.9) are present in several bacteria. Their basic metal site as demonstrated by XRD [64, 65] is a heterodimeric unit formed by

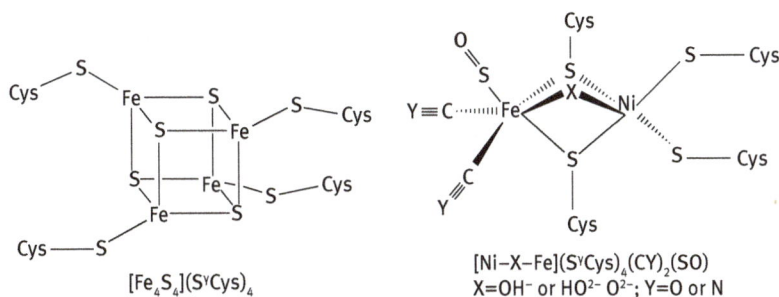

Fig. 13.9: Two different subunits present in FeNi-hydrogenases.

four subunits, three of which are small [Fe] and one contains the bimetallic active center consisting of a dimeric cluster formed by a six-coordinated Fe linked to a penta-coordinated Ni(III) through two cysteine-S and a third ligand whose nature changes with the oxidation state of the metals: in the reduced state, it is an hydride H⁻, while in the oxidized state, it may be either an oxo (O^{2-}) or a sulfide (S^{2-}). It has also been shown that in some microorganisms such as *Desulfomicrobium baculatum*, a S-cysteine is replaced by a Se-cysteine [66], giving place to a trinuclear [FeNiSe]-hydrogenase.

The mechanism of action has been studied by several authors [62, 67], and the implication of a Ni(III)/Ni(II)/Ni(0) system has been proposed. Ni and Fe enzymes apparently have a different role in H_2 production-consumption. In fact, Ni enzymes are more specifically involved in H_2 consumption, while Fe enzymes are more involved in H_2 production.

Ni and Fe are equally implied in the synthesis of the acetyl moiety from CO_2 [22, 68, 69], while Co (as cobalamin) is well known to act as a carrier of methyl groups. Using IR and EPR spectroscopy, the existence of at least seven different forms of the Ni-Fe center was demonstrated. Three S = 1/2, EPR-active states, designated as Ni-A, Ni-B, and Ni-C, were evident [70–79]. Four EPR-silent states identified by IR have been designated as Ni-SU, Ni-SI$_I$, Ni-SI$_{II}$, and Ni-SR (also known as Ni-R). Still, further investigation is needed for the complete elucidation of the structures of the various species, their role in the catalytic cycle, and the details of their interconversion. The Ni-A and Ni-B are not active in the catalysis for H_2 oxidation as they are overoxidized. They can be reactivated by reduction. Ni-C and Ni-R species are believed to be intermediates in the oxidation of H_2. Species designated as Ni-SU, Ni-SI$_I$, and Ni-SI$_{III}$ can be intermediates in the reactivation of the over-oxidized forms of the enzyme, while one of the Ni-SI species is supposed to play a role in the catalytic cycle [64, 80–83].

In the active cycle, nickel changes from EPR active Ni^{III} to Ni^{II} and finally to Ni^{I}, which has not been observed because of its rapid electron transfer. The Ni^{II} forms are themselves high spin as demonstrated by nickel L-edge soft X-ray spectroscopy and density functional theory calculations. Ni-center is the site where external CO binds, as demonstrated by IR [84] and XRD [85] studies on the CO-inhibited forms of [NiFe] H_2-ase derived from *Desulvibrio vulgaris*, with unusual Ni-CO angles of 136.2° and 160.9°.

Models of the active site of [NiFe]H_2-ase have been synthesized as small-molecules [61]. Computational studies on the [NiFe] active site [86–101] support high-spin nickel [102] and suggest that terminal cysteine ligands act as bases in the heterolytic cleavage of dihydrogen through the S-atoms.

13.4.4 Molybdenum-iron-containing N_2-ase

The molybdenum-containing enzyme Mo-N_2-ase is well studied as an active site for dinitrogen and proton reduction. Also known are all-iron, vanadium-containing, and tungnsten-containing enzymes. The Mo-N_2-ase enzyme is composed of two subunits, which are referred to as the iron subunit and the molybdenum-iron subunit [103–111]. The iron subunit contains a single [Fe_4S_4] cluster, which mediates electron transfer to the Fe-Mo-containing subunit. The Fe-Mo subunit contains two 8Fe-7S cluster referred to as P-cluster and two 1Mo7Fe-9S clusters referred to as Fe-Mo cofactors (FeMocos). Metal centers in FeMoco are organized in a bicapped trigonal prism, formed by six iron atoms and capped on opposite sides by an iron and a molybdenum center (Figure 13.10). An interstitial N-atom occupies the center of the trigonal prism. A protein-bound cysteinate ligand binds the capping iron center to the protein. Histidine binds the molybdenum center to the protein; the molybdenum center is further coordinated by a homocitrate ligand. The mechanism of hydrogen production at FeMoco is still not well understood. Only recently, a direct evidence of a hydride ligand bound to FeMoco became available [112]. Alberty, on the basis of thermodynamics, argues that the highly reduced state of FeMoco required for N_2 reduction leads to the incidental production of H_2. Others have suggested that the reductive elimination of H_2 is necessary to produce a more reduced form of the FeMoco [113].

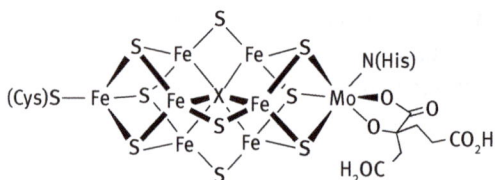

Fig. 13.10: Iron-molybdenum cofactor (FeMoco).

13.5 Methanogenic bacteria

Fatty acids are substrates employed by methanogens, as demonstrated by the fact that they do not accumulate in the digesters where they are produced during the hydrolysis, acidogenesis, and acetogenesis phases. Table 13.4 shows the set of reactions that produce methane. About seven-tenths of the methane produced derive from acetate, whereas the remaining three-tenths result essentially from the hydrogenation of carbon dioxide [48, 114]. These data allow to estimate the biogas potential composition. If biogas is derived only from acetate [Eq. (iii) in Table 13.4], the composition would be 50% methane and 50% carbon dioxide. If 30% of methane derives from the direct reaction of carbon dioxide with hydrogen, the final potential composition is 65% methane and 35% carbon dioxide.

Differences in biogas real composition depend on the process control and the capacity to optimize the various reaction steps. Experiments with labeled carbon have shown that in the acetate molecule, the methyl group gives rise to methane, whereas the carboxylate group produces carbon dioxide, as expected [103]. Methanogenic bacteria may have specific nutritional needs (amino acids, fatty acids, vitamins, metals such as Co, Ni, Mo) [115]. More than 100 methanogenic bacterial species have been isolated; some of the most important are listed in Table 13.5 [116].

Some species have been adapted both to mesophilic and thermophilic conditions. Methanogenic bacteria have been isolated in very different environments and conditions: anaerobic sediments (both fresh and sea water), pulp of trees, flooded lands, man and animal fecal excrements and digestive tracts, and hot springs with temperature up to 85°C [109, 117–119]. Methanogenic bacteria can be categorized into two groups:

– *Methanosarcina* and *Methanotrix*, which belong to the Acetoclasts (or methylotrophic) class and are able to metabolize methanol, methylamine, and, above all, acetate.
– "Hydrogenophils" (or non-methylotrophic), which employ H_2 and CO_2 as substrates for methane production (some may use formate).

Table 13.4: Methanogenesis reactions.

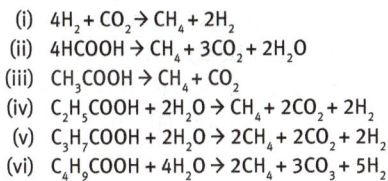

(i)	$4H_2 + CO_2 \rightarrow CH_4 + 2H_2$
(ii)	$4HCOOH \rightarrow CH_4 + 3CO_2 + 2H_2O$
(iii)	$CH_3COOH \rightarrow CH_4 + CO_2$
(iv)	$C_2H_5COOH + 2H_2O \rightarrow CH_4 + 2CO_2 + 2H_2$
(v)	$C_3H_7COOH + 2H_2O \rightarrow 2CH_4 + 2CO_2 + 2H_2$
(vi)	$C_4H_9COOH + 4H_2O \rightarrow 2CH_4 + 3CO_3 + 5H_2$

Table 13.5: Methanogenic bacteria.

Genus and species	Substrates
Methanobacterium formicium DSM 863	H_2-CO_2, formate
Methanobacterium Thermoautrophicum	H_2-CO_2
Methanobacterium bryantii M.O.H.	H_2-CO_2
Methanobacterium wolfei DSM 2970	H_2-CO_2
Methanobrevibacter ruminantium DH1	H_2-CO_2, formate
Methanobrevibacter smithii PS	H_2-CO_2, formate
Methanotermus fervidus DSM 2088	H_2-CO_2
Methanococcus voltae PS	H_2-CO_2, formate
Methanococcus halophilus INMI Z-7982	Methanol, trimethylammine
Methanospirillum hungatei Jf1	H_2-CO_2, formate
Methanomicrobium mobile BP	H_2-CO_2, formate
Methanogenium cariaci JR1	H_2-CO_2, formate
Methanogenium thermophilicum CR1	H_2-CO_2, formate
Methanogenium aggregans MSt	H_2-CO_2, formate
Methanosarcina barkeri MS	H_2-CO_2, methanol trimethylammine, acetate
Methanosarcina thermophila TM-1	Methanol, trimethylammine
Methanoplanus limicola DSM 2279	H_2-CO_2, formate
Methanococcoides methylutens TMA-10	Methanol, trimethylammine
Methanolobus tindarius Tindari 3	Methanol, trimethylammine
Methanotrix soehngenii Opfikon	Acetate
Methanotrix concilii GP6	Acetate
Methanosphaera stadmane MCB-3	Methanol plus H_2

13.5.1 Methanogenesis

Several schemes have been proposed to explain the metabolic path leading to methane. The Barker scheme (1951, Figure 13.11) [120] has a historical interest, as it first proposed that intermediates are bound to carriers (generally marked with X). One of these carriers has been isolated by Mc Bride and Wolfe [121] and shown to be 2-mercaptoethanol sulfonic acid (HS-$CH_2CH_2SO_3^-$), named coenzyme-M: it works during the last phase of the reactions presented in the Barker's scheme. The methane production from the coenzyme-M adduct requires the intervention of the F_{430} coenzyme (Figure 13.11), which has been proposed to have a nickel-tetrapyrrolic active center. Enzymes involved in the synthesis of methane from CO_2 and H_2 are still investigated by several research groups, as their knowledge can bring to the development of new interesting biotechnological applications. Recent studies have shown the implication of Ni [18] in CO_2 reduction to CO and of the Fe-S/Ni protein in the synthesis of the acetyl-moiety from CH_3 and CO [22, 122].

It has been now ascertained that the process goes through the tetrahydrofolate (THF) and CODH cycles (Figure 13.12). In the THF cycle, a molecule of CO_2 is reduced to

$CO_2 + XH \longrightarrow XCOOH$

$\quad\quad\quad\quad\quad |{+2H}$
$\quad\quad\quad\quad\quad |{-H_2O}$

$CH_3OH + XH \quad\quad XCHO$

$\quad\quad\quad\quad\quad\quad\quad |{+2H}$

$\quad\quad\quad\quad -H_2O \;\; XCH_2OH$

$\quad\quad\quad\quad\quad\quad\quad |{+2H}$
$\quad\quad\quad\quad\quad\quad\quad |{-H_2O}$

$CH_3COOH + XH \; \dfrac{-2H}{-CO_2} \; XCH_3$

$\quad\quad\quad\quad\quad\quad\quad |{+2H}$

$\quad\quad\quad\quad\quad\quad XH + CH_4$

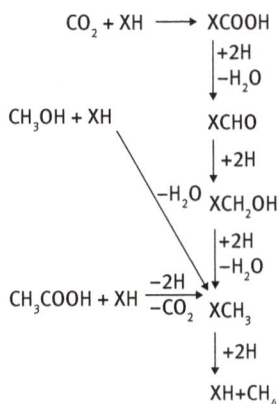

Fig. 13.11: Barker's scheme for methanogenesis.

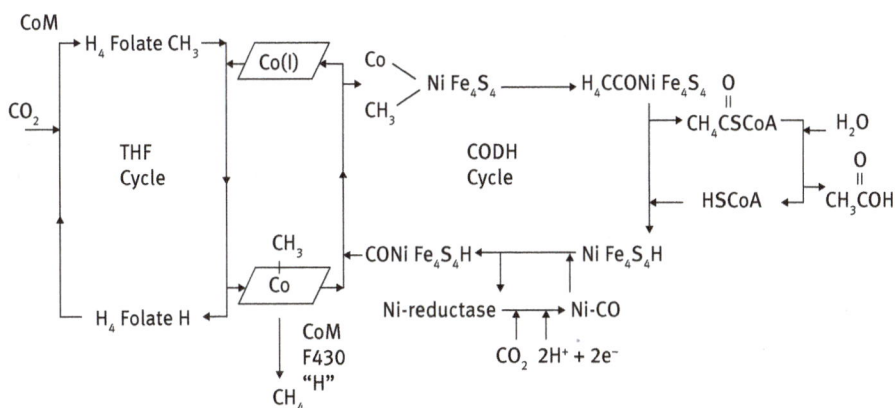

Fig. 13.12: The CODH and THF cycles implied in the formation of biogas.

formate via a formate dehydrogenase enzyme (W dependent), then to formaldehyde by a formaldehyde dehydrogenase enzyme. Formaldehyde is taken by the THF enzyme (a non-metal enzyme) to afford an imino moiety $-N=CH_2$ reduced to a methylamino moiety $-NH-CH_3$. The methyl group is then taken by the cobalamin through a $H-CH_3$ exchange that regenerates the amino group. Vitamin B12 transfers the methyl to the Fe_4S_4-X-Ni enzyme where it is coupled with CO, produced from a second molecule of CO_2 by the action of CODH, to afford the acetyl moiety bonded to Ni.

Acetylcoenzyme (ACoA)-SH takes the acetyl group and forms $CoAS-C(O)CH_3$, which is hydrolyzed to afford back the CoASH moiety and acetic acid CH_3COOH. The latter is decomposed from methanogens into CO_2 and CH_4. However, starting with $2CO_2$ molecules and 8H (or $8H^+$ plus $8e^-$), one gets back a CO_2 molecule and methane plus $2H_2O$ [Eq. (13.8)]

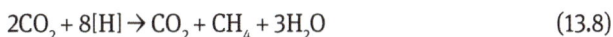

$$2CO_2 + 8[H] \rightarrow CO_2 + CH_4 + 3H_2O \quad\quad\quad (13.8)$$

Thus, the ratio CH_4/CO_2 depends on many factors. It has been demonstrated [123] that the concentration of metal ions such as Fe^{2+}, Ni^{2+}, and Co^{2+} in solution can influence the rate of H_2 formation and the CH_4/CO_2 ratio, most probably through the higher availability of the enzymes of which they are the active centers (see below).

The production of methane from WB is a quite complex process that requires the optimization of several parameters for the production of methane may be effective.

13.5.2 The laboratory equipment for biogas production and system investigation

In the laboratory, biogas production can be investigated using quite simple apparatus such as that represented in Figure 13.13. Using the apparatus shown in Figure 13.13, the production of gas, and its composition, was monitored and the role of Ni, Fe, Co on the CH_4/CO_2 molar ratio was investigated. The analysis of the composition of biogas was carried out by monitoring the H_2, CO_2, and CH_4 production for 8 h after the addition of feed (acetogenesis phase) and over 5 days (methanogenesis phase).

As expected, each metal produced a different effect on the basis of the role of the enzymes in which they are present. Ni(II), Fe(II), and Co(II) were added to the sludge either separately or in combination, always at a subtoxic concentration. The concentration of the metal in solution was increased by 3–4 times with respect to the standard conditions (e.g., 22.4 ppm with respect to 6.5 ppm). The distribution between the liquid and the solid phases was determined by elemental analysis of the liquid and

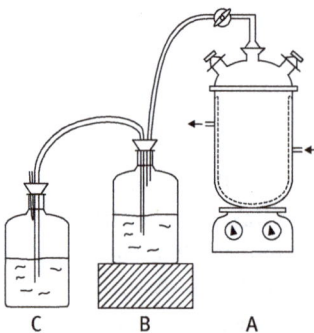

C   B   A

Fig. 13.13: Laboratory equipment for the investigation of the production of biogas. A represents the reactor connected to a thermostat. The stirrer keeps the reaction medium homogeneous. Feed or solution with metals can be added from the side openings on the top. B represents the collector of gas over water (or any other liquid. At $t = 0$, B is completely filled with water, which is then pushed into C by the biogas produced in A). Water is displaced into C, and the volume of produced gas can be measured. A sampling valve (eventually an automatic sampling system connected to a gas chromatograph) allows the withdrawal of gas at the desired time. C represents the water collector; it can be elevated to equalize the level of water in B and C for a correct measure of the volume of produced gas at ambient pressure.

solid phases withdrawn at several reaction times. The metal concentration in solution decreased from 90% of the value read soon after the addition to 70% after 2 weeks. Therefore, during the test at 8 h (which starts soon after the addition of metal and feeding) and 5 days, the bacteria were exposed to the increased metal concentration of nickel available in the solution. The response to such increase is therefore a real cause-effect process. A control reactor (broken line in Figure 13.14a and b) used was fed at time=0 with the same slurry used in the test reactor (continuous line). Figure 13.15a shows that an increase in Ni(II) concentration in the solution causes a decrease in dihydrogen in the gas phase.

This observation agrees well with the role of nickel as part of the active site in H_2-consuming hydrogenases [21]. Therefore, the addition of Ni(II) increases the activity of such H_2-consuming hydrogenases. Conversely, when Fe(II) was added, the increase in H_2 production in the batch reactor was observed within 2.5 h after feeding (Figure 13.14b), in comparison with the control, which had a lower H_2 production. Also, these data agree with the role of iron in H_2-evolving hydrogenases, which promote the formation of H_2 [23].

Noteworthy, the addition of cobalt did not cause any variation in H_2 production at any time after the start of the fermentation. Such metal-dependent effects were evident only when fresh feed was added to a not too much aged system. Figure 13.15 shows the composition of biogas produced after the addition of Ni(II): an increase in the amount of methane in the gas phase was observed with an increase in the CH_4/CO_2 molar ratio from 3 to 4.5. However, Ni(II) caused a more effective conversion of CO_2 into methane.

The observed effects linked to an increased concentration of the metals match the role of metals in enzymes. Among the three metal ions, nickel thus has the most spectacular effect [117]. These results suggest that the controlled addition of Fe, Ni, and Co could be beneficial for improving the methanation process of waste. When metals are

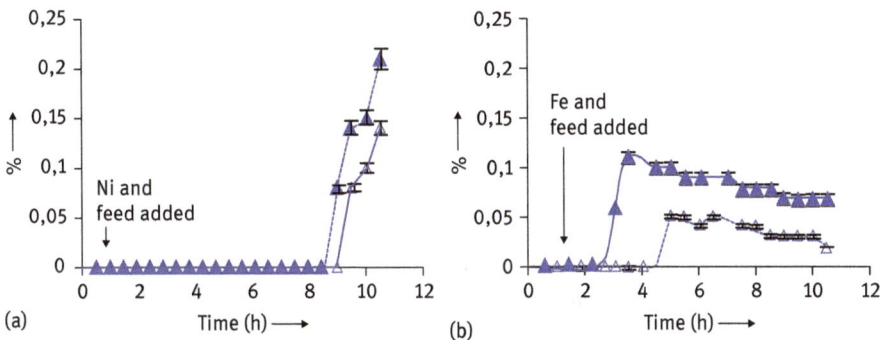

Fig. 13.14: Influence of Ni(II) and Fe(II) on the dihydrogen production during the acidogenesis phase in the anaerobic fermentation of FVG. The broken line represents the control; the continuous line, the reactor where Ni and Fe solutions were added.

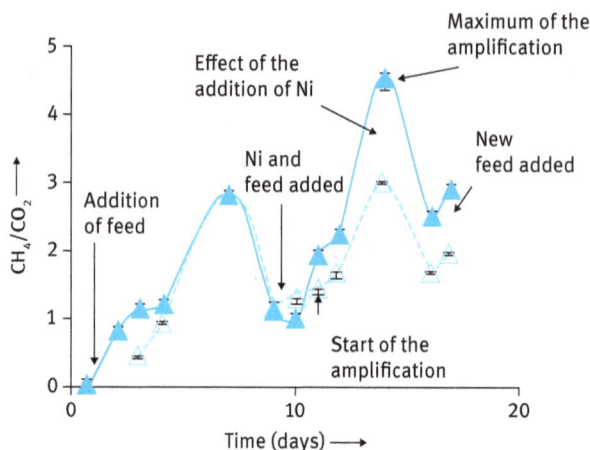

Fig. 13.15: Effect of the addition of Ni(II) on the production of biogas: an increase in the CH_4/CO_2 molar ratio (continuous line) with respect to the control (broken line) is evident.

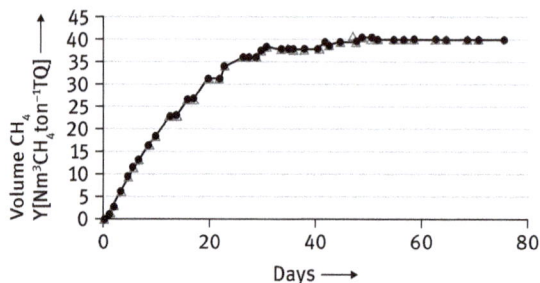

Fig. 13.16: Production of CH_4 with time in a single-feed reactor.

not added, the production of methane continues with a smooth growth up to ca. 40 days when it reaches a plateau (Figure 13.16).

13.6 Industrial reactors

As reported above, the anaerobic digestion can be considered as a sequence of three main degradation processes, namely (i) hydrolysis, (ii) acidogenesis, and (iii) methanogenesis. Each has its own velocity that depends on the nature of the substrate, the biomass properties, and the mass factor or available amount. For a good methanogenic process, it is important to maintain the specific velocity k for the three defined phases as expressed by the relation: $k_{hydrolysis} < k_{acidogenesis} < k_{methanogenesis}$, while the methanogenic process is usually the rate-limiting step. Under the condition above, any accumulation of the substrate is avoided. This becomes particularly true with substrates that have a low degradation rate or may cause important variation of pH, such as fatty acids.

Eq. (13.9) gives the specific rate of the biogas formation, where k is the specific velocity of biogas production, k_{max} is the maximum specific velocity of degradation of the substrate, S is the concentration of the substrate, and k_s is the concentration of the substrate when the specific velocity is half of the maximum:

$$k = k_{max} \frac{S}{S + k_s} X \qquad (13.9)$$

$$\mu = \mu_{max} \frac{S}{S + k_s} \qquad (13.10)$$

$$k_E = k_{max} \frac{S}{S + k_M} \qquad (13.11)$$

Eq. (13.9) is derived from the Monod equation for the growth of microorganisms on a given substrate [Eq. (13.10)], which, on turn, is an adaptation of the well-known Michaelis-Menten equation that gives the hydrolysis or, in general, the conversion of a substrate [Eq. (13.11)] under the action of enzyme E.

In Eq. (13.10), μ is the specific (1 M concentration of the substrate) growth rate, μ_{max} is the maximum specific growth rate, S is the concentration of the limiting substrate for growth, and k_s is the value of S for $m/m_{max} = 0.5$. In Eq. (13.11), E is the enzyme concentration, k_{max} is the maximum hydrolytic rate, and k_M is the Michaelis-Menten constant or the value of the concentration of the substrate at which the reaction rate is half of the maximum. By applying Eq. (13.9) to the three steps of the digestion process, it will be possible define which is the slower and, consequently, which will be the maximum rate of the entire process.

Among the factors that play a key role in biogas production, the quality of feed is prominent. Very often, digestors fed with agricultural biomasses receive a feed with much variable composition, depending on availability and cost. In such cases, an average velocity of digestion can be calculated that may vary over a large interval. Consequently, the digestor must be designed using some "flexibility" criteria and considering some buffer parameters.

The outside working temperature is a parameter that must be adapted to bacterial pools used (psychrophilic, mesophilic, or thermophilic) and influences the dimension of the plant [124]. The concentration of ammonia must be taken under control for a good balance between nitrogen necessary for bacterial growth and the best COD/N/P ratio, which must be 350/7/1 for reactors with a high loading and 1000/7/1 for low loading digestors, avoiding any excess that would disfavor several bioprocesses [125].

$$[NH_3] = \frac{[NH_3] + [NH_4^+]}{1 + \frac{[H^+]}{ka}} \qquad (13.12)$$

As the temperature influences the solubility of gases in a liquid, it will determine the concentration of key gases in solution (CO_2, NH_3, and H_2S).

Plants can be built as a single stage or multiple stages. The choice is dictated by the necessity of providing a better condition to each of the three phases of biogas formation mentioned above. In fact, hydrolysis and acidogenesis require higher temperatures and lower pH than methanogenesis.

Using a one-stage reactor, good average conditions for all the three steps (i.e., hydrolysis, acidogenesis, and methanogenesis) must be implemented that will negatively influence the production of biogas. In a multistage reactor, instead, it is possible to better control the individual conditions and make them most suitable for each phase within an increase in methane production.

The residence time of the bacterial pool in the reactor (sludge age) is also a key issue: it must be longer than the duplication time of microorganisms so to avoid wash out of the biomass from the reactor. In continuously stirred tank reactors (CSTR), such time is equal to the hydraulic retention time (HRT). In reactors working with wastewaters, it is possible to retain the sludge, increasing its age, and release water.

The feeding rate is another key factor: the quantity of feed introduced in the digestor must be always lower than its maximum capacity of degradation in order to avoid accumulation of substrate that will cause inefficiency in biogas production and even deactivation of the digester. All the above considerations, when merged into a plant design, bring about two main types of digestors: one based on the HRT (mainly used for solid biomass digestion) and another one based on the kinetics (used for water treatment).

It is noteworthy that the maximum quantity of biogas produced affects the specific biogas production (Figure 13.17) [118].

Table 13.6 lists some of the key properties of various biomasses commonly available for biogas production. It is possible to see how great is the variability of the composition of the substrate (C/N ratio), dry matter (DM) content, and solids (VS): all such

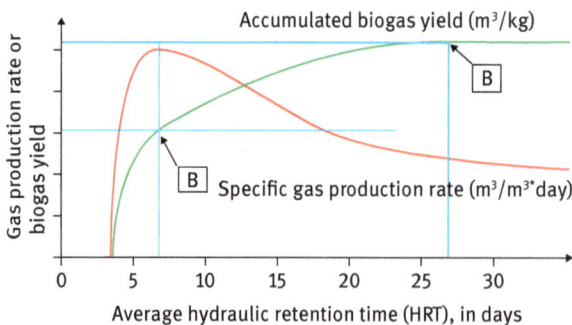

Fig. 13.17: Reciprocal influence of the specific and cumulative rate of biogas production. Increasing the cumulative production reduces the specific gas production.

Table 13.6: Characteristics of common available feedstock for biogas production [126].

Type of feedstock	Organic content	C/N ratio	DM (%)	VS (% of DM)	Biogas yield (m³/kg VS)	Unwanted physical impurities	Other unwanted matters
Pig slurry	Carbohydrates, proteins, lipids	3–10	3–8	70–80	0.25–0.50	Wood shavings, bristles, water, sand, cords, straw	Antibiotics, disinfectants
Cattle slurry	Carbohydrates, proteins, lipids	6–20	5–12	80	0.20–0.30	Bristles, soil, water, straw, wood	Antibiotics, disinfectants, NH_4
Poultry slurry	Carbohydrates, proteins, lipids	3–10	10–30	80	0.35–0.60	Grit, sand, feathers	Antibiotics, disinfectants, NH_4
Stomach/ intestine content	Carbohydrates, proteins, lipids	3–5	15	80	0.40–0.68	Animal tissues	Antibiotics, disinfectants
Whey	75–80% lactose 20–25% protein	–	8–12	90	0.35–0.80	Transportation impurities	
Concentrated whey	75–80% lactose 20–25% protein	–	20–25	90	0.80–0.95	Transportation impurities	
Flotation sludge	65–70% protein 30–35% lipids	–				Animal tissues	Heavy metal disinfectants, organic pollutants
Fermentation slops	Carbohydrates	4–10	1–5	80–95	0.35–0.78	Non-degradable fruit remains	
Straw	Carbohydrates, lipids	80–100	70–90	80–90	0.15–0.35	Sand, grit	
Garden wastes		100–150	60–70	90	0.20–0.50	Soil, cellulosic components	Pesticides
Grass		12–25	30–25	90	0.55	Grit	Pesticides
Grass silage		10–25	15–25	90	0.56	Grit	
Fruit wastes		35		75	0.25–0.50		
Fish oil	30–35% lipids	–					
Soya oil/ margarine	90% vegetable oil	–					
Alcohol	40% alcohol	–					
Food remains			10	80	0.50–0.60	Bones, plastic	Disinfectants
Organic household waste							Heavy metals, organic pollutants
Sewage sludge							Heavy metals, organic pollutants

parameters are obviously linked to the biogas production that eventually spans from 0.15 to 0.95 m³/kg. This has an enormous impact on the economics of the process. Moreover, biomasses that are poor generators of biogas will necessitate an energy input, which is against the objective of biogas production that targets waste reduction and net energy production.

The HRT sensibly varies with the nature of the substrate used. However, for an organic matter that is very easily biodegradable and for high loading rate reactors, the HRT can be set at 6–12 h. Conversely, when pig manure is used, the HRT must be increased to around 15–20 days, and for cow manure, HRT must be increased to 40–50 days. Energy crops demand even longer HRT, reaching the limit of 60–90 days.

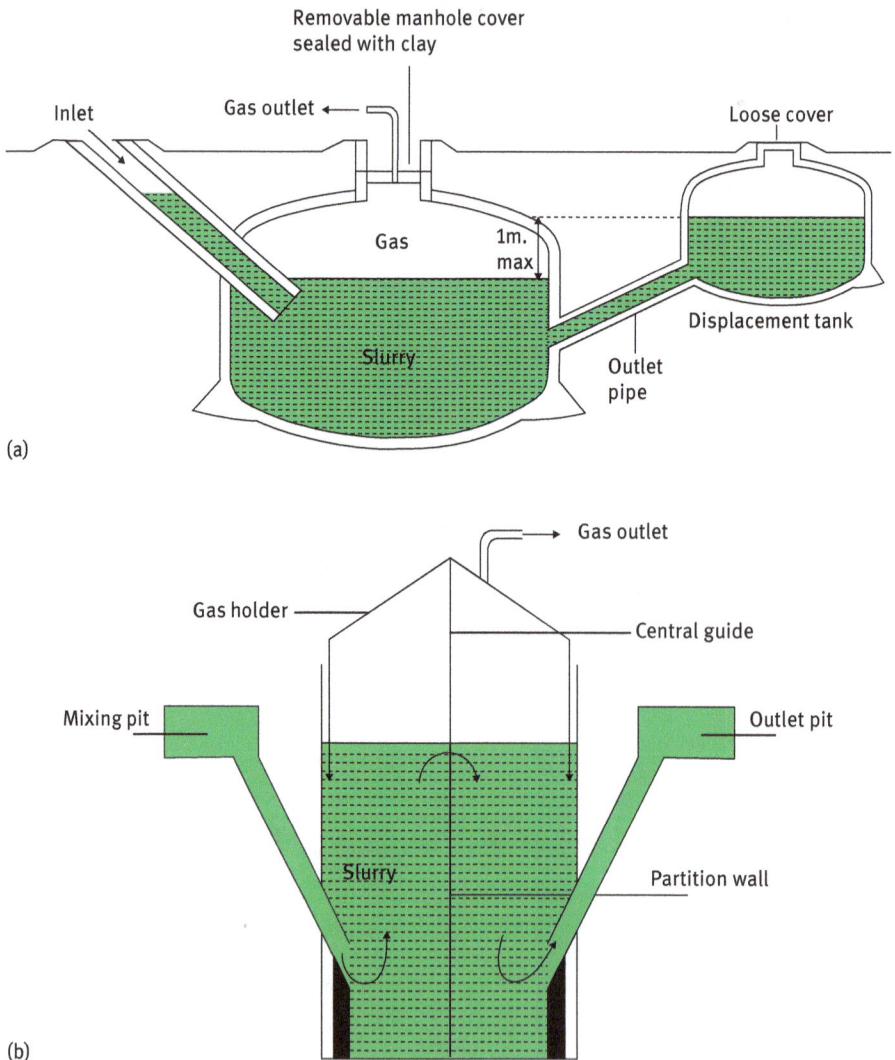

(a)

(b)

Fig. 13.18: (a) Chinese- and (b) Indian-type reactors [127].

13.6.1 Types of reactors and configuration of biogas plants

Digestors are characterized by a quite different complexity, depending on local conditions such as ambient temperature, kind of available biomass, investment capacity, etc. Figure 13.18a shows a Chinese digestor, while an Indian digestor is shown in Figure 13.18b.

These are very basic, simple concept reactors, but much more sophisticated ones are available on the market (see below).

Digestors can be classified according to several properties or working conditions, such as load of solid matter, working temperature, HRT, SRT, continuity or discontinuity of the plant operation, etc.

13.6.1.1 Solid content
If the solid content is considered, digestors can be classified as (i) wet reactors (the solid content is less than 15%), (ii) semiwet reactors (with a solid content between 15% and 20%), and (iii) dry reactor (having a content of solid higher than 20%).

13.6.1.2 Temperature
Based on the working temperature, it is possible to define the following general ranges of temperature. Psychrophilic conditions are relevant to a working temperature of ca. 20°C. Reactors working in such conditions can be used in relatively different climates up to warm climates with the main advantage of avoiding the need for any temperature control system. They can also work in cold climates with a low heat input and without using any heat exchange system. Mesophilic conditions are the mostly used ones worldwide. The process is quite stable and the productivity in biogas is high; the optimal temperature range is 35–37°C. Thermophilic micoorganisms require a temperature of at least 55°C up to 90°C. Under such conditions, the efficiency of the process is higher, with faster kinetics, but the control is much more difficult and large energy input is necessary.

13.6.1.3 Liquid and solid retention time
HRT and SRT issues have been discussed above.

13.6.1.4 Continuity of the process
Digestors can operate on a continuous or discontinuous base. "Batch" and "fill-and-draw" digestors are examples of the latter class. The batch reactors work on a very simple technology: they are filled and closed, the biogas production starts after a lag time and continues until biodegradable matter is present in the reactor. A single

reactor may be used, or even better, a sequence of reactors that are loaded at different times so that there is a continuous production of biogas, even when a reactor has to be stopped for downloading and reloading operations (Figure 13.19).

In this setup, the liquid phase, if it has the correct pH and N-content, can be recycled.

Such kind of reactors require a large feed availability at any loading procedure: it is not very suited for use in the conversion of agricultural wet biomass (like manure or other seasonal biomass). In the above case, fill-and-draw reactors are used instead, in which the biomass is fed discontinuously during a long period of time (3–7 month) and the gas is continuously collected. An obvious drawback is that the production of biogas may vary during the same period according with the amount of biomass fed into the reactor. Lagoons represent a typical implementation of such concept of low-cost reactors (Figure 13.20).

Conversely, continuous reactors are continuously fed while the solid material (digestate) and biogas are continuously withdrawn. They can have cylindrical horizontal, vertical, inclined, egg-shaped setups; they can be upflow or downflow, mixed or not, single stage or multistage, single phase or multiphase.

CSTR represents the most common type of reactor used for the treatment of wet biomass and wastewater under anaerobic digestion conditions. Such reactors work with a solid content ranging from 2% to 15% under thermophilic or mesophilic

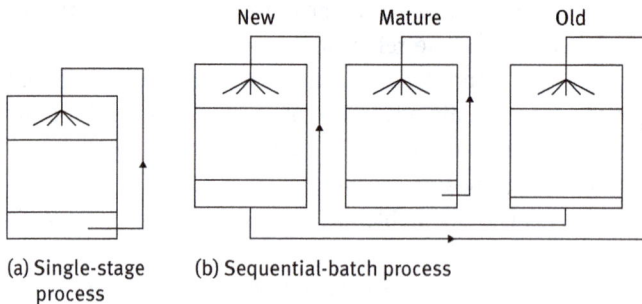

(a) Single-stage process   (b) Sequential-batch process

Fig. 13.19: Scheme of the management of a batch reactor in single stage and in sequence [128].

Fig. 13.20: Biogas collection system from lagoons [129].

conditions. The microorganism biomass and the sludge are continuously (or semi-continuously) stirred using different devices, including endless screws, recirculation pumps, and gas bubbling, for a short (high solid content) or long periods (wastewater treatment).

A particular plug-flow reactor is used in the DRANCO process (Figure 13.21). It is a dry, vertical, downflow reactor developed for the digestion of organic waste source – separated as organic fraction municipal solid waste, thickened sludge, or organic industrial wastes. This process can work at a solids concentration up to 45–50% [130] in the feed represented by solid particles with a size <40 mm that move by gravity from the top to the bottom of the reactor. They are mixed in a proportion of 1 t of fresh waste with 6 t of digested waste as inoculum. The reactor can operate in mesophilic or thermophilic conditions with an HRT of 15–20 days producing 100–200 N m³ biogas per ton of waste.

13.6.2 Biogas from wastewaters

Anaerobic digestion finds a large application in the treatment of wastewater, whether agricultural, municipal, or industrial (provided that species toxic to the microorganisms are not present). It is a quite common treatment method because of its effectiveness in treating wastewater with high organic load and because it gives some economic advantages in terms of biogas production and utilization, either locally or by immission into existing gas ducts. As already mentioned, due to the slow rate of such processes, it is necessary to decouple the biomass retention time (SRT) from that of water (HRT): for this reason, CSTRs cannot be used in this treatment. Attempts have been made to improve the biomass retention time, as shown in Table 13.7.

Quite sophisticated reactors have been developed to solve the BRT-WRT issue. An evolution of the CSTR reactor is the anaerobic contactor process, which is a CSTR with a settling tank usually preceded by a degassifier that enables the removal of biogas

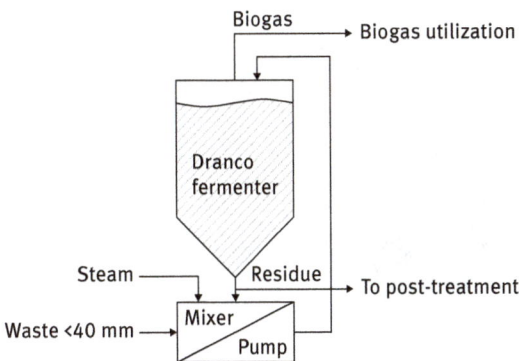

Fig. 13.21: Scheme of the DRANCO process [130].

Table 13.7: Technologies for improving the biomass retention time over water retention time.

Technology	Biomass retention mechanisms	Reactor type
Biomass immobilization in attached growth systems	Anaerobes growth attached on a support media (e.g., plastic, gravel, sand, and activated carbon, plastic foams) to form biofilm	AF, rotating anaerobic contactor, expanded bed reactor FBR
Granulation and flock formation	Anaerobic microorganisms growth in agglomerate to form granules or flocks that settle in the bioreactor	Upflow anaerobic sludge blanket reactor, static granular bed reactor, anaerobic-sequencing batch reactor, anaerobic baffled reactor
Biomass recycling	Suspended biomass is settled in a separate settler and then recycled back to the reactor	Anaerobic contact reactor
Biomass retention	Microfiltration membrane is integrated into an anaerobic reactor to retains biomass	Anaerobic membrane bioreactor

and reduces the buoyancy of the solid particles, favoring their settling velocity: the settled biomass is then recycled back to maintain longer SRT than in CSTR. This technology is particularly useful for high-load suspended solids in wastewater streams.

Anaerobic filters (AFs) filled with packing materials were also used to have a system that retains the biomass inside the reactor preventing its wash out (Figure 13.22).

The internal circulation (IC) reactor has been studied for the use with very high strength waste streams. It is divided horizontally in two parts by a first gas-liquid-solid (GLS) separator, while a second GLS is placed on the top of the reactor (Figure 13.23). This reactor is substantially a vertical tower of 16 to 28 m height with a diameter from 1.5 to 15 m.

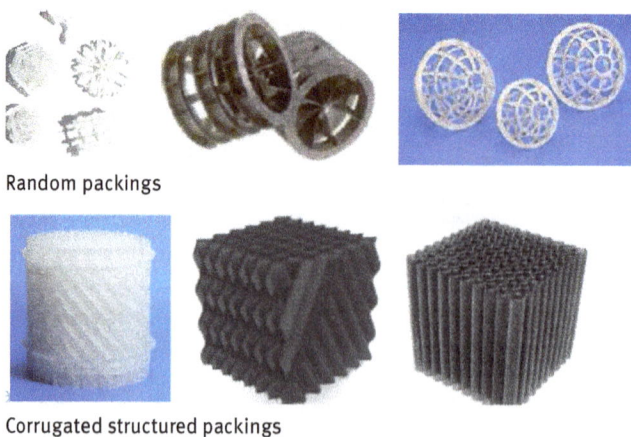

Random packings

Corrugated structured packings

Fig. 13.22: Different types of packing material for AFs.

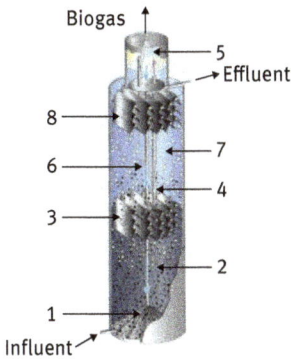

Fig. 13.23: IC reactor: 1, distribution system; 2, expanded bed compartment; 3, first separator; 4, riser; 5, degassing tank; 6, downer; 7, polishing compartment; 8, second separator.

We also mention the expanded granular sludge bed or fluidized bed reactors (FBRs) that use recirculation systems to expand the sludge bed formed by granules or by inert support material where microorganisms can growth.

Anaerobic membrane bioreactors have been studied and applied to improve the solid retention but especially to obtain a clarified effluent [131, 132].

The membrane can be planar or tubular and can be placed inside the reactor or externally. Despite interesting performances, drawbacks are represented by the fouling of membranes, which require frequent cleaning cycles, along with decreased membrane life and increased OPEX.

Bibliography

[1] Hobson, P. N., Bousfield, S., Summer, R., Anaerobic digestion of organic matter, CRC Crit Rev Environ Control **4** (1974) 131–191.

[2] Bernard, S., Gray, F., Aerobic digestion of pharmaceutical and domestic wastewater sludges at ambient temperature, Water Res **34** (2000) 725–734.

[3] Weemaes, M., Grootaerd, H., Simoens, F., Verstraete, W., Anaerobic digestion of ozonized biosolids, Water Res **34** (2000) 2330–2336.

[4] International Energy Agency (IEA) bioenergy, annual report 1994:1.

[5] Farina, R., Spagni, A., From lab-scale to full-scale biogas plants, in Aresta, M., Dibendetto, A., Dumeignil F., editors, Biorefinery: From Biomass to Chemicals and Fuels, Berlin: de Gruyter, 2012, 405–435.

[6] Fan, L. T., Lee, T. H., Kinetic studies of enzymatic hydrolysis of insoluble cellulose: derivation of a mechanistic kinetic model, Biotechnol Bioeng **25** (1983) 2707–2733.

[7] Sleat, R., Math, R., Hydrolytic bacteria, in Chynoweth, P., Isaacson, R., editors, Anaerobic Digestion of Biomass, Elsevier, London, 1987, 15–33.

[8] Maki, L. R., Experiments on the microbiology of cellulose decomposition in a municipal sewage plant, Antoine von Leeuwenhoek **20** (1954) 185–200.

[9] Greenwood, C. T., Starch and glycogen, in Pigman, W., Horton, D., editors, The Carbohydrates, Chemistry and Biochemistry, Academic Press, New York and London, 1970, 417–513.

[10] Hungate, R. E., Microbial ecology of the rumen, Bacteriol Rev **24** (1960) 353–364.

[11] Chynowethh, D. P., Mah, R. A., Volatile acid fermentation in sludge digestion, Adv Chem Ser Am Chem Soc **105** (1971) 41–54.

[12] Heuhelekian, H., Mueller, P., Transformation of some lipids in anaerobic sludge digestion, Sewage Indust Wastes **30** (1985) 1108–1120.

[13] Zehnder, A. J. B., Ecology of methane formation, in Mitchell, R., editor, Water Pollution Microbiology, Wiley & Sons, New York, 1978, 349–376.

[14] Zeikus, J. G., Microbes in their natural environments, in Slater, J. H., Whittenbury, J. W., editors, Cambridge University Press, Cambridge UK, 1983, 423–462. **259** (1984) 7045–7055.

[15] Schink, B., Environmental microbiology of anaerobes, in Zehnder, A. J. B., editor, Principles and Limits of Anaerobic Degradation: Environmental and Technological Aspect, Wiley & Sons, New York, 1983, 1466–1473.

[16] Thauer, R. K., Jungermann, K., Decker, K., Energy conservation in chemothrophic anaerobic bacteria, Bacteriol Rev **41** (1977) 100–180.

[17] Walsh, C. T., Orme-Johnson, W. H., Nickel enzymes, Biochemistry **26** (1987) 4901–4906.

[18] Aresta, M., Quaranta, E., Tommasi, I., Reduction of co-ordinate carbon dioxide to carbon monoxide via protonation by thiols and other Bronsted acids promoted by Ni-system: a contribution to the understanding of the mode of action of the enzyme carbon monoxide dehydrogenase, J Chem Soc Chem Commun (1988) 450–451.

[19] Ellefson, W. L., Wolfe, R. S., Component C of the methylreductase system of methanobacterium, J Biol Chem **256** (1981) 4259–4262.

[20] Rouviere, P. E., Escalante-Semerena, J. C., Wolfe, R. S., Component A2 of the methylcoenzyme M methylreductase system from *Methanobacterium thermoautotrophicum*, J Bacteriol **162** (1985) 61–66.

[21] Albracht, S. P. J., Nickel hydrogenases: in search of the active site, Biochem Biophys Acta **1188** (1994) 167–204.

[22] Tommasi, I., Aresta, M., Giannoccaro, P., Quaranta, E., Fragale, C., Bioinorganic chemistry of nickel and carbon dioxide: an Ni complex behaving as a model system for carbon monoxide dehydrogenase enzyme, Inorg Chim Acta **272** (1998) 38–42.

[23] Adams, M. W. W., The structure and mechanism of iron-hydrogenases, Biochem Biophys Acta **1020** (1990) 115–145.

[24] Hippler, B., Thauer, R. K., The energy conserving methyltetrahydromethanopterin:coenzyme M methyltransferase complex from methanogenic archaea: function of the subunit MtrH, FEBS Lett **449** (1999) 165–168.

[25] Schonheit, P., Mool, J., Thauer, R. K., Nickel, cobalt, and molybdenum requirement for growth of Methanobacterium thermoautotrophicum, Arch Microbiol **123** (1979) 105–107.

[26] Noike, T., Endo, G., Yagushi, J., Matsumoto, J., Characteristics of carbohydrate degradation and the rate-limiting step in anaerobic digestion, Biotechnol Bioeng **27** (1985) 1482–1489.

[27] Chandler, J. A., Jewell, W. S., Gosset, J. M., Van Soest, P., Robertson, J., Predicting methane fermentation biodegradability, Biotechnol Bioeng Symp **10** (1980) 93–107.

[28] Boone, D. R., Terminal reactions in the anaerobic digestion of animal waste, Appl Environ Microbiol **43** (1982) 57–64.

[29] Orpin, C. G., Letcher, A. J., Utilization of cellulose, starch, xylan, and other hemicelluloses for growth by the rumen phycomycete *Neocallimastix frontalis,* Curr Microbiol **3** (1979) 121–124.

[30] Montgomery, L., Macy, J. M., Characterization of rat cecum cellulolytic bacteria, Appl Environ Microbiol **44** (1982) 1435–1443.

[31] Davies, M. E., Cellulolytic bacteria isolated from the large intestine of the horse, J Appl Bacteriol **27** (1964) 373–378.

[32] Dehority, B. A., Cellulolytic cocci isolated from the cecum of guinea pigs (Cavia porcellus), Appl Environ Microbiol **33** (1977) 1278–1283.

[33] Madden, R. H., Bryder, M. J., Poole, N. J., Isolation and characterization of an anaerobic, cellulolytic bacterium, *Clostridium papyrosolvens* sp. nov., Int J System Bacteriol **32** (1982) 87–91.

[34] Skinner, F. A., The isolation of anaerobic cellulose-decomposing bacteria from soil, J Gen Microbiol **22** (1960) 53–534.

[35] Hobson, P. N., Continuous culture of some anaerobic and facultative anaerobic rumen bacteria, J Gen Microbiol **38** (1965) 167–180.

[36] Hishinuma, F., Kanegasaki, S., Takahashi, H., Ruminal fermentation and sugar concentrations, a model experiment with *Selenomonas ruminantium*, Agric Biol Chem **32** (1968) 1327–1330.

[37] Hobson, P. N., Summers, R., The continuous culture of anaerobic bacteria, J Gen Microbiol **47** (1967) 53–65.

[38] Henderson, C., Hobson, P. N., Summers, R., The production of amylase, protease and lipolytic enzymes by two species of anaerobic rumen bacteria, in Continuous Cultivation Of Microorganism, Proceedings of the 4th Symposium, Academia, Prague, 1969, 189.

[39] Dean, A. C. R., Influence of environment on the control of enzyme synthesis, in Dean, A. C. R., Pirt, S. J., Tempest, D. W., editors, Environmental Control of Cell Synthesis and Function, Academic Press, London, 1972, 245.

[40] Bull, A. T., Environmental factors influencing the synthesis and excretion of exocellular macromolecules, in Dean, A. C. R., Pirt, S. J., Tempest, D. W., editors, Environmental Control of Cell Synthesis and Function, Academic Press, London, 1972, 261.

[41] Demain, A. L., Cellular and environmental factors affecting the synthesis and excretion of metabolites, in Dean, A. C. R., Pirt, S. J., Tempest, D. W., editors, Environmental Control of Cell Synthesis and Function, Academic Press, London, 1972, 365.

[42] Brown, C. M., Stanley, S. O., Environment-mediated changes in the cellular content of the pool content of the pool constituents and their associated changes in cell physiology, in Dean, A. C. R., Pirt, S. J., Tempest, D. W., editors, Environmental Control of Cell Synthesis and Function, Academic Press, London, 1972, 345.

[43] Holme, T., Influence of the environment on the content and composition of bacterial envelopes, in Dean, A. C. R., Pirt, S. J., Tempest, D. W., editors, Environmental Control of Cell Synthesis and Function, Academic Press, London, 1972, 391.

[44] Davies, H. C., Rudd, J. H., Influence of environment on growth and cellular content of group A *Haemolytic streptococci* in continuous culture, in Dean, A. C. R., Pirt, S. J., Tempest, D. W., editors, Environmental Control of Cell Synthesis and Function, Academic Press, London, 1972, 401.

[45] Wolin, M. J., Hydrogen transfer in microbial communities, in Bull, A. T., Slater, J. H., editors, Microbial Interaction and Communities, Academic Press, New York and London, 1982, 323–256.

[46] Smith, P. H., Mah, R. H., Kinetics of acetate metabolism during sludge digestion, Appl Microbiol **14** (1966) 368–371.

[47] Lorowitz, W. H., Bryant, M. P., Methanogenic stearate enrichment cultures, Abstr Am Soc Microbiol Annu Meeting **148** (1983).

[48] Boone, D. R., Mah, R. H., Transitional bacteria, in Chynoweth, P., Isaacson, R., editors, Anaerobic Digestion of Biomass, Elsevier, London, 1987, 35–48.

[49] McInerney, M.-J., Bryant, M. P., Hespell, R. B., Casterton, J. W., Syntrophomonas wolfei gen. nov. sp. nov., an anaerobic, syntrophic, fatty acid-oxidizing bacterium, Appl Environ Microbiol **41** (1981) 1029–1039.

[50] Beaty, P. S., McInerney, M. J., Isolation of *Syntrophomonas wolfei* on crotonate, Abstr Ann Meeting Am Soc Microbiol (1986) I132.

[51] Boone, D. R., Bryant, M. P., Propionate-degrading bacterium *Syntrophobacter wolinii* sp., nov. gen. nov. from methanogenic ecosystem, Appl Environ Microbiol **40** (1981) 626–632.

[52] Henson, J. M., Smith, P. H., Isolation of a butyrate-utilizing bacterium in co-culture with methanobacterium thermoautrophicum from thermophilic digester, Appl Environ Microbiol **49** (1985) 1461–1466.

[53] Pierik, A. J., Hulstein, M., Hangen, W. R., Albracht, S. P. J., A low-spin iron with CN and CO as intrinsic ligands forms the core of the active site in [Fe]-hydrogenases, Eur J Biochem **258** (1998) 572–578.

[54] Nicolet, Y., Piras, C., Legrand, P., Hatchikian, E. C., Fontecilla-Camps, J. C., *Desulfovibrio desulfuricans* iron hydrogenase: the structure shows unusual coordination to an active site Fe binuclear center, Structure **7** (1999) 13–23.

[55] De Lacey, A. L., Stadler, C., Cavazza, C., Hatchikian, E. C., Fernandez, V. M., FTIR characterization of the active site of the Fe-hydrogenase from *Desulfovibrio desulfuricans*, J Am Chem Soc **122** (2000) 11232–11233.

[56] Chen, Z., Lemon, B. J., Huang, S., Swartz, D. J., Peters, J. W., Bagley, K. A., Infrared studies of the CO-inhibited form of the Fe-only hydrogenase from *Clostridium pasteurianum I*: examination of its light sensitivity at cryogenic temperatures, Biochemistry **41** (2002) 2036–2043.

[57] Adams, M. W. W., Mortenson, I. E., The physical and catalytic properties of hydrogenase II of *Clostridium pasteurianum*. A comparison with hydrogenase I, J Biol Chem **259** (1984) 7045–7055.

[58] Adams, M. W. W., The mechanisms of H_2 activation and CO binding by hydrogenase I and hydrogenase II of *Clostridium pasteurianum*, J Biol Chem **262** (1987) 15054–15061.

[59] Zambrano, I. C., Kowal, A. T., Mortenson, L. E., Adams, M. W. W., Johnson, M. K., Magnetic circular dichroism and electron paramagnetic resonance studies of hydrogenases I and II from *Clostridium pasteurianum*, J Biol Chem **264** (1989) 20974–20983.

[60] Bennett, B., Lemon, B. J., Peters, J. W., Reversible carbon monoxide binding and inhibition at the active Site of the Fe-only hydrogenase, Biochemistry **39** (2000) 7455–7460.

[61] Georgekaki, I. P., Darensbourg, M. Y., in McCleverty, J. A., Mayer, T. J., editors, Comprehensive Coordination Chemistry II, Elsevier, New York, 2004, 549–568.

[62] Dance, I., Structural variability of the active site of Fe-only hydrogenase and its hydrogenated forms, Chem Commun **17** (1999) 1655–1656.

[63] Nicolet, Y., De Lacey, A. L., Vernede, X., Fernandez, V. M., Hatchikian, E. C., Fontecilla-Camps, J. C., Crystallographic and FTIR spectroscopic evidence of changes in Fe coordination upon reduction of the active site of the Fe-only hydrogenase from *Desulfovibrio desulfuricans*, J Am Chem Soc **123** (2001) 1596–1601.

[64] Volbeda, A., Charon, M. H., Piras, C., Hatchikian, E. C., Frey, M., Fontecilla-Camps, J. C., Crystal structure of the nickel-iron hydrogenase from *Desulfovibrio gigas*, Nature **373** (1995) 580–587.

[65] Volbeda, A., Garcin, E., Piras, C., de Lacey, A. L., Fernandez, V. M., Hatchikian, E. C., Frey, M., Fontecilla-Camps, J. C., Structure of the [NiFe] hydrogenase active site: evidence for biologically uncommon Fe ligands, J Am Chem Soc **118** (1996) 12989–12996.

[66] Garcin, E., Vernede, X., Hatchikian, E. C., Volbeda, A., Frey, M., Fontecilla-Camps, J. C., The crystal structure of a reduced [NiFeSe] hydrogenase provides an image of the activated catalytic center, Structure **7** (1999) 557–566.

[67] Higuchi, Y., Ogata, H., Miki, K., Yasuoka, N., Yagi, T., Research article Removal of the bridging ligand atom at the Ni-Fe active site of [NiFe] hydrogenase upon reduction with H_2, as revealed by X-ray structure analysis at 1.4 Å resolution, Structure **7** (1999) 549–556.

[68] Hu, Z., Spangler, N. J., Anderson, M. E., Xia, J., Ludden, P. W., Lindahl, P. A., Munck, E., Nature of the C-cluster in Ni-containing carbon monoxide dehydrogenases, J Am Chem Soc **118** (1996) 830–845.

[69] Aresta, M., Narracci, M., Dibenedetto, A., Tommasi, I., 223th ACS Meeting, Inorganic Chemistry Division, Boston, MA, August 18–22, 2002, Abstract 61.

[70] Albracht, S. P. J., The use of electron-paramagnetic-resonance spectroscopy to establish the properties of nickel and the iron-sulphur cluster in hydrogenase from *Chromatium vinosum*, J Biochem Soc Trans **13** (1985) 582–585.

[71] De Lacey, A. L., Hatchikian, E. C., Volbeda, A., Frey, M., Fontecilla-Camps, J. C., Fernandez, V. M., Infrared-spectroelectrochemical characterization of the [NiFe] hydrogenase of *Desulfovibrio gigas*, J Am Chem Soc **119** (1997) 7181–7189.

[72] De Lacey, A. L., Stadler, C., Fernandez, V. M., Hatchikian, E. C., Fan, H. J., Li, S., Hall, M. B., IR spectroelectrochemical study of the binding of carbon monoxide to the active site of *Desulfovibrio fructosovorans* Ni-Fe hydrogenase, J Biol Inorg Chem **7** (2002) 318–326.

[73] Fernandez, V. M., Hatchinkian, E. C., Cammack, R., Properties and reactivation of two different deactivated forms of *Desulfovibrio gigas* hydrogenase, Biochim Biophys Acta **812** (1985) 69–79.

[74] Cammack, R., Fernandez, V. M., Schneider, K., Activation and active sites of nickel-containing hydrogenases, Biochimie **68** (1986) 85–91.

[75] Fernandez, V. M., Hatchinkian, E. C., Patil, D., Cammack, R., ESR-detectable Nickel and iron-sulphur centers in relation to the reversible activation of *Desulfovibrio gigas* hydrogenase, Biochim Biophys Acta **883** (1986) 145–154.

[76] Maroney, M. J., Pressler, M. A., Mirza, S. A., Shaukat, A., Whitehead, J. P., Gurbiel, R. J., Hoffman, R. J., Insights into the role of nickel in hydrogenase, Adv Chem Ser Mech Bioinorg Chem **246** (1995) 21–60.

[77] Dole, F., Fournel, A., Magro, V., Hatchinkian, E. C., Bertrand, P., Guigliarelli, B., Nature and electronic structure of the Ni-X dinuclear center of *Desulfovibrio gigas* hydrogenase. Implications for the enzymatic mechanism, Biochemistry **36** (1997) 7847–7854.

[78] Bleijlenivens, B., Faber, B. W., Albracht, S. P. J., The [NiFe] hydrogenase from *Allochromatium vinosum* studied in EPR-detectable states: H/D exchange experiments that yield new information about the structure of the active site, J Biol Inorg Chem **6** (2001) 763–769.

[79] Vincent, K. A., Cracknell, J. A., Parking, A., Armstrong, F. A., Hydrogen cycling by enzymes: electrocatalysis and implications for future energy technology, Dalton Trans **21** (2005) 3397–3403.

[80] Bagley, K. A., Van Garderen, C. J., Chen, M., Duin EC, Albracht, S. P. J., Woodruff, W. H., Infrared studies on the interaction of carbon monoxide with divalent nickel in hydrogenase from *Chromatium vinosum*, Biochemistry (1994) 9229–9936.

[81] Bagley, K. A., Duin, E. C., Roseboom, W., Albracht SPJ, Woodruff, W. H., Infrared-detectable group senses changes in charge density on the nickel center in hydrogenase from *Chromatium vinosum*, Biochemistry **34** (1995) 5527–5535.

[82] Van de Spek, T. M., Arendsen, A. F., Happe, R., Yun, S., Bagley, K. A., Stukens, D. J., Hagen, W. R., Albracht, S., P. J., Similarities in the architecture of the active sites of Ni-hydrogenases and Fe-hydrogenases detected by means of infrared spectroscopy, Eur J Biochem **237** (1996) 629–634.

[83] Berlier, Y., Fauque, G. D., Legall, J., Choi, E., Peck, H. D, Jr., Lespinat, P. A., Inhibition studies of three classes of Desulfovibrio hydrogenase: application to the further characterization of the multiple hydrogenases found in *Desulfovibrio vulgaris Hildenborough*, Biochem Biophys Res Commun **146** (1987) 147–153.

[84] Bleijlevens, B., Van Broekhuizen, F. A., De Lacey, A. L., Roseboom, W., Fernandez, V. M., Albracht, S. P. J., The activation of the [NiFe]-hydrogenase from *Allochromatium vinosum*. An infrared spectro-electrochemical study, J Biol Inorg Chem **9** (2004) 743–752.

[85] Ogata, H., Mizoguchi, Y., Mizuno, N., Miki, K., Adachi, S. I., Yasuoka, N., Yagi, T., Yamauchi, O., Hirota, S., Higuchi, Y., Structural studies of the carbon monoxide complex of [NiFe] hydrogenase from *Desulfovibrio vulgaris*: suggestion for the initial activation site for dihydrogen, J Am Chem Soc **124** (2002) 11628–11635.

[86] De Gioia, L., Fantucci, P., Guigliarelli, B., Bertrand, P., Ab initio investigation of the structural and electronic differences between active-site models of [NiFe] and [NiFeSe] hydrogenases, Int J Quantum Chem **73** (1999) 187–195.

[87] Siegbahn, P. E. M., Blomberg, M. R. A., Pavlov, M. W., Crabtree, R. H., The mechanism of the Ni-Fe hydrogenases: a quantum chemical perspective, J Biol Inorg Chem **6** (2001) 460–466.

[88] De Gioia, L., Fantucci, P., Guigliarelli, B., Bertrand, P., Ni-Fe hydrogenases: a density functional theory study of active site models, Inorg Chem **38** (1999) 2658–2662.

[89] Bruschi, M., Gioia, L., Zampella, G., Reither, M., Fantucci, P., Stein, M., A theoretical study of spin states in Ni-S4 complexes and models of the [NiFe] hydrogenase active site, J Biol Inorg Chem **9** (2004) 873–884.

[90] Niu, S., Thomson, L. M., Hall, M. B., Theoretical characterization of the reaction intermediates in a model of the Nickel–Iron hydrogenase of *Desulfovibrio gigas*, J Am Chem Soc **121** (1999) 4000–4007.

[91] Fan, H. J., Hall, M. B., Recent theoretical predictions of the active site for the observed forms in the catalytic cycle of Ni-Fe hydrogenase, J Biol Inorg Chem **6** (2001) 467–473.

[92] Li, S., Hall, M. B., Modeling the active sites of metalloenzymes 4. Predictions of the unready states of [NiFe] *Desulfovibrio gigas* hydrogenase from density functional theory, Inorg Chem **40** (2001) 18–24.

[93] Niu, S., Hall, M. B., Modeling the active sites in metalloenzymes 5. The heterolytic bond cleavage of H_2 in the [NiFe] hydrogenase of *Desulfovibrio gigas* by a nucleophilic addition mechanism, Inorg Chem **40** (2001) 6201–6203.

[94] Stein, M., Van Lenthe, E., Baerends, E. J., Lubitz, W., Relativistic DFT calculations of the paramagnetic intermediates of [NiFe] hydrogenase. Implications for the enzymatic mechanism, J Am Chem Soc **123** (2001) 5839–5840.

[95] Stein, M., Lubitz, W., Quantum chemical calculations of [NiFe] hydrogenase, Curr Opin Chem Biol **6** (2002) 243–249.

[96] Foerster, S., Stein, M., Brecht, M., Ogata, H., Higuchi, Y., Lubitz, W., Single crystal EPR studies of the reduced active site of [NiFe] hydrogenase from *Desulfovibrio vulgaris Miyazaki F*, J Am Chem Soc **125** (2003) 83–93.

[97] Stein, M., Lubitz, W., Relativistic DFT calculation of the reaction cycle intermediates of [NiFe] hydrogenase: a contribution to understanding the enzymatic mechanism, J Inorg Biochem **98** (2004) 862–877.

[98] Van Gastel, M., Stein, M., Brecht, M., Schroeder, O., Lendzian, F., Bittl, R., Ogata, H., Higuchi, Y., Lubitz, W. A single-crystal ENDOR and density functional theory study of the oxidized states of the [NiFe] hydrogenase from *Desulfovibrio vulgaris Miyazaki F*, J Biol Inorg Chem **11** (2006) 41–51.

[99] Pavlov, M., Siegbahn, P. E. M., Blomberg, M. R. A., Crabtree, R. H., Mechanism of H-H activation by nickel-iron hydrogenase, J Am Chem Soc **129** (1998) 548–555.

[100] Pavlov, M., Blomberg, M. R. A., Siegbahn, P. E. M., New aspects of H_2 activation by nickel-iron hydrogenase, Int J Quant Chem **73** (1999) 197–207.

[101] Siegbahn, P. E. M., Proton and electron transfers in [NiFe] hydrogenase, Adv Inorg Chem **56** (2004) 101–125.

[102] Fan, H. J., Hall, M. B., High-Spin Ni(II), a surprisingly good structural model for [NiFe] hydrogenase, J Am Chem Soc **124** (2002) 394–395.

[103] Kim, J., Rees, D. C., Crystallographic structure and functional implications of the nitrogenase molybdenum–iron protein from Azotobacter vinelandii, Nature **360** (1992) 553–560.

[104] Peters, J. W., Stowell, M. H., Soltis, S. M., Finnegan, M. G., Johnson, M. K., Rees, D. C., Redox-dependent structural changes in the nitrogenase P-cluster, Biochemistry **36** (1997) 1181–1187.

[105] Mayer, S. M., Lawson, D. M., Gormal, C. A., Roe, S. M., Smith, B. E., New insights into structure-function relationships in nitrogenase: a 1.6 Å resolution X-ray crystallographic study of *Klebsiella pneumoniae* MoFe-protein, J Mol Biol **292** (1999) 871–891.

[106] Einsle, O., Tezcan, F. A., Andrade, S. L. A., Schmid, B., Yoshida, M., Howard, J. B., Rees, D. C., Nitrogenase MoFe-protein at 1.16 Å resolution: a central ligand in the FeMo-Cofactor, Science **297** (2002) 1696–1700.

[107] Kim, J., Rees, D. C., Structural models for the metal centers in the nitrogenase molybdenum-iron protein, Science **257** (1992) 1677–1682.

[108] Chan, M. K., Kim, J., Rees, D. C., The nitrogenase FeMo-cofactor and P- cluster pair: 2.2 Å resolution structures, Science **260** (1993) 792–794.

[109] Kim, J., Woo, D., Rees, D. C., X-ray crystal structure of the nitrogenase molybdenum-iron protein from Clostridium pasteurianum at 3.0-Å resolution, Biochemistry **32** (1993) 7104–7115.

[110] Bolin, J. T., Campobasso, N., Muchmore, S. W., Morgan, T. V., Mortenson, L. E., The structure and environment of the metal cluster in the nitrogenase MoFe protein from Clostridium pasteurianum, in Stiefel, E. I., Coucouvanis, D., Newton, W. E., editors, Molybdenum Enzymes, Cofactors and Model Systems, American Chemical Society, Washington, DC, 1993, 186–195.

[111] Peters, J. C., Mehn, M. P., Bio-organometallic approaches to nitrogen fixation chemistry, in activation of small molecules, in Activation of Small Molecules: Organometallic and Bioinorganic Perspectives, W.B. Tolmon, Wiley-VCH Verlag GmbH & Co. KGaA, Weinheim, Germany, 2006, 81–120.

[112] Igarashi, R. Y., Laryukhin, M., Dos, S. antos, P. C., Lee, H. I., Dean, D. R., Seefeldt, L. C., Hoffman, B. M., Intermediates trapped during nitrogenase reduction of N≡N, CH_3-N=NH, and H_2N-NH_2, J Am Chem Soc **127** (2005) 6231–6241.

[113] Orgo, S., Kure, B., Nakai, H., Watanabe, Y., Fukuzumi, S., Why do nitrogenases waste electrons by evolving dihydrogen?, Appl Organomet Chem **18** (2004) 589–594.

[114] Jeris, J. S., McCarty, P. L., The biochemistry of methane fermentation using C_{14} tracers, J Water Pollut Control Fed **37** (1965) 158–192.

[115] Zeikus, J. G., The biology of methanogenic bacteria, Bacterial Rev **41** (1977) 514–541.

[116] Ferguson, T., Mah, R. H., Methanogenic bacteria, in Chynoweth, P., Isaacson, R., editors, Anaerobic Digestion of Biomass, Elsevier, London and New York, 1987, 49–63.

[117] Mah, R. A., Word, D. M., Baresi, L., Glass, T. L., Biogenesis of methane, Ann Rev Microbiol **31** (1977) 309–341.

[118] Stetter, K. O., Thomm, M., Winter, J., Wildgruber, G., Huber, H., Zilling, W., Janecovic, D., König, H., Palm, P., Wonderl, S., Methanothermus fervidus, sp. nov., a novel extremely thermophilic methanogen from an Icelandic hot spring, Zbl Bakt Hyg I Abt Orig **C2** (1981) 166–158.

[119] Miller, T. L., Wolin, M. J., (1983), Oxidation of hydrogen and reduction of methanol to methane in the sole energy source for a methanogen isolated from human feces, J Bacteriol **153** (1983) 1051–1055.

[120] Barker, H. A., Biological formation of methane, in Bacterial Fermentations, Wiley & Sons, New York, 1956, 1–27.

[121] Mc Bride, B. C., Wolfe, R. S., A new coenzyme of methyl transfer, coenzyme M, Biochemistry **10** (1977) 2315–2324.

[122] Ragsdale, S. W., Kumar, M., Nickel-containing carbon monoxide dehydrogenase/acetyl-CoA synthase, Chem Rev **96** (1996) 2515–2536.

[123] Aresta, M., Narracci, M., Tommasi, I., Influence of iron, nickel and cobalt on biogas production during the anaerobic fermentation of fresh residual biomass, Chem Ecol **19** (2003) 451–459.

[124] LfU, *Biogashandbuch Bayern – Materienband*, Bayerisches Landesamt für Umwelt, Augsburg, Germany, 2007.

[125] Angelidaki, I., Ahring, B. K., Anaerobic thermophilic digestion of manure at different ammonia loads: effect of temperature, Water Res **28** (1994) 727–731.

[126] Al Seadi, T., Good Practice in Quality Management of AD Residues from Biogas Production, IEA Bioenergy and AEA Technology Environment, Oxfordshire, UK, 2001.

[127] Gunnerson, C. G., Stuckey, D. C., Anaerobic Digestion: Principles and Practices for Biogas Systems: UNPD Management, World Bank, Washington, DC, 1986.

[128] Wilson, P., Anaerobic Treatment of Agricultural Residues and Wastewaters, Application of High-Rate Reactors, Department of Biotechnology, Media-Tryck, Lund University, Lund, 2004.

[129] CRPA, Energy from Biomass, Il Divulgatore, Centro di Divulgazione Agricola, Italy, 2005 [in Italian].

[130] De Baere, L., The DRANCO technology:a unique digestion technology for solid organic waste, 2010, http://www.ows.be/pages/index.php?menu=85&submenu=129&choose_lang=EN.

[131] Spagni, A., Casu, S., Crispino, N. A., Farina, R., Mattioli, D., Filterability in a submerged anaerobic membrane bioreactor, Desalination **250** (2010) 787–792.

[132] Casu, S., Crispino, A. N., Farina, R., Mattioli, D., Ferraris, M., Spagni, A., Wastewater treatment in a submerged anaerobic membrane bioreactor, J Environ Sci Health A Toxic Hazard Subst Environ Eng **47** (2012) 204–209.

Index

www.ingramcontent.com/pod-product-compliance
Lightning Source LLC
Chambersburg PA
CBHW081046220326
41598CB00038B/6997